Surface Chemistry of Biological Systems

ADVANCES IN EXPERIMENTAL MEDICINE AND BIOLOGY

Volume 1
THE RETICULOENDOTHELIAL SYSTEM AND ATHEROSCLEROSIS
Edited by N. R. Di Luzio and R. Paoletti • 1967

Volume 2
PHARMACOLOGY OF HORMONAL POLYPEPTIDES AND PROTEINS
Edited by N. Back, L. Martini, and R. Paoletti • 1968

Volume 3
GERM-FREE BIOLOGY — EXPERIMENTAL AND CLINICAL ASPECTS
Edited by E. A. Mirand and N. Back • 1969

Volume 4
DRUGS AFFECTING LIPID METABOLISM
Edited by W. L. Holmes, L. A. Carlson, and R. Paoletti • 1969

Volume 5
LYMPHATIC TISSUE AND GERMINAL CENTERS IN IMMUNE RESPONSE
Edited by L. Fiore-Donati and M. G. Hanna, Jr. • 1969

Volume 6
RED CELL METABOLISM AND FUNCTION
Edited by George J. Brewer • 1970

Volume 7
SURFACE CHEMISTRY OF BIOLOGICAL SYSTEMS
Edited by Martin Blank • 1970

Surface Chemistry of Biological Systems

Proceedings of the American Chemical Society Symposium on
Surface Chemistry of Biological Systems held in New York
City September 11-12, 1969

Edited by

Martin Blank

Department of Physiology
College of Physicians and Surgeons
Columbia University, New York

PLENUM PRESS • NEW YORK–LONDON • 1970

Library of Congress Catalog Card Number 70-110799

Softcover reprint of the hardcover 1st edition 1970
A Division of Plenum Publishing Corporation
227 West 17th Street, New York, N.Y. 10011
United Kingdom edition published by Plenum Press, London
A Division of Plenum Publishing Corporation, Ltd.
Donington House, 30 Norfolk Street, London W.C.2, England

ISBN-13: 978-1-4615-9007-1   e-ISBN-13: 978-1-4615-9005-1
DOI: 10.1007/ 978-1-4615-9005-1

PREFACE

     This volume of Advances in Experimental Medicine and Biology
is based on an American Chemical Society Symposium entitled:
"Surface Chemistry of Biological Systems", which took place in New
York on September 11-12, 1969. Thanks to the special photo offset
process used by the publishers, the papers are appearing very soon
after their presentation, and at a lower cost than usual. These
advantages are appreciated by the scientific community.

     As the title of the volume indicates we have attempted to
bring the scientific approach and techniques of surface chemistry
to the complex problems of biological systems. Two previous sym-
posia in this field have been published, one in the Journal of
Colloid and Interface Science (24:1-127, 1967) and the other in
the Journal of General Physiology (52:187S-252S, 1968). The
previous publication outlets, a chemical and a biological journal,
help to emphasize the interdisciplinary nature of the material and
also the appropriateness of the choice of Advances in Experimental
Medicine and Biology for the current symposium.

     Surface chemistry continues to be useful and productive in
the direct study of biological systems, and also indirectly as a
context for the consideration of biological problems. There has
been considerable progress in the last few years, and this volume
is composed of research contributions by leading workers on a
variety of problems in this general area. Along with the recent
results of research on specific problems, each paper includes an
up-to-date survey of the general field highlighting those areas in
which there has been considerable activity in the last few years,
and also indicating some of the directions in which future
research will continue.

     The papers can be grouped into three categories:

1-   The first group deals with model systems, such as monolayers
     bilayers and dispersions, where the emphasis is on inter-
     actions between the substances themselves and also between
     the substances and other materials in bulk phases. These
     papers are directed primarily at the problems of membrane
     stability, and in particular the role of lipids.

2-   The second group deals with the composition of the surfaces
      of natural systems and the various interactions that can
      result between the surfaces. These papers consider such
      problems as the chemical composition of a simple membrane,
      the substances present at the outer surfaces of various cells,
      the lung surfactant system, etc.

3-   The third group deals with various aspects of transport
      across surfaces and the factors that influence transport.
      These papers emphasize the specialized mechanisms that are
      characteristic of transport across natural membranes, e.g.
      gastric secretion, pinocytosis in ameba, sugar transport in
      red blood cells. In the case of sugar transport it is of
      some interest to note that some components of the red cell
      which are believed to be "carriers" of sugars in membranes
      can be shown to act as carriers in a simple interfacial
      system.

      There is some overlap between the groups, which is to be ex-
pected in an active field, but the groupings help one to see the
trends in the general areas. For example, work is continuing on
the composition and surface structure of membranes and there is a
growing interest in the study of interactions in and between surfaces. There is also renewed awareness of the changes that occur
in natural surfaces with time and as a result of the processes
under study. This latter factor is particularly important in the
study of transport processes where one must now begin to consider
that membranes may change (in degree of hydration, charge, etc.)
during a process and that average properties, e.g. permeability
constants, may prove to be inadequate. Finally, one must note the
advances that are being made in the study of medical problems,
such as the characterization and the mechanism of deposition of
lipids in atheromatous plaques, the development of compatible
materials for intravascular prostheses, the characterization of
the lung surfactant system, etc.

      Although this is a rather small collection of papers, I think
the reader will find that they cover a wide variety of topics and
that they serve to crystallize the many recent advances in this
active field of research.

                          Martin Blank
                          Department of Physiology
                          Columbia University
                          College of Physicians and Surgeons

                          October 1969

LIST OF PARTICIPANTS

Morris B. Abramson
The Saul R. Korey Department of Neurology
Albert Einstein College of Medicine
Bronx, New York

R. E. Baier
Applied Physics Department
Cornell Aeronautical Laboratory
Cornell University
Buffalo, New York

Jesse M. Berkowitz
Division of Gastroenterology
Department of Medicine
Meadowbrook Hospital
East Meadow, New York

Joel L. Bert
Department of Mechanical Engineering
University of California
Berkeley, California

Michael Borowitz
Cardiovascular Research Institute and
Department of Medicine
University of California Medical Center
San Francisco, California

Philip W. Brandt
Columbia University
New York, New York

Giuseppe Colacicco
York College
City University of New York
Flushing, New York

R. C. Dutton
Laboratory of Technical Development
National Heart Institute
Bethesda, Maryland

M. T. A. Evans
Unilever Research Laboratory
Colworth/Welwyn
The Frythe, Welwyn
Herts, England

Irving Fatt
Department of Mechanical Engineering
University of California
Berkeley, California

M. A. Frommer
Polymer Department
Weizmann Institute of Science
Rehovoth, Israel

Morton Galdston
Associate Professor of Medicine
New York University School of Medicine
New York, New York

Jon Goerke
Cardiovascular Research Institute and
Department of Medicine
University of California Medical Center
San Francisco, California

V. L. Gott
Department of Surgery
Johns Hopkins Hospital
Baltimore, Maryland

Helen H. Harper
Cardiovascular Research Institute and
Department of Medicine
University of California Medical Center
San Francisco, California

Klaus B. Hendil
Carlsberg Laboratory
Carlsbergvey, Valby, Denmark

L. Irons
Unilever Research Laboratory
Colworth/Welwyn
The Frythe
Welwyn, Herts, England

Robert Katzman
The Saul R. Korey Department of Neurology
Albert Einstein College of Medicine
Bronx, New York

A. Khaïat
Polymer Department
Weizmann Institute of Science
Rehovoth, Israel

Christian Mathot
Rockefeller University
New York, New York

E. Mayhew
Department of Experimental Pathology
Roswell Park Memorial Institute
Buffalo, New York

I. R. Miller
Polymer Department
Weizmann Institute of Science
Rehovoth, Israel

J. Mitchell
Unilever Research Laboratory
Colworth/Welwyn
The Frythe
Welwyn, Herts, England

Thomas Jenner Moore
St. Luke's Hospital Center and
College of Physicians and Surgeons
New York, New York

P. R. Mussellwhite
Unilever Research Laboratory
Colworth/Welwyn
The Frythe
Welwyn, Herts, England

P. A. Myers
New England Institute for Medical Research
Ridgefield, Connecticut

Martin S. Nachbar
Department of Medicine
New York University School of Medicine
New York, New York

Shinpei Ohki
Department of Biophysical Sciences
State University of New York at Buffalo
Buffalo, New York

Demetrios Papahadjopoulos
Department of Experimental Pathology
Roswell Park Memorial Institute
Buffalo, New York

Melvin Praissman
The Department of Physiology
Mount Sinai School of Medicine
New York, New York

Alexandre Rothen
Rockefeller University
New York, New York

Milton R. J. Salton
Department of Microbiology
New York University School of Medicine
New York, New York

Emile M. Scarpelli
Department of Pediatrics
Albert Einstein College of Medicine
New York, New York

Dinesh O. Shah
Surface Chemistry Laboratory
Marine Biology Division
Lamont-Doherty Geological Observatory
Columbia University
Palisades, New York

Donald M. Small
Biophysics Unit
Department of Medicine
Boston University School of Medicine
Boston, Massachusetts

H. Ti Tien
Department of Biophysics
Michigan State University
East Lansing, Michigan

D. J. Wilkins
New England Institute for Medical Research
Ridgefield, Connecticut

L. Weiss
Department of Experimental Pathology
Roswell Park Memorial Institute
Buffalo, New York

CONTENTS

The Effect of the Modification of Protein Structure on
the Properties of Proteins Spread and
Adsorbed at the Air-Water Interface   1
M. T. A. Evans, J. Mitchell, P. R. Mussellwhite, and
L. Irons

The Interaction of Calcium with Monolayers of Stearic
and Oleic Acid   23
J. Goerke, H. H. Harper, and M. Borowitz

Studies of Thermal Transitions of Phospholipids in Water:
Effect of Chain Length and Polar Groups of
Single Lipids and Mixtures   37
M. B. Abramson

The Physical State of Lipids of Biological Importance:
Cholesteryl Esters, Cholesterol, Triglyceride .   55
D. M. Small

The Effect of Hydrocarbon Configuration and Cholesterol
on Interactions of Choline Phospholipids
with Sulfatide   85
M. B. Abramson and R. Katzman

Lipid-Polymer Interaction in Monolayers: Effect of
Conformation of Poly-L-Lysine on Stearic
Acid Monolayers 101
D. O. Shah

Interactions of DNA with Positively Charged Monolayers . . . 119
M. A. Frommer, I. R. Miller, and A. Khaiat

The Effect of Modifiers on the Intrinsic Properties of
            Bilayer Lipid Membranes (BLM)   135
    H. Ti Tien

Asymmetric Phospholipid Membranes: Effect of pH and Ca^{2+} . .   155
    S. Ohki and D. Papahadjopoulos

Dissociation of Functional Markers in Bacterial Membranes . .   175
    M. S. Nachbar and M. R. J. Salton

RNA in the Cell Periphery   191
    E. Mayhew and L. Weiss

Immunological Reactions Carried Out at a Liquid-Solid
            Interface with the Help of a Weak Electric
            Current   209
    A. Rothen and C. Mathot

Electrophoresis and Adsorption Studies of Proteins and
            Their Derivatives on Colloids and Cells   217
    D. J. Wilkins and P. A. Myers

Surface Chemical Features of the Blood Vessel Walls and
            of Synthetic Materials Exhibiting
            Thromboresistance   235
    R. E. Baier, R. C. Dutton, and V. L. Gott

Lipid-Protein Association in Lung Surfactant   261
    M. Galdston and D. O. Shah

Absence of Lipoprotein in Pulmonary Surfactants   275
    E. M. Scarpelli and G. Colacicco

Relation of Water Transport to Water Content in Swelling
            Biological Membranes   287
    J. L. Bert and I. Fatt

Kinetic and Equilibrium Behavior of Simple Sugars in a
            Water-Butanol-Lipid System   295
    T. J. Moore

Use of Synthetic Membrane Models in the Study of Gastric
            Secretory Processes   309
    J. M. Berkowitz and M. Praissman

Properties of the Plasma Membrane of Amoeba   323
    P. W. Brandt and K. B. Hendil

Index   337

THE EFFECT OF THE MODIFICATION OF PROTEIN STRUCTURE ON THE PROPERTIES

OF PROTEINS SPREAD AND ADSORBED AT THE AIR-WATER INTERFACE

M.T.A. Evans, J. Mitchell, P.R. Mussellwhite, L. Irons

Unilever Research Laboratory Colworth/Welwyn, The Frythe

Welwyn, Herts, England

INTRODUCTION

The results of experimental biology clearly show that living systems contain complex organisations of membranes, organelles, solid-gel, solid-liquid and liquid-gel interfaces. Here proteins play key roles, the complete understanding of which requires a knowledge of their structures and interactions with surrounding substances, and a determination of how these factors govern functional relationships. With problems of this difficulty, it is usual to isolate individual components and try to understand their behaviour in simple, model systems. A classic example of this sort of approach is to be found in the study of the surface chemistry of proteins, where a vast amount of work has been concerned with the behaviour of one protein at an interface. We will examine some of the important conclusions of such work here, and in particular, we propose to analyse the relationship between the native structure of a protein and its interfacial configuration and properties.

The tremendous advance in the understanding of protein structure in solution and the crystalline state during the last two decades unfortunately has not been parallelled by a similar increase in the knowledge of their interfacial configuration. Protein chemistry in general has benefited enormously from the development and application of powerful physical methods, while the limited number of experimental techniques which are available severely restricts the study of proteins at interfaces. Present knowledge stems mainly from the measurement of surface pressure, surface potential and the rheological properties of spread and adsorbed films at the air-water and oil-water interfaces, although some work has been carried out at solid-liquid interfaces. The

1

state of current research is displayed and discussed in the recent reviews of Loeb[1] and James and Augenstein[2]. Certain generally acceptable conclusions have been drawn from this extensive work on the interfacial properties of proteins. For instance, it is thought that at the air–water interface, and probably also at the oil–water interface, a protein can form two types of film:-

(a) A dilute film, in which all the molecules are in the same extensively unfolded state.
(b) A concentrated film, which may contain only native and unfolde molecules, or molecules in many different degrees of unfolding[2].

Three main factors will dictate whether a protein forms a dilute or concentrated film at an interface. They are:-

(i) The decrease in surface free energy which will result if the protein unfolds.
(ii) The forces which will act to maintain the protein in its native configuration at the interface.
(iii) The surface pressure against which the molecule has to expand in order to unfold.

General considerations such as these may now be applied to particular cases. We shall show that the characteristics of certain proteins at the air–water interface can be related to differences in their tertiary structures. To illustrate this we will refer to the surface properties of β–casein, bovine serum albumin and lysozyme.

The influence of primary and quaternary structure on surface properties was investigated by the technique of selective chemical modification. For this, some derivatives of β–casein were studied at the air–water interface. In these molecules, a controlled variation of charge and hydrophobic character enabled an estimate to be made of the effect of aggregation and electrostatic repulsion on the interfacial properties of the protein.

EXPERIMENTAL
Materials
Lysozyme (3 x crystallised)was obtained from Calbiochem, Los Angeles. Crystalline bovine serum albumin was supplied by the Armour Pharmaceutical Company, Eastbourne, England. Acid casein was prepared from bulk Ayrshire milk by acidification to pH4.6 at 25°. Crude β–casein was isolated from the acid casein by urea fractionation[3], and subsequently purified by chromatography on DEAE cellulose[4].

For this study the succinyl, acetyl, n-butyryl, n-hexanoyl and n-decanoyl derivatives of β–casein were prepared. Most of these derivatives have been described by Hoagland[5,6]. Satisfactory compounds were prepared by reacting β-casein with the appropriate anhydride in either phosphate buffer (2% Na_2HPO_4, pH8.0) or dimethyl sulphoxide. Exhaustive dialysis against distilled water was necessary to remove excess reagents. The derivatives were isolated by lyophilisation.

Methods

The purity of the β-casein and its derivatives was examined by gel electrophoresis.

Polyacrylamide gel electrophoresis was performed using a 10% gel slab in Tris-EDTA borate buffer at pH8.5, 400v and 50mA.

Starch-gel electrophoresis was carried out in urea-Tris-citrate buffer pH8.6, with a 12% gel[7].

The extent of modification of β-casein was measured by the loss of ninhydrin colour in the derivatives compared to β-casein[8]. Substitution of protein hydroxyl groups was investigated by an alkaline hydroxylamine procedure[9].

Experimental conditions for the preparation of the β-casein derivatives were chosen so that in all cases, substitution of available amino groups was greater than 90% by the ninhydrin reaction. The amino groups involved were principally the ε-NH_2 of the lysines and the terminal amino group of the protein.

It was found that with the method of modification described, some 15-30% of available serine, threonine and tyrosine hydroxyl groups were also substituted.

Protein concentrations were measured by a semimicro Kjeldhal method.

Spreading Experiments at the Air-Water Interface

The relationship between surface concentration (c) and surface pressure (π) was determined in two ways.

(1) A pressure-area (π-A) isotherm was obtained by reducing the surface area occupied by a fixed mass of protein.

(2) The pressure-concentration (π-C) isotherm was found by spreading an increasing amount of protein onto a fixed surface area.

A Langmuir-Adam surface balance with a fused silica trough and teflon barriers was used for these measurements. The spreading area was 331 cm^2. The force on the mica float was measured with a torsion wire system which enabled the surface pressure to be determined to an accuracy of ±0.1 dyne cm^{-1}. The protein was spread on phosphate or glycine buffer substrates, which were maintained at $25^\circ \pm 0.5^\circ$. The buffers were as follows:-

1.   5.65mM Na_2HPO_4, 3.05mM NaH_2PO_4, 0.08M NaCl, pH7.0, I = 0.1.
2.   0.115M glycine in 0.1M sodium hydroxide, pH10.0, I = 0.1.
3.   0.76mM glycine in 12.4mM sodium hydroxide, 0.088M sodium chloride, pH11.7, I = 0.1.

The buffers were made up from spectroscopic grade sodium chloride and analytical grade salts. Water for surface work was double distilled, deionised and redistilled from alkaline permanganate

For the π-A isotherms, approximately 1.5 x 10^{-2}mg of protein was spread from a 0.03% solution in the substrate buffer onto an area of 331 cm^2. The spreading method was similar to that of Trurnit[10]. A clean glass rod was wetted for 15 mm along its length by gently dipping it into the subphase. It was then clamped with 5mm of its length dipping below the interface. Protein solution from an Agla glass micrometer syringe was run onto the glass rod 10mm above the surface. The rod was then removed from the surface and the film allowed to stand for 15 minutes. For lysozyme, a period of 3 hours was allowed for equilibration. After this the film was compressed and the corresponding area measured. Measurements were made continuously up to a pressure of 1 dyne cm^{-1}. At higher pressures, 10-20 minute intervals were allowed between readings, so that the film could equilibrate.

π-C Isotherms were obtained by spreading successive aliquots of from 0.5-2.0 x 10^{-2}mg of protein onto a constant area of 331cm^2. After the addition of each aliquot, the film was allowed to equilibrate as before, then the surface pressure was measured. The spreading solution and technique were as previously described.

Compressibility data were obtained from the pressure-area isotherms by plotting $-\frac{1}{A}\frac{dA}{d\pi}$ against A.

Adsorption Experiments at the Air-Water Interface

The (π-t) isotherm, i.e. the change of surface tension with time as protein adsorbed to an initially clean surface was measured using a Wilhelmy plate suspended from a du Nüöy torsion head. The volume of buffer solution in a circular trough of surface area 47cm^2 was reduced to 100 ml by suction at the surface. 10ml of subphase were removed and 10 ml of protein solution (0.01%) injected slowly under the surface, so that the two solutions gently mixed. Measurements of surface pressure were then made over a period of 2-3 hours.

I. THE RELATIONSHIP BETWEEN NATIVE PROTEIN STRUCTURE AND INTERFACIAL CONFIGURATION

The wide variety of surface behaviour exhibited by proteins at the air-water interface is well illustrated by the properties of lysozyme, bovine serum albumin and β-casein. The π-A and π-C isotherms of the three proteins are displayed in Figure 1.

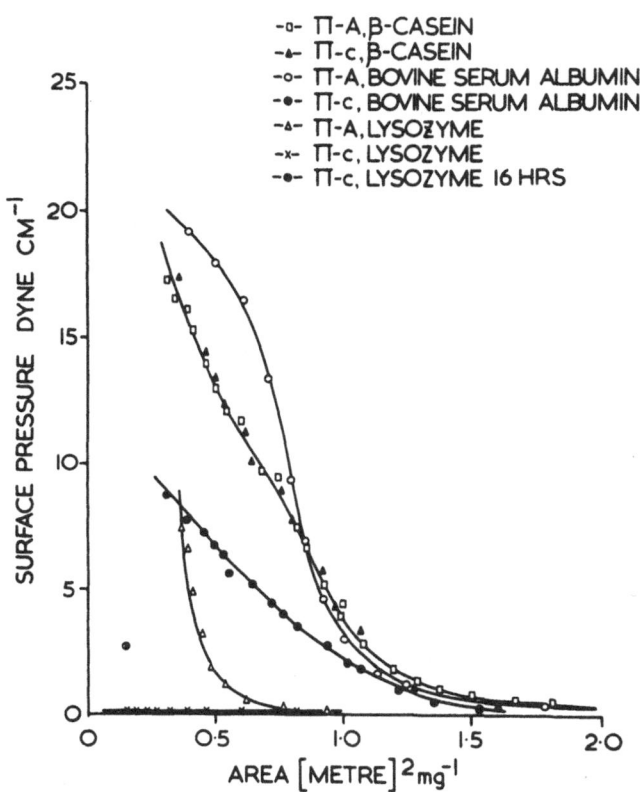

FIGURE 1:   π-A and π-C isotherms of β-casein, bovine serum albumin
            and lysozyme on phosphate buffer (pH7) subphase at 25°.

At minimum compressibility, the π-A isotherms of bovine serum
albumin and β-casein have areas of 0.85 and 0.90m^2 mg^{-1} respectively.
Such areas are regarded as being characteristic of dilute films[11].
Furthermore, the limiting area of the π-A isotherm for bovine serum
albumin differs very little from those obtained when the protein is
spread on concentrated salt solutions, or when spreading solvents
are used[12,13]. Both of these methods favour the formation of dilute
films. It is, therefore, reasonable to suppose that the π-A films
of β-casein and bovine serum albumin are dilute films containing
only unfolded molecules.

     On the other hand, even when a film of lysozyme was left for
two hours to equilibrate before compression, the subsequent π-A

isotherm only gave an area at minimum compressibility of
approximately $0.35 m^2 mg^{-1}$. This is consistent with previous work
on lysozyme[14], where it was found that the protein could not be
spread completely on substrates of low salt concentration. The
small limiting area found for lysozyme probably reflects the presence
of a concentrated rather than a dilute film at the air–water
interface[15], together with an appreciable loss of protein to the
subphase[14].

Considerable differences are evident in the π-C isotherms of
these proteins (Figure 1). The π-C isotherm of β-casein is in
agreement with its π-A isotherm, which strongly suggests that the
π-C isotherm of β-casein is also formed by a dilute film[16].

Negligible surface pressures were recorded when the standard
equilibration conditions were used for a π-C isotherm of lysozyme.
However, after sixteen hours at a nominal surface area of $0.14 m^2 mg^{-1}$,
a pressure of 2.7 dyne cm^{-1} was attained. Bovine serum albumin
showed an intermediate behaviour. The π-A and π-C isotherms agreed
up to a surface pressure of about 2 dyne cm^{-1}, but above this there
was a marked divergence[17]. Films of pepsin[18] and trypsin[19] exhibit
π-A and π-C isotherms with similar properties, although these were
not formed under the same spreading conditions as the protein films
we describe. Enzyme activity can be recovered from the π-C films
of pepsin and trypsin, and this is regarded as evidence for the
presence of native or incompletely unfolded molecules[2]. By analogy,
it is therefore probable that bovine serum albumin also forms a concentrated π-C film.

For lysozyme however, the π-C film generates so little surface
pressure that both the surface concentration and hence the condition
of the molecules at the interface are uncertain. However, other
facts we will present make it reasonable to assume that lysozyme
may also form concentrated films at the air–water interface.

The π-t isotherms of bovine serum albumin, lysozyme and β-
casein are shown in Figure 2. These approximately reflected the
maximum pressures attained in the π-C films. This was to be
expected, since the adsorption of an increasing amount of protein
to a constant surface area closely resembles the method of formation
of a π-C film. At this concentration, β-casein lowers the surface
tension more than bovine serum albumin, while lysozyme has little
effect. Yamashita and Bull[20] have recently shown that adsorbed
films of lysozyme contain native or near native molecules. The
long term ageing effects shown by the gradual change in surface
pressure in adsorbed films of lysozyme and bovine serum albumin
(Figure 2) may then be attributed to the slow unfolding and re-
orientation of native molecules in the adsorbed surface film.
This is a consequence of the non-equilibrium nature of concentrated
films, for since a protein can give both a dilute and a concentrated

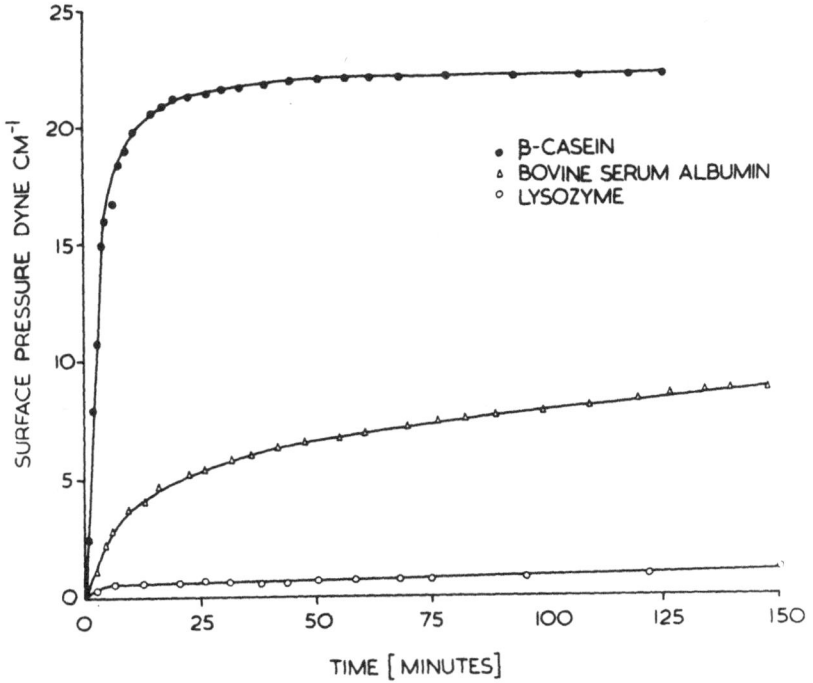

FIGURE 2: π–t Isotherms of β–casein, bovine serum albumin
and lysozyme at the air/phosphate buffer (pH7)
interface at 25° 0.001% protein concentration
in bulk phase. After 6 hours a surface pressure
of 3.3 dyne cm⁻¹ was obtained for lysozyme.

film at the same concentration and on the same subphase, it follows
that one type of film is metastable. As it is possible to convert
a concentrated film to a dilute film by expansion, but impossible
to perform the reverse process, then the dilute film is stable. In
this context it is noteworthy that β–casein π–t films show no
appreciable ageing effects.

Our conclusions concerning the types of films formed by these
proteins are summarised in Table I, together with some structural
parameters of the molecules.

We have already outlined the criteria which govern whether a
molecule will readily unfold at an interface. One of these was the

TABLE I

	β-Casein	Bovine serum albumin	Lysozyme
π-A film	Dilute	Dilute	Concentrated
π-t or π-C film	Dilute	Concentrated	Concentrated
Molecular weight %age α-helix	25,000[21] 10[23]	65,000[22] 47[24]	14,300[22] 23[24]
S-S bond content per molecule	0[25]	17[22]	4[22]
Hydrophobicity	1320[26]	1120[27]	970[27]

surface pressure against which the protein has to spread. A consideration of this factor reveals that unfolding will be favoured in a spread π-A film as opposed to a spread π-C film, since in the former all molecules will enter a surface which is at low or zero surface pressure. In the latter, successive aliquots of protein are required to spread against the increasing surface pressure of molecules already present. Therefore, for many globular proteins of which bovine serum albumin is a typical example, adsorbed films will be of a concentrated form, and will bear little relationship to the dilute π-A spread film except at very low surface pressures.

The lack of structural barriers to unfolding of the molecule was suggested as an explanation for the pronounced surface activity of high density apolipoprotein[28]. The differing surface activity of β-casein, bovine serum albumin and lysozyme can also be attributed to a variation in this factor. An understanding of this necessitates a brief consideration of the structure of these proteins. β-Casein is now thought to be a predominantly random coil molecule with no disulphide bonds, while both lysozyme and bovine serum albumin are globular proteins containing secondary structures, their polypeptide chains being crosslinked by disulphide bonds (Table I).

It is not obvious from Table I why bovine serum albumin should be less resistant to surface denaturation than lysozyme, since it contains about the same number of disulphide bonds per residue as lysozyme, and a greater helical content. However, studies on the denaturation of these proteins in bulk solution suggest that the barriers to unfolding in lysozyme are greater than in bovine serum albumin. For example, lysozyme at neutral pH is very stable to heat[29], while bovine serum albumin is not[30]. Bovine serum albumin is denatured more readily by moderate concentrations of guanidine hydrochloride than lysozyme[31] and unfolds at a pH slightly below its isoelectric point[30]. Thus one of the main factors determining the surface behaviour of these two proteins appears to be the strength of the forces maintaining tertiary structure. If these forces are weak, then the probability of unfolding is high. β-Casein, which has little tertiary structure, consequently has high surface activity. When tertiary structural content is low or absent, then secondary structure and intramolecular cohesion should become important. In this respect, the work of Malcolm[32] and Loeb and Baier[33] suggests that even in dilute films proteins and polypeptides may maintain their secondary structure.

If helical regions are not disrupted at the surface, the activation energy for unfolding will not be related to secondary structure. Proteins usually contain less secondary structure than synthetic polypeptides, this structure being distributed in discontinuous segments throughout the polypeptide chain. It is therefore quite feasible for major reorientations of protein poly-peptide chains to occur at the surface without disrupting the helical portions.

Solubility may be critical in determining the surface isotherm behaviour, though not necessarily the type of film that is formed. If a molecule does not unfold readily when subjected to surface forces, it will not stay long at the interface and will tend to dissolve back into the subphase. With low solubility, this tendency will be reduced, so that the molecules remain longer at the interface. The probability of unfolding then becomes greater, as desorption is less likely. Solubility considerations will, therefore, affect the time a molecule remains in the interface, but not the probability per unit time of its unfolding there. This is a property of the molecule itself.

In Table I we have included the hydrophobicity parameter derived by Bigelow[27]. This has been calculated for many proteins on the basis of their defined apolar amino acid content. It might appear from the table that proteins with a high content of apolar amino acids are more efficient at lowering the surface tension than hydrophilic proteins, provided that tertiary structural constraints are low. This has been suggested as an additional reason for the marked surface activity of high density apolipoprotein[28]. To examine this sort of effect we have systematically altered the hydrophobicity and charge of β-casein by selective chemical modification.

II. THE PROPERTIES OF SOME DERIVATIVES OF β-CASEIN AT THE AIR-WATER INTERFACE

The π-A isotherms of β-casein and its acyl derivatives are shown in Figure 3. An interesting feature of these isotherms is the expansion apparent at moderate and high surface pressures as the length of the n-alkyl side chain on the modifying group increases. Plots of compressibility against area for acetyl, n-butyryl and native β-casein (Figure 4) gave bimodal patterns rather than the

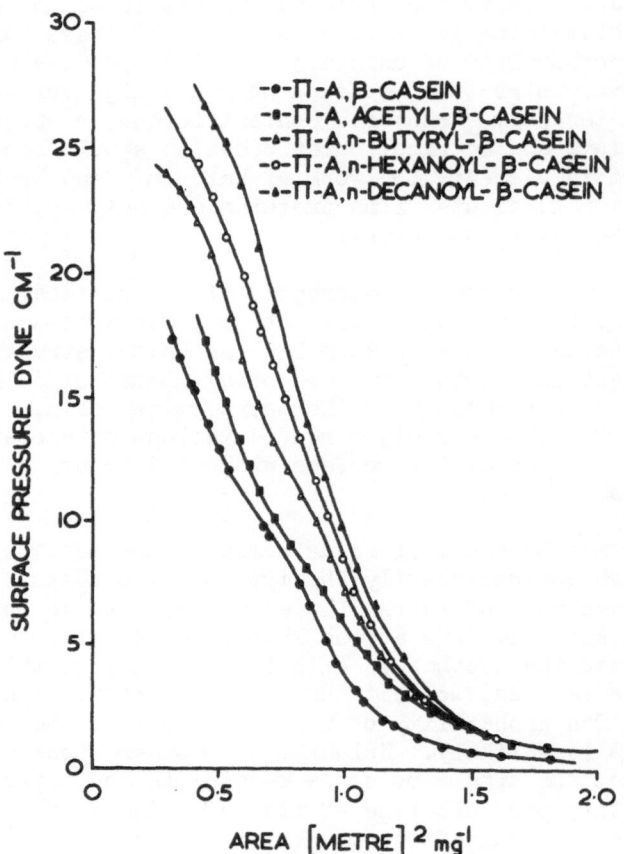

FIGURE 3:   π-A isotherms of β-casein and its acyl derivatives on
            phosphate buffer (pH7) subphase at 25°.

single minimum normally obtained from a protein. This was caused
by an irregularity in the π-A isotherms between $0.5-1.0 m^2 mg^{-1}$.
These effects were reproducible and appeared to be unique to
β-casein and its short chain acyl derivatives. Such behaviour has
not been observed in other proteins studied under comparable compression conditions. However, n-hexanoyl and n-decanoyl β-caseins
did not show this effect.

      In Figure 5, the π-A isotherm of succinyl β-casein at pH7.0 is
compared with those of β-casein at pH 10.0 and pH 11.7. Assuming a
high degree of substitution of the available lysine amino groups,
the charge on the succinyl derivative will be approximately three
times that of β-casein, which at pH7.0 possesses a net negative

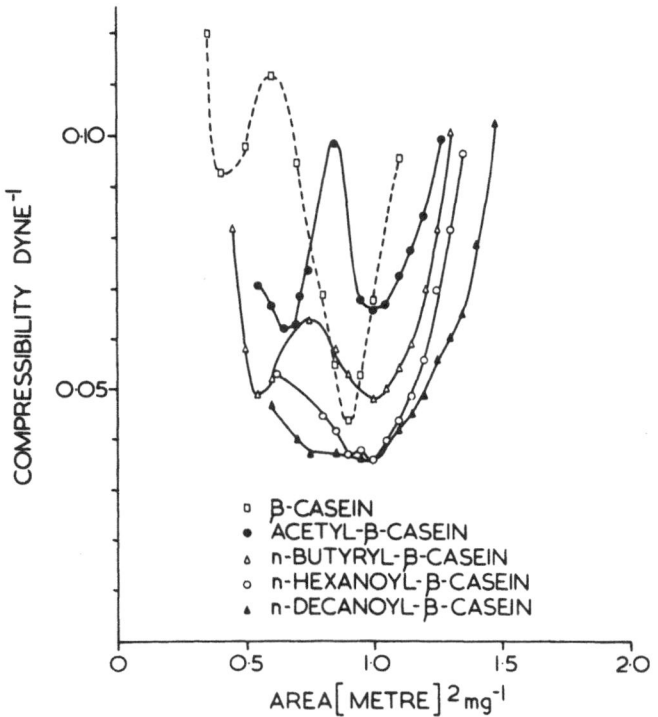

FIGURE 4: Compressibilities of β-casein and its acyl derivatives
on phosphate buffer (pH7) subphase at 25°.

charge of eleven units[34]. Similarly the negative charge on the acyl
derivatives will be about twice that of β-casein at pH7.0. This is
because in the first case the lysine will bear a negative instead of
a positive charge, while in the second the positive charge is
neutralised by the modification.

Although the collapse pressures shown by the π-A isotherms of the
acyl derivatives increase considerably with the chain length of the
modifying group (Figure 3), the π-t isotherms do not show
correspondingly large differences. It is noteworthy however, that
the slow long-term changes in surface tension evident in the π-t
films of n-butyryl, n-hexanoyl and n-decanoyl β-caseins are larger
than those shown by native β-casein and the other derivatives.

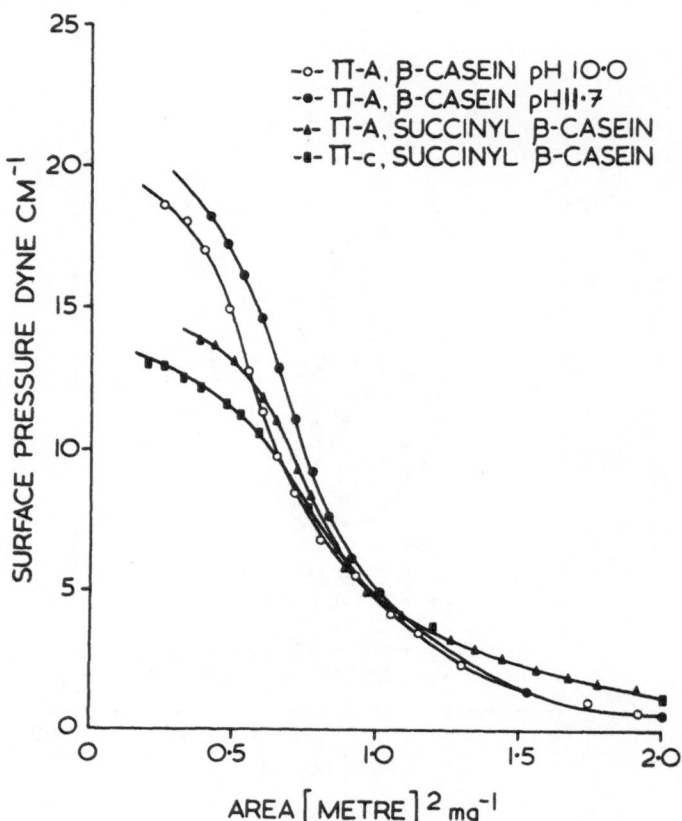

FIGURE 5:   π-A isotherms of β-casein on glycine buffer (pH10.0 and
             11.7) subphases at 25°. π-A and π-C isotherms of
             succinyl-β-casein on phosphate buffer (pH7) subphase at
             25°.

     In Figure 6 the π-C and π-A isotherms of β-casein and its
acyl derivatives are compared. The π-C isotherm of succinyl
β-casein is included, with its π-A isotherm in Figure 5. The π-A
and π-C isotherms of β-casein and its acetyl derivative were
nearly coincident over the whole pressure range measured, while
at low or moderate pressures the same was true for the other
acylated derivatives. For these however, divergence occurred at
high pressure.

     There are two possible explanations for the incremental shift
of the acyl π-A isotherms with the chain length of the modifying
group. Either the extra hydrophobic character imparted by the
hydrocarbon side chains acts to stabilise segments of the polypeptide

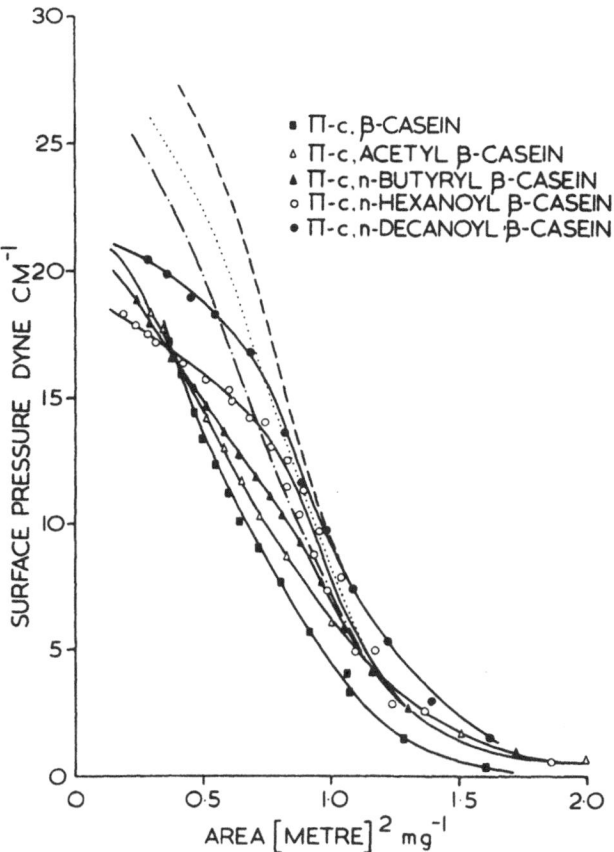

FIGURE 6: π-C Isotherms of β-casein and its acyl derivatives on
    phosphate buffer (pH7) subphase at 25°. The π-A isotherms of
    n-butyryl, n-hexanoyl and n-decanoyl β-caseins are shown as
    dotted lines.

chain against collapse, or the area occupied by the modified lysine
residues at the interface is related to the side chain length. The
first alternative seems to fit the data satisfactorily, since the
expansion of the π-A isotherms of the acyl derivatives is
accompanied by a decrease in compressibility, which may be
associated with an increase in resistance to collapse (Figure 4).
The π-A isotherm of β- casein is more compressible than that of a
typical globular protein such as bovine serum albumin (Figure 1).
This is because of the low degree of interaction between the amino
acid residues in the film, and is in accord with the properties of
a molecular structure seen as a randomly coiled thread with no
covalent and little noncovalent intrachain constraint.

Modification might be expected to reduce the degree of collapse by making the energy of desorption of individual lysine residues larger and by increasing the extent of both inter and intramolecular non-polar interactions. In support of this, Hoagland[6] has found that there is an increase in aggregate size in the bulk phase as the chain length of the modifying group in the acyl derivatives increases. High resolution n.m.r. studies also show evidence of line broadening in the n-butyryl, n-hexanoyl and n-decanoyl derivatives consistent with the presence of strong hydrophobic interactions[55].

It is difficult to determine whether the expansion of the π-A isotherms of the acyl derivatives is in any way due to a change in the area occupied by the various modified lysine residues at the interface. The basic problem is the definition of the area of close packing of a very compressible protein film, and the accurate measurement of its change. If this were feasible, it would be a simple matter to calculate the extra contribution of area per residue supplied by modified lysine groups whose side chains were oriented horizontally in the interface, and compare this with the observed alteration in area of close packing. If, however, collapse is entirely the determining factor in the differences between the π-A isotherms, then this must begin at areas greater than the first minimum in the compressibility curve ($\sim 1.0 m^2 mg^{-1}$).

Chemical modifications of this nature carried out on globular proteins exhibiting well defined limiting areas may well provide more accurate information as to the orientations of the modified lysines and hence to the secondary structure present in the film. We are currently investigating this possibility.

Bigelow[27] has extended the work of Tanford[36] and produced a quantitative basis for assessing the hydrophobicity of amino acids and their side chains. In this respect β-casein has an unusual distribution of amino acid content, possessing 99 residues of high hydrophobicity, between 1.5 and 3.0 k. cal/residue on Bigelow's scale, only 35 residues of moderate hydrophobicity (between 0 and 1.5 k cal/residue) and 75 polar amino acids with hydrophobicity defined as zero. Langmuir and Waugh[37] have suggested that when protein films were compressed, polar amino acids collapsed out at low surface pressure and apolar residues at high pressure. Thus when a β-casein film is compressed, it follows that there could be a range of pressures corresponding to the region of collapse of the residues of moderate hydrophobicity; since these are few there will be a considerable change in surface pressure over a small area in the isotherm. This could be the reason for the irregularities in the π-A isotherms of β-casein and its short chain acyl derivatives. In the n-hexanoyl and n-decanoyl β-casein isotherms, where these irregularities are not observed, the degree of interaction between residues is much larger so that the idea of considering collapse in terms of hydrophobicities of individual amino acids may not be valid. The order of amino acids in the sequence is an additional factor to be considered.

The influence of charge on the π-A isotherm is most marked at high areas. In this region the isotherms of the acyl derivatives coincide, and resemble that of β-casein at high pH (Figures 3,5). Films of both the succinyl and acyl derivatives exert a higher pressure at high areas than does the native protein.

Payens[38] has derived an equation for the electrostatic contribution to the film pressure of a monolayer of charge density σ esu cm^{-2} on a substrate of equivalent electrolyte concentration C (equation 1).

$$(1) \quad \pi_{el} = 6.1 \ C^{\frac{1}{2}} \left[\cosh \left\{ \text{arc sinh } 28.10^{-6} \sigma C^{-\frac{1}{2}} \right\} -1 \right]$$

Figure 7 shows the experimental and calculated portions of the π-A isotherms of β-casein and its derivatives at high areas. The assumption is made that the difference between the β-casein isotherm and the others is solely due to charge effects. The charge on β-casein at pH7.0 was taken as 11 negative units, and those on the acyl and succinyl derivatives as 19 and 27 negative units respectively. These latter figures allow for slightly less than 100% substitution, as well as incomplete ionisation at pH7.0. Agreement between the calculated and experimental isotherms was fairly satisfactory, so that it is probable that the increase in pressure exhibited by the modified caseins at high area can be ascribed entirely to electrostatic effects. From Figure 6 it also seems that the charge on β-casein at high pH closely resembles that on the acyl derivatives.

Although at areas lower than about 1.3m^2 mg^{-1}, effects other than electrostatic forces become important in determining the π-A isotherm, the lower collapse pressures of the β-casein films at high pH may also be related to the higher charge preventing cohesive interactions between the amino acid residues. The low collapse pressure of the succinyl β-casein film (Figure 6) may also be attributed to this, since the negative charge here is also greatly increased. In addition, the modified lysines are given increased polar character by the substituent carboxyl group.

Since there is a clear analogy between the method of formation of adsorbed and π-C spread films, we have made a quantitative comparison between adsorption pressures in τ-t isotherms and the pressures obtained from π-C isotherms. This can be done for the β-casein derivatives by assuming that the rate of arrival of molecules at an interface in an adsorption experiment is given by an irreversible diffusion equation (equation 2)[39].

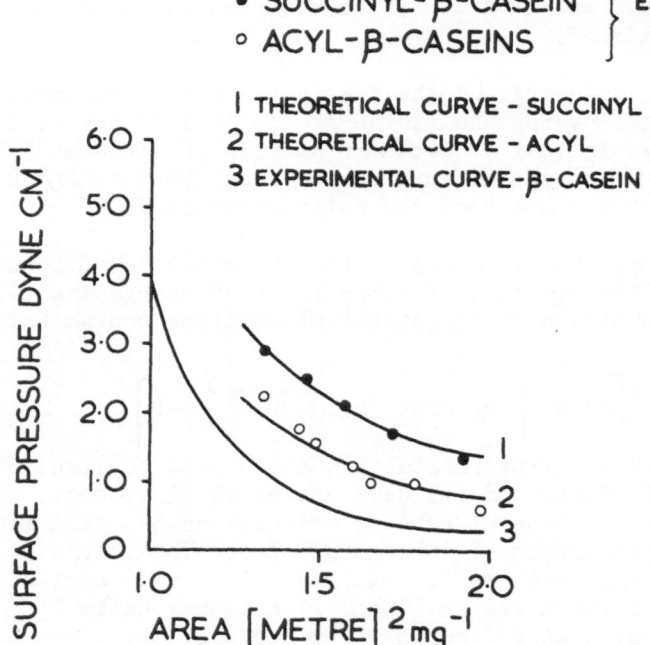

FIGURE 7: Experimental and calculated portions of the π-A
          isotherms of β-casein, succinyl β-casein and acyl
          β-caseins on phosphate buffer (pH7) subphase at 25°.

$$(2) \quad n = 2n_o \sqrt{\frac{Dt}{\pi_c}}$$

(n is the surface concentration of the material after time t, D is
the diffusion coefficient, n_o the bulk concentration of the protein
and π_c = 3.142.)

The relationship between the surface pressure and surface concentration in the calculated π-t isotherms is derived from the π-C
data. In Figures 8 and 9 the experimental isotherms are compared
with the curves calculated on this basis. It can be seen that
agreement between calculation and experiment is fairly good for
most of the modified proteins. This would lend some support to the
idea that there is a similarity between π-C and π-t films. From
the work of MacRitchie and Alexander[40] it might be considered dubious
to assume that the rate of adsorption was determined only by diffusion.
However, these workers were concerned with globular proteins, while
β-casein is disordered. It has been shown recently[41] that the rate

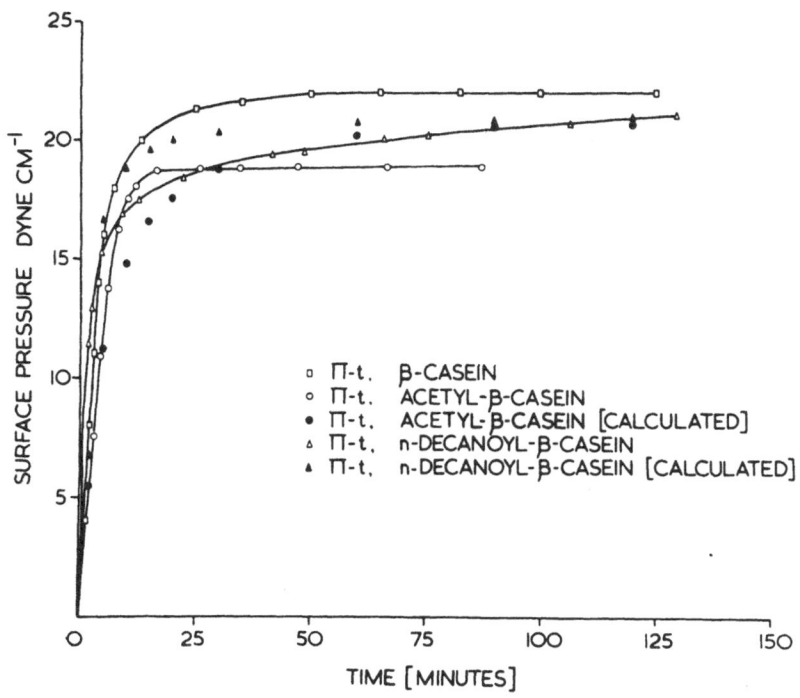

FIGURE 8: π-t Isotherms of β-casein, acetyl β-casein and n-decanoyl
β-casein at the air/phosphate buffer (pH7) interface at
25° 0.001% protein concentration in bulk phase.
Calculated and experimental points are shown.

of adsorption of disordered proteins to an interface more nearly
corresponds to that predicted for an irreversible diffusion process.

The difference between the π-A and π-C isotherms of n-butyryl,
n-hexanoyl and n-decanoyl β-caseins at high surface pressures may be
due to the highly aggregated nature of these proteins. Previous
work[16] has shown that the divergence between the π-A and π-C
isotherms of K-casein at high pressures was due to the presence of
disulphide bonds which help to maintain the highly aggregated nature
of this protein. In ultracentrifuge studies we have demonstrated
that n-butyryl, n-hexanoyl and n-decanoyl β-caseins form stable
polydisperse aggregates which do not dissociate on dilution[35],
while β-casein, though forming a strongly aggregating system,
dissociates into monomers at low temperatures and on dilution[21].
However, aggregation cannot be the reason for the divergence of the
π-A and π-C isotherms of succinyl-β-casein at high pressures since
this is a monomeric species at 25°[5]. Here there is possibly some
loss to the subphase from the π-C film by collapse due to its
enhanced polar character.

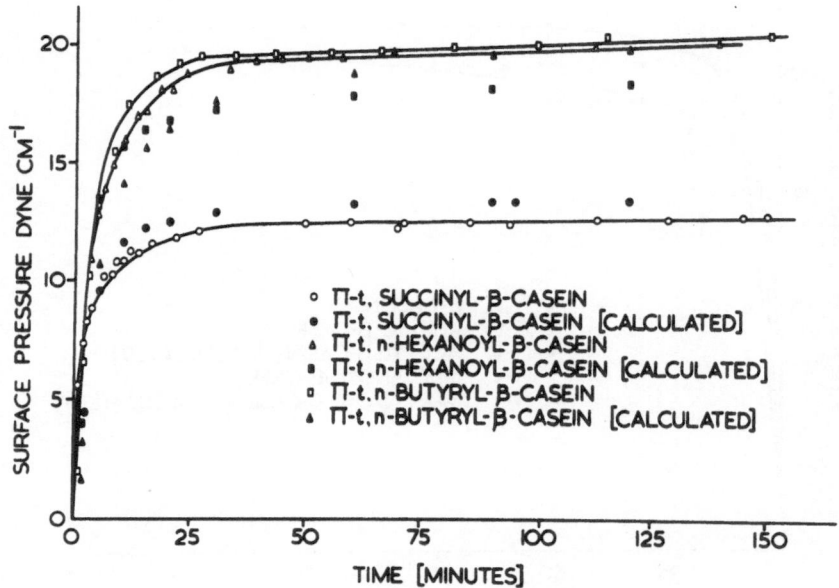

FIGURE 9:   π-t Isotherms of succinyl n-butyryl and n-hexanoyl
            β-caseins at the air/phosphate buffer (pH7) interface
            at 25°. 0.001% protein concentration in bulk phase.
            Calculated and experimental points are shown.

        In the π-A film, it appears that the aggregates in the acyl
derivatives will unfold at the surface, since if this were not so
the π-A isotherms of these proteins would be less expanded than
that of β-casein. This situation does in fact occur between monomer
and aggregated γ-globulin[42].

        However, when n-butyryl, n-hexanoyl and n-decanoyl β-caseins
are spread against high surface pressures, the aggregates appear to
resist unfolding, so that the π-C and adsorbed films are of the
concentrated rather than the dilute type. The greater, though
gradual changes in the surface tension of adsorbed π-t films of
n-butyryl, n-hexanoyl and n-decanoyl β-caseins when compared with
acetyl, succinyl and native β-casein tend to confirm this.

        It will be useful at this point to recapitulate some of the
main conclusions about the relationship between surface properties
and protein structure.

(a) A disordered protein is more surface active than a globular protein, and will be more likely to form a dilute film. It will adsorb faster to an interface, unfold more easily, and cause a larger change in interfacial tension. Disorder in this context implies a lack of intrachain restraints, not necessarily an absence of secondary structure.

(b) We have demonstrated the influence of primary and quaternary structure on the surface properties of β-casein, by changing the chemical composition and aggregation behaviour of the protein. Caseins modified with long chain acyl groups exhibited expanded π-A isotherms, characterised by high collapse pressures. Strong apolar interactions probably occur in these films. In contrast, the π-A isotherm of succinyl β-casein shows no expansion and a low collapse pressure. Proteins which are strongly aggregated will be more likely to form concentrated films than will monomeric species.

It has been suggested that proteins present at biological interfaces possess unique surface properties[28]. These might possibly be similar to β-casein in that they would have few tertiary structural restraints rather than a globular conformation. Such proteins, because of their high surface activity, would be the most likely to be found at, and remain at an interface. The higher collapse pressures found for the π-A isotherms of β-caseins modified with long chain acyl groups is reminiscent of the increase in pressure above the collapse pressure of the individual components, found when lipids[38,43] or detergents[44] interact with proteins at the air-water interface. It has been proposed that the properties of these complexes are due to the intimate association of lipids and proteins in surface aggregates or mosaics. Such micellar protein aggregates may also occur in the n-hexanoyl and n-decanoyl films at high pressures in the air-water interface. We consider that certain chemically modified proteins at interfaces may display properties closely analogous to those of lipid-protein complexes, so that a study of these single protein models could perhaps clarify the understanding of more involved systems, where such structures and interactions are now thought to be present[45].

ACKNOWLEDGMENTS
   The authors thank Mr. Donald Adams and Mr. Peter Miller for expert technical assistance at all stages of this work. Thanks are also due to Miss J. Greatorex for preparation of the manuscript.

REFERENCES

1. G.I. Loeb, Surface Chemistry of Proteins and Polypeptides, U.S. Naval Research Laboratory, Report 6381 (1965).

2. L.K. James and L.G. Augenstein, Advan. Enzymol., 28 (1966) 1.

3. R. Aschaffenberg, J. Dairy Res., 30 (1963) 259.

4. M.P. Thompson, J. Dairy Sci., 49 (1966) 792.

5. P.D. Hoagland, J. Dairy Sci., 49 (1966) 783.

6. P.D. Hoagland, Biochemistry, 7 (1968) 2542.

7. R.G. Wake and R.L. Baldwin, Biochim. Biophys. Acta, 47 (1961) 225.

8. E.W. Yemm and E.C. Cocking, Analyst, 80 (1955) 209.

9. A. Gounaris and.G.E. Perlmann, J. Biol. Chem., 242 (1967) 2739

10. H.J. Trurnit, J. Colloid Sci., 15 (1960) 1.

11. H.B. Bull, Adv. Protein Chem., 3 (1947) 110.

12. P. Joos. Mededel Koninkl Vlaam. Acad. Welenschap Belg., 30 (1968) 3.

13. J.T. Pearson and A.E. Alexander, J. Colloid Interface Sci., 27 (1968) 53.

14. T. Yamashita and H.B. Bull., J. Colloid and Interface Sci., 24 (1967) 310.

15. K. Hamaguchi, Bull. Chem. Soc. Japan., 31 (1958) 123.

16. J. Mitchell, L. Irons and G.J. Palmer, Proc. 5th Intern. Congr. Surface Activity, Barcelona, 1968, in press.

17. P.R. Mussellwhite and G.J. Palmer, J. Colloid Interface Sci., 28 (1968) 168.

18. L.G. Augenstein and B.R. Ray, J. Phys. Chem., 61, (1957) 1385.

19. D.F. Cheeseman and H. Schuller, J. Colloid Sci., 9, (1954) 113.

20. T. Yamashita and H.B. Bull, J. Colloid Interface Sci., 27 (1968) 19.

21. T.A.J. Payens and B.W. van Markwijk, Biochim. Biophys. Acta, 71 (1963) 517.

22. B.E. Davidson and F.J.R. Hird, Biochem. J., 104 (1967) 473.

23. T.T. Herskovits, Biochemistry, 5 (1966) 1018.

24. J.A. Gordon, J. Biol. Chem., 243 (1968) 4615.

25. R. Pion, J. Garnier, B. Dumas, P.J. de Koning and P.J. van Rooyen, Biochem. Biophys. Res. Commun., 20 (1965) 246.

26. R.J. Hill and R.G. Wake, Nature, 221 (1969) 635.

27. C.C. Bigelow, J. Theoret. Biol., 16 (1967) 187.

28. G. Camejo, G. Colacicco and M. Rapport, J. Lipid Res. 9, (1968) 562).

29. S. Beychok and R.C. Warner, J. Am. Chem. Soc., 81 (1959) 1892.

30. J.F. Foster, in F.W. Putnam, The Plasma Proteins, Vol. 1., Academic Press, New York, 1960, p.179.

31. B. Jirgensons, Arch. Biochem. Biophys., 39 (1952) 261.

32. B.R. Malcolm, Surface Activity and the Microbial Cell, Society of Chemical Industry, London, 1965, p.102.

33. G.I. Loeb and R.E. Baier, J. Colloid Interface Sci., 27 (1968) 38.

34. T.L. McMeekin, The Proteins, 2 (1954) 389.

35. M.T.A. Evans, and L. Irons, to be published.

36. C. Tanford, J. Am. Chem. Soc., 84 (1962) 4240.

37. I. Langmuir and D.F. Waugh, J. Am. Chem. Soc., 62 (1940) 2771.

38. T.A.J. Payens, Biochim. Biophys. Acta, 38 (1960) 539.

39. H.J. Trurnit, Arch. Biochem. Biophys., 51 (1954) 176.

40. F. MacRitchie and A.E. Alexander, J. Colloid Sci., 18 (1963) 453

41. J. Mitchell, L. Irons and G.J. Palmer, Biochim. Biophys. Acta., submitted for publication.

42. M. Demeny, S. Kochwa and H. Sobotka, J. Colloid Interface Sci., 22 (1966) 144.

43. G. Colaccico, _J. Colloid Interface Sci._ 29 (1969) 346.

44. I. Blei, Molecular Association in Biological and Related
     Systems, Advances in Chemistry series 84 (1968) 149.

45. J.M. Steim, Molecular Association in Biological and Related
     Systems, Advances in Chemistry series 84, (1968) 259.

THE INTERACTION OF CALCIUM WITH MONOLAYERS OF STEARIC AND OLEIC ACID[1]

Jon Goerke[2], Helen H. Harper, and Michael Borowitz

Cardiovascular Research Institute and Department of
Medicine, University of California Medical Center,
San Francisco, California 94122

ABSTRACT

The interaction of monolayers of stearate and oleate with
Ca^{++} and H^+ have been studied in a Tris-buffered system where
trans-surface potential (ΔV) and surface accumulation of Ca^{45}
per mole lipid (θ) have been studied. ΔV-area data suggest that
the orientation of the carbonyl dipole is the same for both
lipids when the monolayers are close-packed. θ is higher for
stearate than for oleate at bulk calcium concentrations in the
range 10^{-7} to 10^{-4} M, but for both lipids it is far greater than
would be expected from a simple mass-law relationship using the
bulk affinity constant. Both the titration curves and the Ca^{45}
accumulation data can be fitted reasonably well by a Donnan model
which takes monolayer charge into account.

INTRODUCTION

The surface chemical properties of fatty acid monolayers
have been the subject of many studies, largely because these
compounds can be obtained in reasonably pure form and are among

1. Supported in part by USPHS grants HE-06285, HE-5251 and FR-00122
2. American Heart Association Established Investigator 1967-1972

the simplest agents known to lower air/water surface tension.
This simplicity, while somewhat deceiving, has often been exploited
so as to verify a new equation of state for the surface. Another
group of studies, this among them, seeks to use such monomolecular
arrays of charges as models of cell membranes, in the hope that
ionic behavior near these structures will approximate that near the
cell. The analysis that follows draws heavily on Danielli (1,2,3)
but could as easily have followed the Stern treatment (4).

<h2 style="text-align:center">METHODS</h2>

Air/water surface tension, surface area, trans-surface
electrical potential (ΔV) and surface beta-radioactivity were
monitored in a Langmuir-Wilhelmy surface balance at 24.5 $\pm$ 1°C
with Ca^{45} in the subphase and either stearic or oleic acid monolayers on the surface (fig. 1).

Stearic acid, obtained from Applied Science, had a melting
point of 70.0 - 70.1°C and showed only trace impurities on gas-

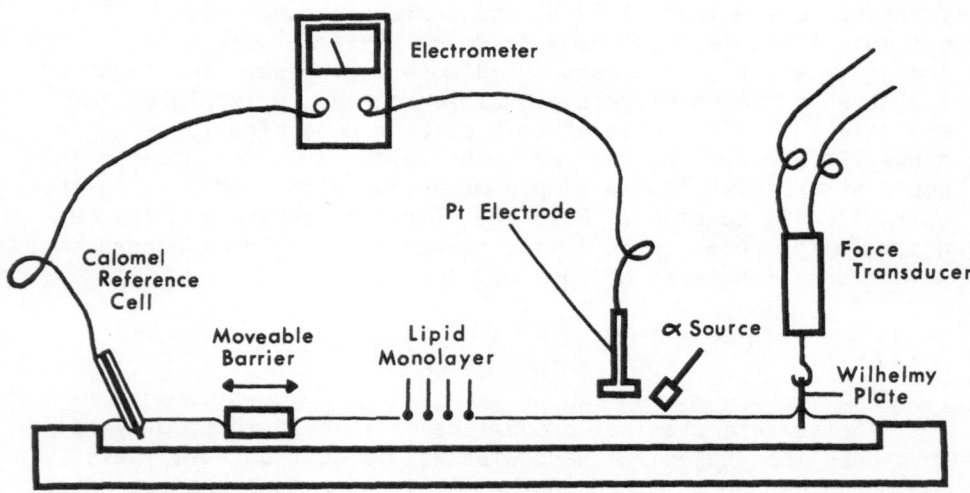

Figure 1. Langmuir-Wilhelmy surface balance.
The Geiger tube for Ca^{45} measurement is not shown, nor is the
clean surface "island" for V_o measurement.

liquid chromatography (GLC). Oleic acid, melting point 16.4°C, was obtained from Applied Science and had less than 1% impurities by GLC. These lipids were delivered to the air-water surface in 20 to 30 μg amounts from redistilled Hexane solution using calibrated 30 to 50 μl pipettes. pH was set to 6.80 $\pm$ .02 or as otherwise indicated using 2 mM Tris buffer (International Chemical and Nuclear Company) titrated with HCl or KOH. Water for solutions was doubly or triply distilled after passage thru a mixed ion-exchange resin.

Ca^{45}, obtained as the chloride from New England Nuclear Corporation at specific activities varying from 11 to 25 $c \cdot g^{-1}$, was reputedly 99% pure. Cold $CaCl_2 \cdot 2H_2O$ (Baker Analyzed Reagent) was used to vary the subphase Ca concentration. Chemical calcium from both sources was checked by flame photometry and the specific activity of the Ca^{45} was reassayed by the method of Goldstein (5).

The trough for the Langmuir-Wilhelmy surface balance was milled from white virgin teflon securely mounted on 0.5 inch stainless steel leaving inner dimensions of 40 × 10.1 × 0.5 cm. A tightly fitting teflon barrier was motor driven at linear rates of 0.1 $cm \cdot sec^{-1}$ between measurements for the radioactivity series, and at rates of 0.05 $cm \cdot sec^{-1}$ (stearate) and 0.1 $cm \cdot sec^{-1}$ (oleate) for the pH potential-area isotherms. The entire trough and measurement assembly was contained within an electrically shielded enclosure. Barrier position was monitored with a potentiometer linked to the barrier mechanism.

Surface tension was measured by the Wilhelmy method using a roughened platinum plate 4 × 0.5 × 0.015 cm flamed just before use. Force on the plate was determined by a force transducer (Statham model G10B) with amplifier (Sanborn model 311A).

A $Radium^{226}$ ionizing electrode (5 μC, U.S. Radium Corp.) was used solely as a removeable alpha-radiation source to ionize the air gap between the surface and a well aged platinum electrode connected to an electrometer amplifier (Keithley model 603). A calomel cell at the other end of the trough acted as the reference electrode. Source and electrode were incorporated in a teflon carriage which could be positioned via a teflon slide either over the experimental solution surface or over an area of the surface isolated by a teflon ring. This latter area could easily be cleaned by aspiration, and surface potential measurements taken over it were used as reference potentials, V_o. The trans-surface potential difference (ΔV) due to the monolayer was then determined by subtracting this clean surface reading from that obtained over the monolayer. V_o was found to vary by as much as 10 mV during a day for aged platinum electrodes and far more for newly made ones.

Prior to each experiment the solution surface was cleaned by aspiration until ΔV was less than 5 mV. The potential titration curves (fig. 3) represent single experiments at each pH.

A Geiger tube (Amperex #18536) and ratemeter (Tracerlab model SC-34) were used to monitor surface beta-radioactivity due to Ca^{45} for 20 minute periods at different values of surface tension. The tube was calibrated for Ca^{45} by summing the readings taken from the surface of a 10 x 10 cm lucite plate which had been sprayed with a Ca^{45} solution and allowed to dry. The plate surface Ca^{45} was subsequently eluted and counted in a liquid scintillation counter to assay the amount of isotope involved. Because of its density, lucite was assumed to give a back-scattering effect close to that of water. It was found that the readings were substantially independent of distance from 0.3 to 0.7 cm over the planar source although tube-to-surface distance was kept constant in all experiments. Surface accumulation of Ca was determined from the difference in counts between the clean solution surface and the monolayer covered surface.

Replicate 50 μl samples of the mixed subphase were counted in a beta-scintillation counter (Packard model 3003) and used with bare surface radiation counts to determine Geiger tube efficiency for each experiment. Counting vials contained PPO and POPOP in 10 ml toluene, 5 ml absolute ethanol, and 50 μl 1N HCl.

Two separate Ca^{45} experiments were performed at each pCa and the results averaged except for those stearate experiments (fig. 4) employing a concentration range of 4 to 8 Gibbs (4 to 8 x 10^{-10} Moles·cm^{-2}) where single experiments were done. In these last experiments, Ca^{45} uptake was measured at 4 to 6 positions over the lipid covered surface and the average of these readings was used in the calculations.

The four electrical outputs developed as above were appropriately scaled with voltage dividers and led into four channels of a 10 inch multi-point chart recorder (Leeds and Northrup Speedomax W) with sampling time of one second per point. Thus each channel was sampled once every four seconds.

Surface pressure (π) (dyne·cm^{-1}) for any given monolayer was taken to be the difference between the surface tension of the clean air-water interface (γ_o) and the tension of the interface with a monolayer present (γ_m):

$$\pi = \gamma_o - \gamma_m$$

When an attempt was made to hold monolayers for 20 minutes at pressures exceeding 5 dyne·cm^{-1} for oleate and 20 dyne·cm^{-1} for stearate it was found necessary to decrease the trough area

periodically. This was interpreted as a movement of lipid out of
the film either into solution, onto the teflon walls and barrier
or into a folded configuration. In order to compensate for this
phenomenon separate π - A isotherms were obtained at successively
faster barrier speeds, starting with 0.025 cm·sec^{-1}. Several
isotherms at the slowest speed giving constant patterns were then
averaged to produce reference stearate and oleate isotherms for the
full range of calcium concentrations employed. The values of π
chosen for obtaining experimental data (1, 2, 5, 10, 20 and 30
dyne·cm^{-1}) were then referred to these isotherms to obtain the
surface area per molecule of the lipid species. Oleate molecular
areas determined in the above manner were used to calculate the
upper of the lower pair of curves in figure 5. The lowest oleate
curve was calculated using the measured area and weight of lipid
applied.

RESULTS

Representative surface tension-area data (fig. 2) has been
plotted as π versus molecular cross-sectional area. The smallest
area obtainable for oleate just prior to film collapse is seen to
exceed that for stearate, presumably reflecting both the inclina-
tion of the oleate hydrocarbon chain enforced by its 9-cis double
bond and the tendency of such a rotating structure to sweep out
larger areas. While there were minor changes in oleate isotherms
with increasing bulk calcium (Ca_b), there was a rather marked
condensation of stearate isotherms in the 1-20 dyne·cm^{-1} region as
has been reported by other authors (6). Figure 2 shows two typical
isotherms obtained at pCa_b of 7 and 3.

Trans-surface potential (ΔV) measurements have been expressed
in terms of the effective surface dipole moment. ΔV is commonly
thought to be the sum of a number of components (3), the most
important of which are due to the average electrostatic potential
(Ψ_o) and the average vertical component of the permanent molecular
dipole (μ)

$$\Delta V = \Psi_o + \frac{12 \pi \mu}{A} \qquad (1)$$

In equation 1 mixed units have been used for convenience: ΔV and
Ψ_o will be in millivolts if μ is expressed in milliDebyes and A in
$Å^2$. After multiplying both sides by $A/12\pi$ we equate the left hand
side with μ'.

$$\mu' = \frac{\Delta V \, A}{12 \pi} = \frac{\Psi_o A}{12 \pi} + \mu \qquad (2)$$

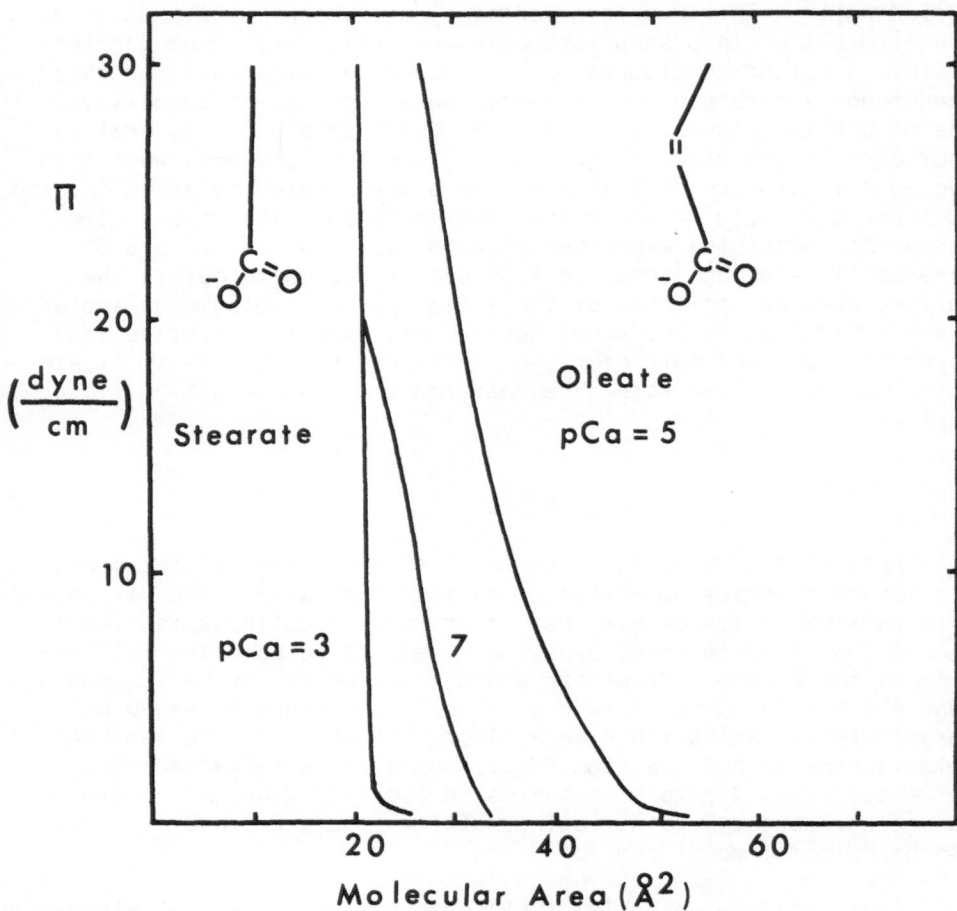

Figure 2.   Pressure-Area isotherms.
The larger molecular area of oleate is evident as is the lowering
of π in stearate monolayers by the addition of Ca.

        μ ought to be nearly identical for both lipids, hence at low
pH where ionization is suppressed, stearate and oleate values for
μ' should be identical if polar group orientation is the same.
Figure 3 shows titration data for the range pH 2-11 at high surface pressures where the molecular orientations would be best
defined.   At the acid end of both data sets μ' approaches +200 mD
which agrees well with the figure of 210 mD given by Goddard and
Ackilli (6) for stearate.

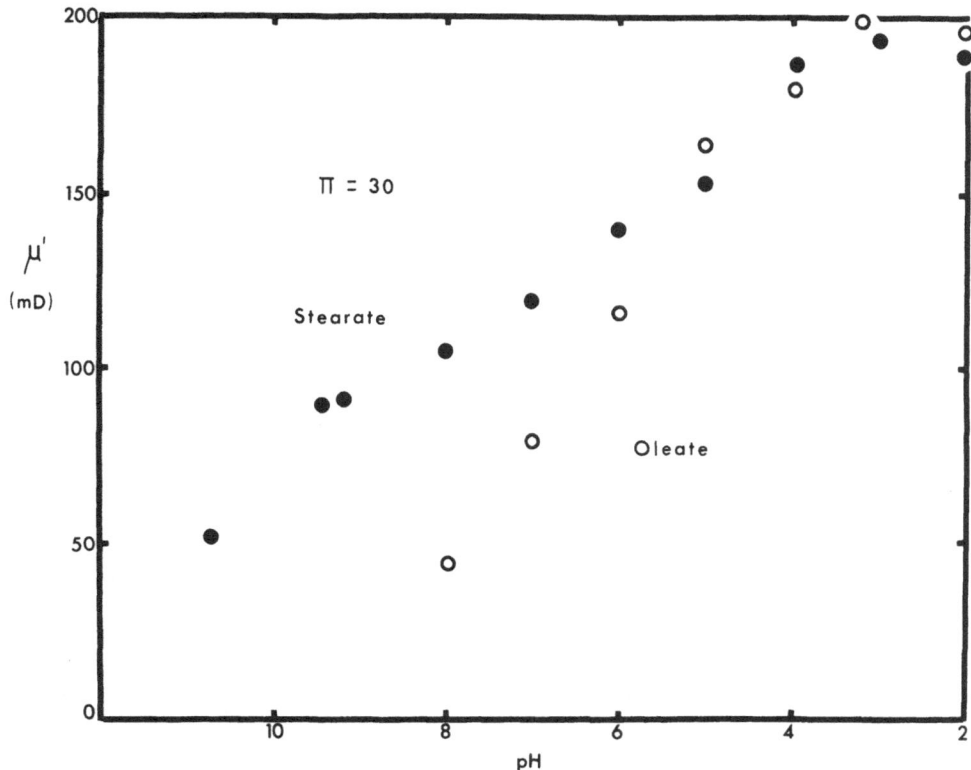

Figure 3.   Potential curves.

$u' \equiv \dfrac{\Delta V}{12} \dfrac{A}{\pi}$.   2 mM Tris buffer was titrated with KOH or HCl.

Equal values at pH 2 presumably reflect the vertical component
of permanent dipole terms.

At higher pH with increased ionization μ' decreases and for
oleate seems to become negative in this system. If the value of
200 mD at pH 2 were due solely to the rather strong carbonyl dipole
(2300 mD), then its oxygen would appear to be inclined downwards
approximately 5° from the horizontal. The rapid decline of stearate
ΔV at pH somewhat greater than 9 was also observed by Goddard and
Ackilli and by Spink (7).

Data from Ca^{45} uptake measurements has been expressed using
the ratio (θ) of excess surface Ca per unit area to the total
applied lipid per surface area. The effect of merely increasing
surface lipid concentration is thereby normalized out. Figure 4 is
a plot of θ versus lipid concentration (L_s) in Gibbs lipid, and

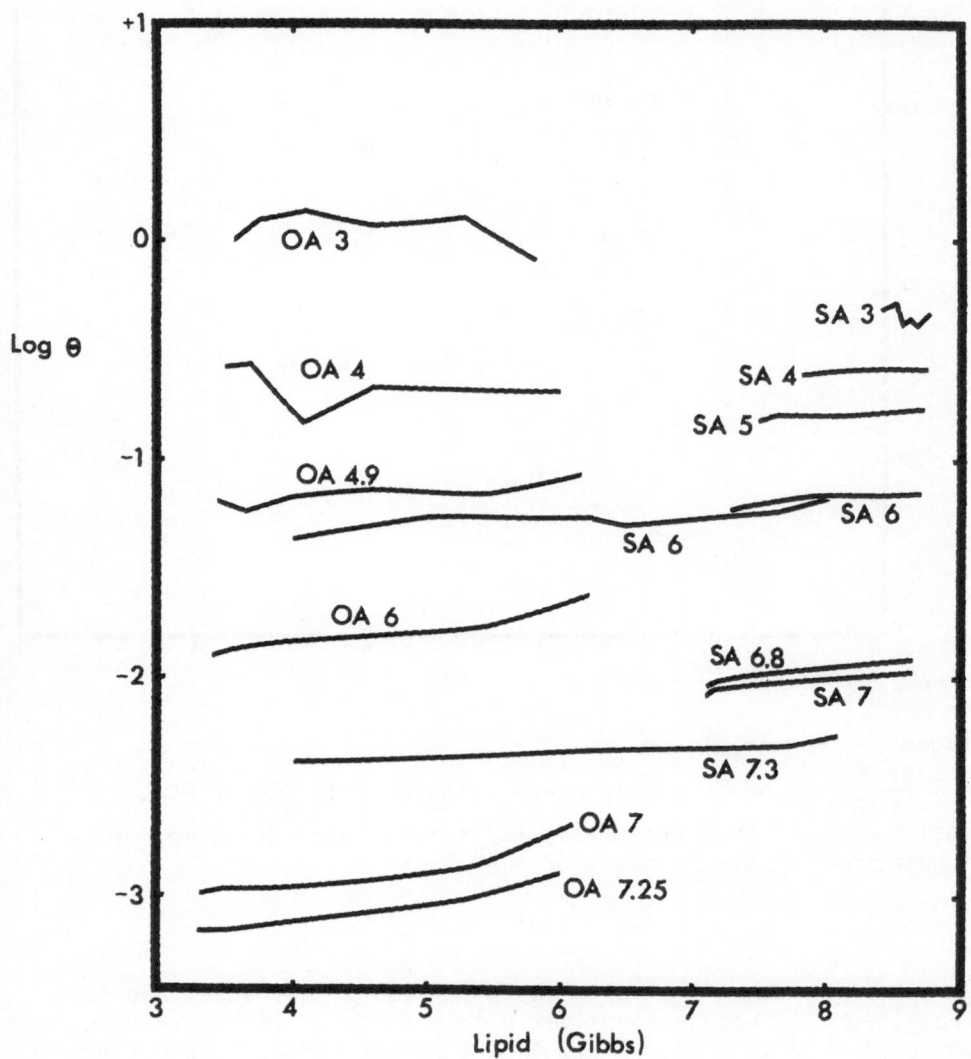

Figure 4. Log (calcium/lipid surface ratio) versus surface lipid concentration. OA = oleic acid, SA = stearic acid, numerical values adjacent to curves are bulk pCa. θ = 0.5 seems to be the limiting value for stearate.

shows the strong dependence of θ on pCa for both stearate and oleate. In the range of L_s studied there was a tendency for θ to increase as the monolayers became more concentrated. At comparable values of L_s and for the lower range of calcium concentrations, θ was appreciably higher for stearate than for oleate. At higher calcium concentrations however, the reverse appeared to be true, although the dispersion of oleate data at the low specific activities used in these runs renders this finding less significant.

In those stearate experiments where Ca^{45} uptake was measured at low L_s there was a considerable variation of counting rates: at 4 Gibbs the area most remote from the spreading region had no increase in activity whatsoever whereas another area had 5 times the clean solution count. Since the physical diameter of the Geiger window was 2 cm and the pattern was stable for over an hour, this was taken as evidence of the presence of large "islands" of stearate in the surface. The effect was still present but markedly diminished at $\pi = 5$ dyne·cm^{-1}.

Figure 5 is a graph of log (θ/(1-θ)) versus pCa_b for stearate and oleate near the lower limits of their cross-sectional areas. Curves at other surface pressures were similarly shaped.

DISCUSSION

These experiments illustrate the markedly increased affinity of fatty acids for calcium and hydrogen ions when these lipids are arrayed in monolayers. Figure 6 reproduces the stearate potential data with superimposed theoretical curves. The steeper right-hand curve, reminiscent of bulk titration curves, is an attempt to predict μ' from the bulk dissociation constant for long chain fatty acids, ignoring any effect the monolayer field might have on counter-ion distribution. Using a K_a of 1.1×10^{-5} (8) to determine surface charge density σ in esu·cm^{-2} one can calculate ψ_0 in statvolts from

$$\psi_0 = \frac{2\ kT}{Ze}\ \sinh^{-1}\left(\frac{\sigma}{c_i}T/2 \left(\frac{500\ \pi}{DRT}\right)^{1/2}\right)\ (3)$$

which is the potential in the plane of monolayer head groups relative to the bulk phase as developed by Gouy (9) and Chapman (10). c_i is the ionic strength in moles·l^{-1}, e is the electronic charge and the other terms have their usual meanings. μ' was calculated from equation 2 using μ = 194 mD, the maximum value obtained with this data set. It should be noted that the Gouy-Chapman relation has been used solely to relate surface charge to potential and that the usual Boltzmann counter-ion double layer distribution has been specifically omitted. This simple approach to the problem indicates that either our value K_a is a gross overestimate or our analysis has excluded an important force peculiar to the arrangement

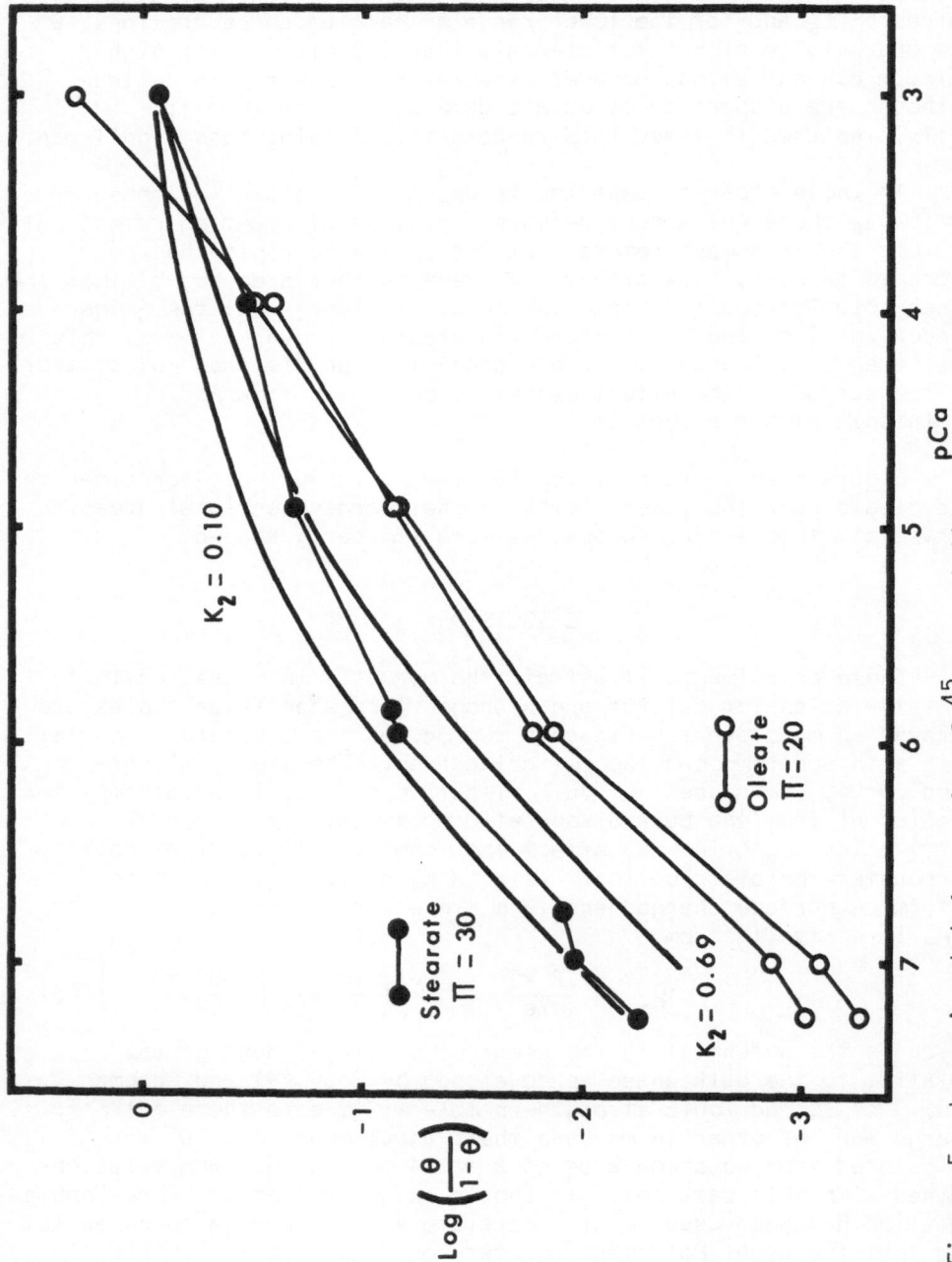

Figure 5. Langmuir plot for surface Ca^{45} accumulation.

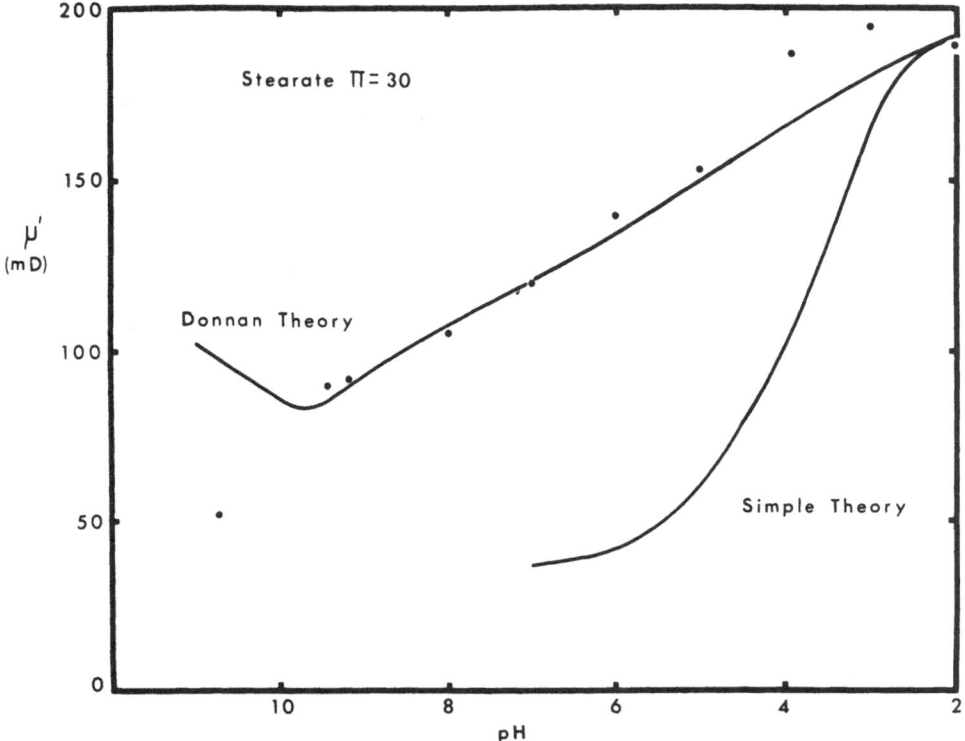

Figure 6. Stearate potential titration.
Curve at right uses Gouy potential Ψ_o, ignoring Boltzmann
distribution of counter-ions. Curve at left uses Donnan potential
and counter-ion distribution.

of carboxyl groups in a monolayer. The importance of such a monolayer array of charge can be shown by including it in a model to
predict surface potential or μ'. In the subsequent Donnan analysis
the solution is considered to have two phases, bulk and surface,
with the lipid head groups restricted to the surface phase. The
thickness of this latter phase is taken arbitrarily as the Debye-
Hückel distance $1/\varkappa$ with

$$\varkappa^2 = \frac{4\pi e^2}{kT}\; c_i$$

where c_i is the bulk ionic strength. In the pH range of these
experiments $1/\varkappa$ varied from 28 to 438 Å. Using the usual Donnan
relations as in equation 4, surface phase electroneutrality and
the K_a given above for lipid and surface phase hydrogen ions

$$\frac{Tris_s^+}{Tris_b^+} = \frac{OH_b^-}{OH_s^-} = \left(\frac{Ca_s^{++}}{Ca_b^{++}}\right)^{1/2} \quad etc. \quad (4)$$

(H_s^+), one obtains a third degree equation in H_s^+ which can be
solved numerically using different values of bulk hydrogen (H_b^+)
and L_s. L_s was set equal to 5.3 Gibbs for a representative close-
packed oleate curve and to 8.3 Gibbs for stearate. With an
assumed maximum of $\mu' = 194$ mD the smooth curve at the left of
figure 6 is a good representation of the data, particularly in the
midrange. The same simple and Donnan analyses for oleate are shown
in figure 7. The fit is less good here, perhaps due to the
instability of oleate monolayers at alkaline pH or more probably to
the change of μ with film expansion.

The Donnan treatment can also be applied to the interaction
of Ca with lipid monolayers. Considering first a simpler analysis
which ignores surface electrostatic forces, one can calculate an
apparent affinity constant K for Ca and lipid from:

$$Ca^{++} + L^- \rightleftharpoons CaL^+$$

Using $\theta = CaL/L_T$ (where CaL is the total excess of calcium per unit
area in the surface as measured, say, by the Geiger tube, and L_T is
the total lipid per unit area) one can easily show that

$$\log\left(\frac{\theta}{1-\theta}\right) = \log K - pCa$$

By fitting the data of figure 5 from pCa 7 to 6 to this relation
one obtains for oleate $K_{OA} = 1 \times 10^4$, and for stearate $K_{SA} = 8 \times 10^4$.
From other considerations Danielli (2) gives the comparable bulk
lipid-calcium affinity constant as 1.45. Again, this disparity can
be largely explained by considering the "surface" calcium in a
Donnan model to be composed of an ionized surface phase component
(Ca_s^{++}) and a lipid-bound component (CaL^+) having an affinity constant $K = 1.45$ in equilibrium with Ca_s^{++}. With the same considerations as above for the hydrogen binding case including the same
value of K_a one can obtain a fourth degree equation in Ca_s^{++}.
Solving this numerically yields the lower smooth curve in figure 5
where $Ca_s^{++} + CaL^+$ has been used to calculate the surface excess of
Ca and hence θ. K_2 values given in figure 5 represent dissociation
of CaL^+ and are the reciprocals of the above affinity constants.
By arbitrarily increasing K to 10, one obtains the upper smooth curve
which fits the extremes of the stearate data better.

Nowhere in the foregoing analysis has there been a suggestion
as to why stearate can attract more Ca to the surface than can
oleate. The data at hand do not allow strict determination of

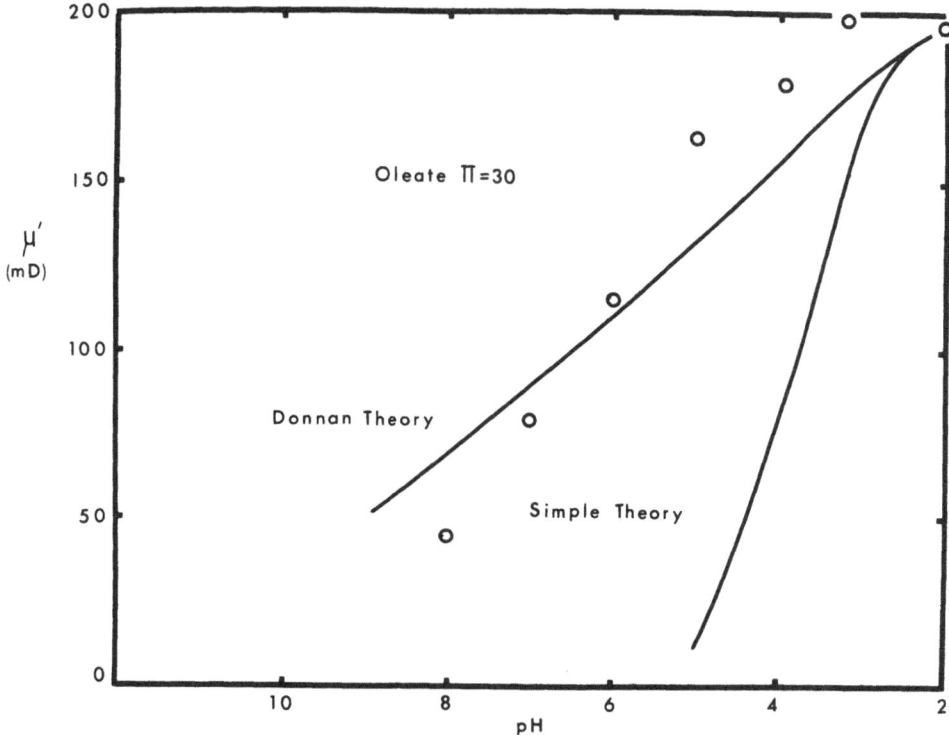

Figure 7. Oleate potential titration.
Curves calculated as for Figure 6.

μ at pH 6.8 for either acid, and hence a difference in ψ_o cannot
be invoked as the cause. The presence of islands would be
expected to enhance surface accumulation, but of course only at
low L_s. At pCa 6 for example, one can predict only a 30% enhancement of θ in a Donnan calculation. Perhaps there is a real increase
in binding through coordination as suggested by Deamer and
Cornwell (11).

These demonstrations of the power of negative surfaces are
quite relevant to biology, where for instance cell membranes and
intracellular membranes composed of charged phospholipids, fatty
acids and proteins abut the aqueous environment. Any consideration
of the interaction of charged species in the aqueous phase with
a membrane containing charged groups must therefore consider the
effect of charge array in addition to the usual group affinity
constants.

We wish to express our thanks to Professor John A. Clements for his interest and advice, and to Miss Sarah Jones and Mrs. Lynne Kalnasy for technical assistance.

BIBLIOGRAPHY

1.   Danielli, J.F., Proc. Roy. Soc. Lond., Ser. B, 1937, 122, 155.
2.   Danielli, J.F., J. Exptl. Biol., 1944, 20, 167.
3.   Danielli, J.F. and J.T. Davies, Adv. in Enzymology, 1951, 11, 35.
4.   Stern, O., Z. Elektrochem., 1924, 30, 508.
5.   Goldstein, G., Nucleonics, 1965, 23, 67.
6.   Goddard, E.D., and J.A. Ackilli, J. Coll. Sci., 1963, 18, 585.
7.   Spink, J.A., J. Coll. Sci., 1963, 18, 512.
8.   Garvin, J.E., and M.L. Karnovsky, J. Biol. Chem., 1956, 221, 211.
9.   Gouy, M., J. de Phys., Ser. 4, 1910, 9, 457.
10.   Chapman, D.L., Phil. Mag., 1913, 25, 475.
11.   Deamer, D.W., and D.G. Cornwell, Biochim. Biophys. Acta, 1966, 116, 555.

STUDIES OF THERMAL TRANSITIONS OF PHOSPHOLIPIDS IN WATER: EFFECT

OF CHAIN LENGTH AND POLAR GROUPS OF SINGLE LIPIDS AND MIXTURES[*]

Morris B. Abramson

The Saul R. Korey Department of Neurology, Albert

Einstein College of Medicine, Bronx, New York

ABSTRACT

Several phospholipids and mixtures of phospholipids in water-rich systems were studied by differential thermal analysis and polarized light microscopy. Sharp transition temperatures were observed for all lipids which were single molecular types. This transition is associated with the melting of the hydrocarbon chains and formation of the liquid crystalline form. For mixtures of lecithin of different hydrocarbon chain lengths, the transition temperature range was broadened and increased with the concentration of the long chain component. The transition temperatures of a series of lipids, which had the same hydrocarbon chains but different polar groups, phosphatidyl ethanolamine, N,N-dimethyl phosphatidyl ethanolamine, and lecithin, were lowered as the methylation of the amine increased. Mixtures of lecithin and phosphatidyl ethanolamine with the same hydrocarbon chains showed that the transition temperature of the phosphatidyl ethanolamine is lowered by the presence of lecithin and in some compositions two transition temperatures were observed. The mixing of an acidic lipid, dicetyl phosphoric acid, with lecithin showed that small concentrations of the acid lower the transition temperature of lecithin. At concentrations of the acid above 21 mole percent, the transition temperature of dicetyl phosphoric acid is observed.

[*]This work was supported by Multiple Sclerosis Grant 503A and U.S. Department of the Interior, Office of Saline Water Grant 14-01-0001-1277.

INTRODUCTION

Several recent investigations have indicated the value of
thermal studies as another method for obtaining knowledge of the
ultrastructure of biologic systems. Interesting investigations
of the membranes of <u>Mycoplasma laidlawii</u> were made by Steim and
co-workers (1969) who used differential scanning calorimetry to
show the effect of fatty acid content on the thermal transitions
of these biomembranes. Other thermal studies gave information
on the organization of lipids in ox brain myelin and of the
lipids of human erythrocyte ghosts (Ladbrooke et al., 1968a,b).

The success of these studies rests in good measure on the
establishment of sufficient information concerning the thermal
properties of relatively simple systems containing compounds
important in biologic structures. Some of these properties for
phospholipids are described in other publications (Chapman et al.,
1966, 1967; Ladbrooke et al., 1968c). Since knowledge of the
properties of phospholipids in water-rich systems is of major
significance toward the understanding of biomembranes, the
present paper describes some of the thermal transitions of lipid-
water systems as observed by differential thermal analysis
(d.t.a.) and microscopic studies. Synthetic lipids are used in
which the hydrocarbon chain structures as well as the polar groups
are of one type. These well-defined compounds have characteristic
and sharp transition temperatures which may be used as indicators
to gain an insight into the inter-relationships of lipids in
mixed systems.

EXPERIMENTAL

Materials

The lipids used in this study came from the following
sources: L-2,3-dipalmitoyl lecithin, L-2,3-dipalmitoyl phos-
phatidyl ethanolamine, L-2,3-dipalmitoyl-glycerine-1-phosphoryl
N,N-dimethyl ethanolamine, L-2,3-dihexadecyl-glycerine-1-phos-
phoryl choline, L-2,3-dihexadecyl-glycerine-1-phosphoryl ethanol-
amine, and lysolecithin from Mann Research Laboratories, New
York. Dicetyl phosphoric acid was from K and K Laboratories,
Plainview, New York. L-2,3 dimyristoyl lecithin was a gift from
Professor Eric Baer. Egg lecithin obtained from Sylvana Company,
Milburn, New Jersey was purified by silicic acid chromatography.
All lipids used gave single bands on thin layer chromatograms.

Mixed lipid systems were prepared by weighing the solids into a small test tube so that 7-10 mg total solid was present. This was dissolved in 1 ml chloroform-methanol (2:1 by volume) and mixed thoroughly. The solvent was evaporated in a stream of nitrogen and the tube was then kept in vacuum overnight. Removal of all solvents was checked by reweighing the tube and its contents. Single lipid systems were prepared from 7-8 mg weighed into a small test tube. Dispersions were prepared by adding 0.1 ml redistilled water and crushing the solid particles in water with a small spatula. After standing for 2 hours, the solid was crushed again and the suspension mixed vigorously in a cyclo-mixer. This was repeated after a second 2 hour period. After standing overnight at 4° the systems were mixed again and forced repeatedly through a narrow-tipped pipette until no large aggregates were visible.

Thermal Analysis

Differential thermal analyses were performed using the DuPont 900 Thermal Analyzer. Samples of 25 µl of the mixed suspension containing 7-10% lipid were transferred by pipette to the the 4 mm diameter sample tube. Most of the determinations were at a heating rate of 10°/min and a sensitivity of 0.1°/inch. Powdered silica was used as reference solid. At the completion of the thermal analysis, the system and the heating block were cooled by careful application of dry ice and the heating cycle was repeated to determine reversibility of the changes. In most instances second aliquots of the lipid system were taken and the thermal diagram was repeated. All determinations of single lipid systems were repeated with second preparations of the aqueous dispersion.

Polarized Light Microscopy

Small aliquots of all systems used for d.t.a. studies were viewed in a polarizing light microscope equipped with a Kofler heating stage. The sample on a glass slide was heated from room temperature to 70°C and changes in structure of the lipid aggregates were observed.

RESULTS

In order to interpret the thermal diagrams shown below, several points can be reviewed. An endothermic transition is indicated by a peak in a downward direction. The onset of this transition occurs at the temperature at which the curve shows a

downward departure from the baseline. The transition is complete
when the peak is at its lowest point and absorption of heat by
the change is complete. Baselines which are not horizontal are
common for solids dispersed in liquids. These may result from
shifting of the particles with respect to the thermocouple position,
change in packing of the solid or other mechanical change.

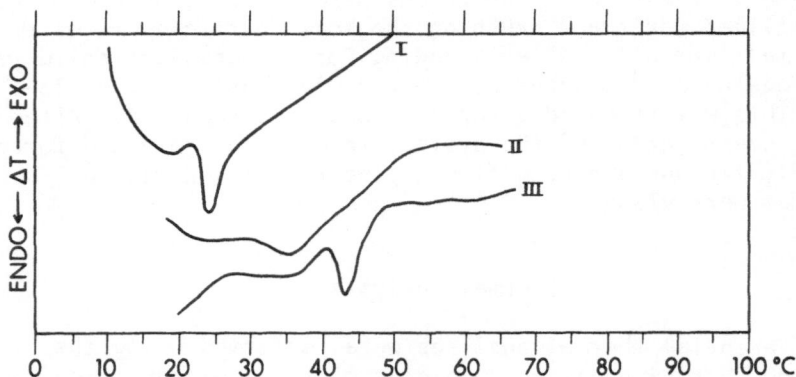

Fig. 1 - D.T.A. diagrams for synthetic lecithins of different
hydrocarbon chain lengths and a mixture of two forms of lecithin.
Systems contained 7-10% lecithin in water. The peak indicates the
transition from the gel to the lamellar liquid-crystalline form.

    I   L-2,3 dimyristoyl lecithin

   II   L-2,3 dimyristoyl lecithin   40 mole %
        L-2,3 dipalmitoyl lecithin   60 mole %

  III L-2,3 dipalmitoyl lecithin

Mixed Hydrocarbon Structures

Thermal diagrams of dimyristoyl lecithin and dipalmitoyl
lecithin both show sharp transitions, the former at 23°-24°, the
latter at 41°-43°. These values are in agreement with the findings of Chapman et al. (1967) who used somewhat more concentrated
aqueous mixtures. These changes are attributed to the transition
from a gel structure to the lamellar or smectic liquid crystal
form. Mixtures of these two forms of lecithin showed transitions
over a temperature range varying roughly from 20° to 39°. In
Figure 1 the d.t.a. diagrams are shown for aqueous systems of
dimyristoyl lecithin, dipalmitoyl lecithin and a mixture of 40
mole percent dimyristoyl and 60 mole percent dipalmitoyl lecithins.
In the latter (Curve II), a broad transition range is observed

from 31-38°. This transition at a temperature closer to that of dipalmitoyl lecithin is consistent with the larger percentage of this component in the mixture. Inspection of this system by polarized light microscopy showed the typical spherical structures observed for lecithin systems. These were highly bi-refringent and did not show any inhomogeneities. The two forms of lecithin appeared to be completely intermixed in these systems and exhibited transition temperatures intermediate between those of the single lipids.

Since egg lecithin is a mixture of compounds that differ in the length and degree of unsaturation of their hydrocarbon chains, it is anticipated that its thermal transitions would take place over a broad temperature range. In Figure 2, Curve I, the d.t.a.

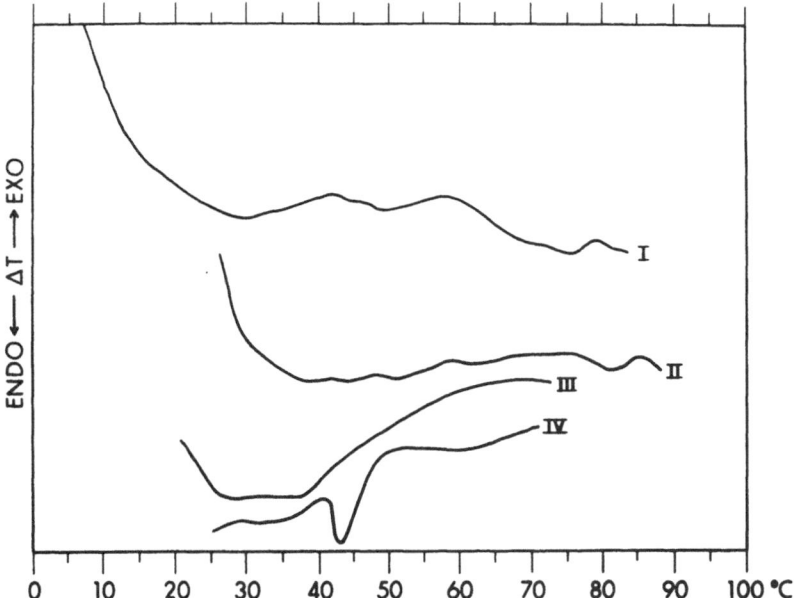

Fig. 2 - D.T.A. diagrams for egg lecithin, dipalmitoyl lecithin and mixtures of these forms of lecithin. Systems contained 7-10% lecithin in water.

I   egg lecithin

II   egg lecithin   63 mole %
     L-2,3 dipalmitoyl lecithin   37 mole %

III egg lecithin   23 mole %
     L-2,3 dipalmitoyl lecithin   77 mole %

IV   L-2,3 dipalmitoyl lecithin

diagram for egg lecithin in water shows an endothermic change
from 10-40° with smaller changes at approximately 50° and 75°. A
mixture of 37 mole percent dipalmitoyl lecithin with egg lecithin
showed the initial transition beginning at about 22° with a smaller
transition at 80°, Curve II. A mixture containing 77 mole percent
dipalmitoyl lecithin showed a much reduced transition range from
roughly 20-40°. The diagrams for egg lecithin alone and those of
egg lecithin with dipalmitoyl lecithin in which the added dipalmitoyl component increased the percent of saturated acyl chains
indicate that the different molecular forms of lecithin are intermixed. When the mixture contains a wide range of hydrocarbon
chain lengths and degrees of unsaturation the transition is over
a broad temperature range. Increasing the presence of one
molecular form then brings the transition temperature closer to
that of the predominant species.

Mixed Lipid Types

In another series of experiments, mixtures were prepared of
dipalmitoyl lecithin and dipalmitoyl phosphatidyl ethanolamine.
These compounds are alike in the length and saturation of the
hydrocarbon chains. They differ only in the nature of the
terminal amino group of the polar moiety of the molecule. The
lecithin contains the tri-methyl amine while diacyl phosphatidyl
ethanolamine contains the primary amine. The thermal diagrams
for three mixtures of these phospholipids are shown in Figure 3,
Curves II, III, IV. These diagrams can be compared with the
ones for the single compounds, dipalmitoyl lecithin, Curve I and
dipalmitoyl phosphatidyl ethanolamine Curve V. In Curve II the
peak associated with the transition of dipalmitoyl lecithin (the
major component in the mixture) at 40° is present and another
smaller and broader peak at 54-56° appears. A mixture of almost
equi-molar amounts of the two compounds (Curve III) shows no
evidence of the peak at 40° but the peak at the higher temperature
is more pronounced and appears at a slightly higher temperature
(52-59°) than in Curve II. With a further increase of the phosphatidyl ethanolamine component to 72 mole percent (Curve IV),
the peak at the higher temperature is sharply defined, ranging
from 53-60.5°. Dipalmitoyl phosphatidyl ethanolamine alone shows
a transition from 62-65°.

Polarized light microscopy of these systems showed two types
of structures. Lecithin alone gave spherical particles with
heavy walls which were highly bi-refringent. Phosphatidyl
ethanolamine consisted of amorphous masses, also bi-refringent.
At the transition temperature these formed myelin figures in a
strikingly sharp transformation. These figures were structures
with parallel walls extruded from the amorphous mass. In some

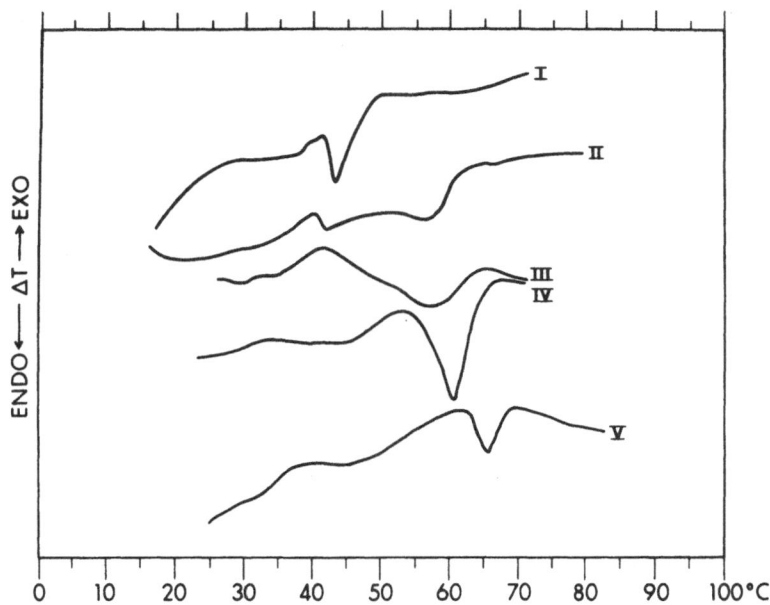

Fig. 3 - D.T.A. diagrams for dipalmitoyl lecithin, dipalmitoyl
phosphatidyl ethanolamine and mixtures of these two lipids showing the effect of different polar groups. Systems contained
7-10% lipids in water.

I	L-2,3 dipalmitoyl lecithin		
II	L-2,3 dipalmitoyl lecithin	66 mole %	
	L-2,3 dipalmitoyl phosphatidyl		
	ethanolamine	34 mole %	
III	L-2,3 dipalmitoyl lecithin	49 mole %	
	L-2,3 dipalmitoyl phosphatidyl		
	ethanolamine	51 mole %	
IV	L-2,3 dipalmitoyl lecithin	28 mole %	
	L-2,3 dipalmitoyl phosphatidyl		
	ethanolamine	72 mole %	
V	L-2,3 dipalmitoyl lecithin		

- - - -

instances stacked arrays of these structures formed. Mixtures
of the two lipids contained both types of aggregates. The temperatures at which they underwent a change differed in the mixed

systems from that observed for the single lipids. At the lower
transition temperature the spherical bodies became deformed in
shape when in contact with other particles and the walls became
thinner.

Addition of Acidic Lipid

The effects of the addition of negatively charged groups of
an acidic component were studied in another series of mixtures.
In these, dicetyl phosphoric acid was mixed with dipalmitoyl
lecithin. Figure 4 shows the thermal diagrams of the single
compounds, dipalmitoyl lecithin and dicetyl phosphoric acid, in
Curves I and VII respectively. In Curves II, III, and IV in
which the mole percent of dicetyl phosphoric acid is 5, 10, and
18%, the transition characteristic of dipalmitoyl lecithin is
present but at a slightly lower temperature beginning at 39⁰. In
Curve IV this transition is broadened and extends beyond 50⁰. A
mixture containing 21 mole percent of the acidic component
(Curve V) does not show the transition for lecithin but a broad
peak beginning at 58⁰ appears. A mixture of 39 mole percent
dicetyl phosphoric acid shows a sharp peak at 69⁰ beginning at
65⁰ as in Curve VI. This does not differ greatly from dicetyl
phosphoric acid alone with a transition beginning at 67⁰ and
exhibiting a peak at 71⁰.

Figure 5 shows thermal diagrams for related compounds which
have specific chemical differences. Curves I, II, and III permit
a comparison of L-2,3-dipalmitoyl lecithin, L-2,3-dipalmitoyl-
glycerine-1-phosphoryl N,N-dimethyl ethanolamine and L-2,3-
dipalmitoyl phosphatidyl ethanolamine in Curves I, II, and III
respectively. Each of these compounds shows a single sharp
transition temperature. The transition in Curve II beginning at
48⁰ and reaching a peak at 50⁰ is somewhat closer to the related
lecithin than to the phosphatidyl ethanolamine. Thermal diagrams
were also obtained for L-2,3-dihexadecyl-glycerine-1-phosphoryl
choline and L-2,3-dihexadecyl-glycerine-1-phosphoryl ethanolamine.
The former shows a sharp peak at 46⁰ beginning at 43⁰ (Curve IV),
the latter has a major peak at 69⁰ beginning at 68⁰ (Curve V).
A smaller thermal transition occurs from 52-56⁰. In both these
compounds the transition to the liquid crystalline form occurs
at a temperature about 3-4 degrees above that for the related
diglyceride compound.

The thermal diagrams of another series of mixtures are shown
in Figure 6. These mixtures are similar to the ones in Figure 3.
They contained dipalmitoyl lecithin mixed with L-2,3-dihexadecyl-
glycerine-1-phosphoryl ethanolamine in place of the L-2,3-
dipalmitoyl phosphatidyl ethanolamine in the mixtures shown in

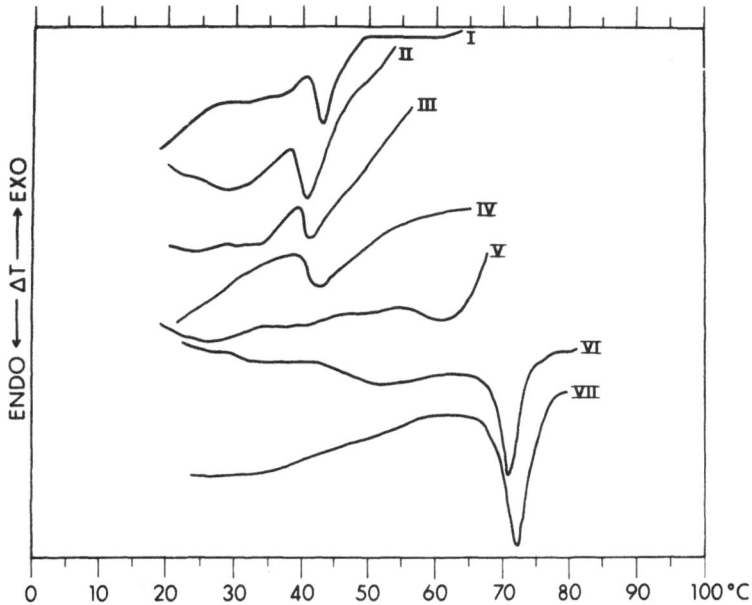

Fig. 4 - D.T.A. diagrams for dipalmitoyl lecithin, dicetyl phosphoric acid and mixtures of these compounds showing the effect of
an acidic component. Systems contained 7-10% lipids in water.

I	L-2,3 dipalmitoyl lecithin		
II	L-2,3 dipalmitoyl lecithin dicetyl phosphoric acid	95 mole % 5 mole %	
III	L-2,3 dipalmitoyl lecithin dicetyl phosphoric acid	90 mole % 10 mole %	
IV	L-2,3 dipalmitoyl lecithin dicetyl phosphoric acid	82 mole % 18 mole %	
V	L-2,3 dipalmitoyl lecithin dicetyl phosphoric acid	79 mole % 21 mole %	
VI	L-2,3 dipalmitoyl lecithin dicetyl phosphoric acid	61 mole % 39 mole %	
VII	dicetyl phosphoric acid		

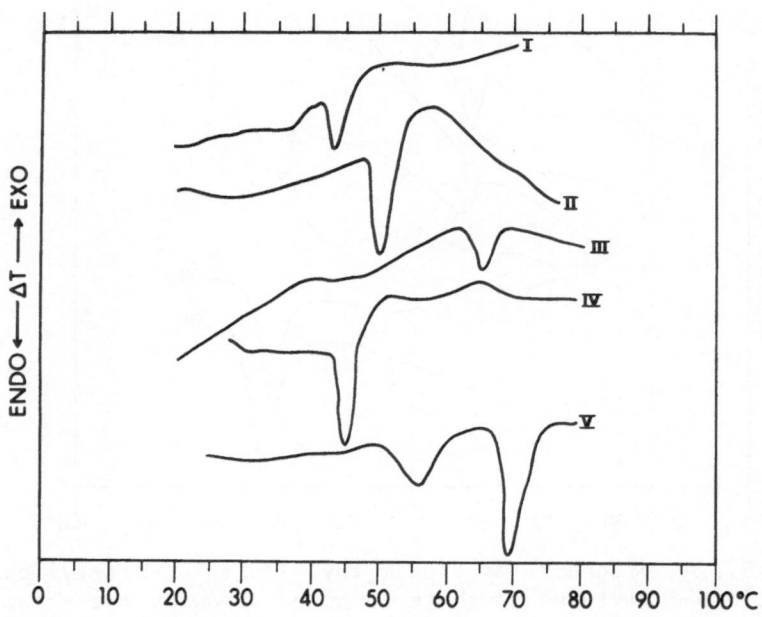

Fig. 5 - D.T.A. diagrams for different phospholipids showing in
I, II and III the effect of methylation of the amine and IV, V
diether analogues of I and III. Systems contained 7-10% lipids
in water.

  I   L-2,3 dipalmitoyl lecithin

  II   L-2,3 dipalmitoyl-glycerine-l-phosphoryl N,N-dimethyl
       ethanolamine

  III L-2,3 dipalmitoyl phosphatidyl ethanolamine

  IV   L-2,3-dihexadecyl-glycerine-l-phosphoryl choline

  V   L-2,3-dihexadecyl-glycerine-l-phosphoryl
       ethanolamine

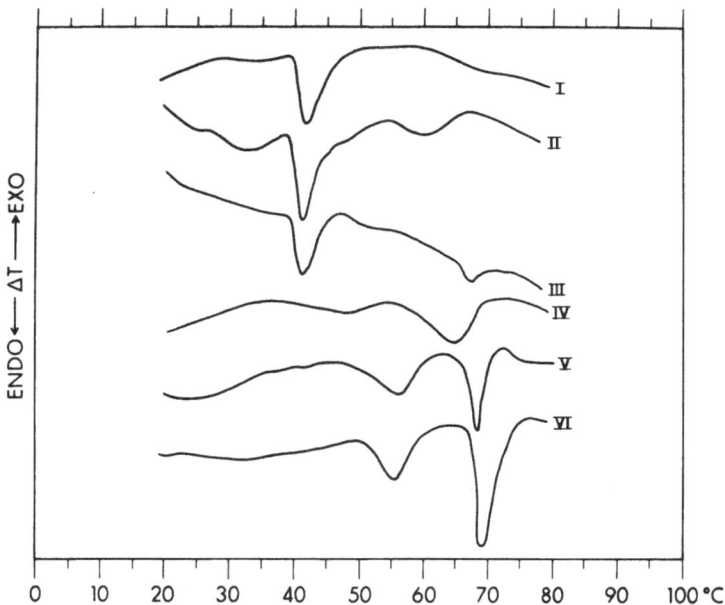

Fig. 6 - D.T.A. diagrams of mixtures of A. L-2,3 dipalmitoyl
lecithin and B. L-2,3 dihexadecyl-glycerine-1-phosphoryl ethanolamine showing the effect of mixtures of two different polar groups
and diether and diester linkage to glycerine. Systems contained
7-10% lipids in water.

I	A - 83 mole % B - 17 mole %	IV	A - 31 mole % B - 69 mole %
II	A - 80 mole % B - 20 mole %	V	A - 16 mole % B - 84 mole %
III	A - 49 mole % B - 51 mole %	VI	B - 100 mole %

- - - -

Figure 3. The peak associated with the transition of dipalmitoyl
lecithin is evident in Curves I, II and III and in which its
concentration decreases from 83 to 49 mole percent. It is not
detected in Curve IV where its concentration is 31%. The transition for L-2,3-dihexadecyl-glycerine-1-phosphoryl ethanolamine
may be seen in all curves except I. In Curve II this is a broad
peak beginning at 56°. In Curve III for the roughly equi-molar
mixture the peak is less broad and begins at 65°. Curves IV and V
show two peaks which approach the two transitions for the single
compound L-2,3-dihexadecyl-glycerine-1-phosphoryl ethanolamine in

sharpness and in temperature. This series compares well with the series shown in Figure 3 in which L-2,3 dipalmitoyl phosphatidyl ethanolamine was present. One difference that may be noted is that the peak for lecithin in the equi-molar mixture is quite definitely present in Figure 6 but not in Figure 3.

Several experiments were performed with mixtures of dipalmitoyl lecithin and lysolecithin which contained 70% C_{16} chains and 30% C_{18} chains. Lysolecithin in water showed no sharp thermal transition from 15-80°C. A mixture containing 9.5 mole percent lysolecithin lowered the transition temperature of dipalmitoyl lecithin 2-3 degrees. A mixture containing 26% lysolecithin did not give the same diagram when reheated. Another mixture containing 34% lysolecithin gave a sharp peak at 50° but this was not repeated on the second heating of the sample. Apparently the changes produced by lysolecithin are more complex than those observed in other mixtures of lipids. These will require further investigation.

TABLE I

Major Transition Temperatures	°C
Single Lipids in Water	
L-2,3-dimyristoyl lecithin	23-24
L-2,3-dipalmitoyl lecithin	41-43
L-2,3-dipalmitoyl-glycerine-1-phosphoryl-N,N-dimethyl ethanolamine	48-50
L-2,3-dipalmitoyl phosphatidyl ethanolamine	62-65
dicetyl phosphoric acid	67-71
L-2,3-dihexadecyl-glycerine-1-phosphoryl choline	43-46
L-2,3-dihexadecyl-glycerine-1-phosphoryl ethanolamine	68-69

DISCUSSION

Chapman and his coworkers have shown (Chapman et al., 1967) that synthetic forms of lecithin mixed with 30% or more by weight of water, underwent a mesomorphic transition at temperatures much below the transition temperatures for the anhydrous or the monohydrate forms of that lecithin. For aqueous systems of dimyristoyl lecithin this transition was observed at 23° and for dipalmitoyl lecithin at 41°. At this transition the lecithin changes from a gel structure to the smectic liquid crystalline form. The structural changes involved are the melting of the hydrocarbon chains and penetration of water between the layers of polar head groups of the molecules. In the smectic state the hydrocarbon groups are in a fluid condition with freedom of movement within the bimolecular layers of the liquid crystal. Movement of the molecules within the plane is conceivable, but movement between planes is not possible because the polar groups are bound to water at the interface and quite possibly by intermolecular bonds established between the polar moieties of neighboring molecules. In this form the phospholipids exist in structures resembling biomembranes. The difference in the sharp transition temperatures encountered with myristoyl lecithin as compared with dipalmitoyl lecithin results from the lower temperature required to disrupt the Van der Waals forces between hydrocarbon chains which decreases with the smaller number of CH_2 groups in dimyristoyl lecithin. Mixtures of dimyristoyl and dipalmitoyl lecithin show a transition that takes place over a broader temperature range intermediate between the transition temperatures of the two single lipids. It, therefore, appears that the two forms of lecithin become intermixed with the resulting hydrocarbon layer composed of a mixture of chains melting at temperatures determined by the mole percent of the components. The polar regions immersed in the aqueous layers are the same for the single forms of lecithin and the mixtures. The thermal diagrams for egg lecithin in Figure 2 shows the broad range of melting of the mixture of hydrocarbon moieties present. Egg lecithin contains a high percentage of unsaturated chains which have lower melting points than the corresponding saturated chains. The transition to the smectic form begins at a low temperature. When dipalmitoyl lecithin is mixed with egg lecithin the transition range becomes narrower and approaches that of dipalmitoyl lecithin as its concentration in the mixture increases. This again points to the mixing of the chains of the different forms of lecithin with a resulting transition temperature dependent upon the composition of the hydrocarbons.

When two different lipid types are mixed, it is quite possible that more complex changes take place. In the mixing process as performed in the experiments reported here, several changes may be inferred. These may be complex and many aspects of them have not yet been fully explored. When a lipid is dissolved in chloroform-methanol or other organic solvent, the lipid is either dispersed at the molecular level or in the form of micelles. The micelles in these instances have the polar groups directed away from the solvent. A mixture of two lipids in such a solvent may contain a mixture of molecules or micelles containing a mixture of molecules

or some mixture of micelles. Furthermore, on evaporation of the
solvent, differences in the solubilities of the lipids in the
solvent may lead to the formation of different forms of solid
deposit as the amount of solvent decreases. In the d.t.a. studies
of mixed lecithins and phosphatidyl ethanolamine, the endothermic
peak characteristic of dipalmitoyl lecithin is seen in Figure 3,
Curve II, where it is present in a concentration of 66 mole percent but is not seen in Curve III at 49%, whereas the peak for
phosphatidyl ethanolamine is seen in Curve II where its concentration is 34 mole percent and more strongly in Curve III at
51%. This may indicate that the structure established by the
phosphoryl ethanolamine in water is more stable than that of phosphorylcholine. The microscopic examination of the systems used
for Curves II and III showed two forms of aggregates, the bi-
refringent spheres characteristic of lecithin systems and amorphous masses which were found with dipalmitoyl phosphatidyl
ethanolamine below the transition temperature. On heating the
mixed system the two types of particles showed changes at
different temperatures. In all the curves for mixtures of the two
lipids, the transition temperature for the dipalmitoyl phosphatidyl ethanolamine is lower than that shown by dipalmitoyl
phosphatidyl ethanolamine alone. This presumably results from the
incorporation of lecithin molecules into the particles of phosphatidyl ethanolamine which is nevertheless predominant in
determining the physical characteristics of the particles when
its concentration is greater than some value between the 34 mole
to 51 mole percentages of Curves II and III.

Although phosphorylcholine lipids and phosphorylethanolamine
lipids are both relatively abundant in biologic membranes, they
are not equivalent or interchangeable, each maintaining its
unique character. An interesting finding related to this is the
work of Tomasz (1968) who showed that ethanolamine can be incorporated in place of choline into the structural components of the
cell wall of pneumococcus with resulting changes in cell characteristics. With ethanolamine the cells remain associated after
cell division and cannot be lysed as readily as cells containing
choline. This difference in the amine groups is further supported by Zull and Hopfinger (1969) who calculated the potential
energy fields about the nitrogen atom in choline and ethanolamine.
They find that in ethanolamine a stronger potential energy field
exists in the region of the nitrogen atom than in choline. This
could result in a stronger interaction with anions in inter-or
intra-molecular bonding. The higher temperature that is found
for the transition to the liquid crystalline form for dipalmitoyl
phosphatidyl ethanolamine as compared with dipalmitoyl lecithin,
and the persistence of the structure characteristic of the
ethanolamine phospholipid over that of the choline phospholipid
as shown by the d.t.a. curves of Figure 3 are consistent with the

greater force field at the positive charged sites of the ethanolamine. In Figure 5, the Curves I, II and III show that in the
dimethylethanolamine phospholipid the steric effect of the two
methyl groups is not as great as the effect in choline and the
forces between molecules is intermediate between that of choline
and ethanolamine as shown by their transition temperatures. In a
parallel sequence in the structure of the pneumococcal cell walls,
incorporation of N,N-dimethylamino ethanolamine results in
organisms more closely resembling the normal one containing
choline than does ethanolamine incorporation (Tomasz, 1968).

The phospholipids of biomembranes are either bi-ionic as
lecithin or phosphatidyl ethanolamine or anionic as the phos-
phoinositides, phosphatidylserine or phosphatidic acid. In recent
studies of the permeabilities of model lipid systems the properties of liposomes have been explored. In these liposomes, the
addition of an acidic lipid to lecithin permits the formation of
lipid aggregates in which membrane-like structures entrap small
volumes of aqueous solution (Bangham et al., 1965). The acidic
component used is often dicetyl phosphoric acid. The thermal
diagrams of Figure 4 show that small concentrations of dicetyl
phosphoric acid are incorporated into the structure of dipalmitoyl
lecithin and a small lowering of the transition temperature of the
lecithin results. When the concentration of the acid reaches
roughly 21 mole percent there is no sharp transition and the
lecithin is in the liquid-crystal state at a temperature below its
normal transition. These experiments do not, however, indicate
whether this change results from the presence of negative phosphate groups which disrupt the network of polar groups of lecithin
in the aqueous boundary layer or whether it stems from the in-
trusion of hydrocarbon chains which are not carboxy acid ester
linked to glycerine. The appearance of a sharp peak at 65° for a
mixture containing 39 mole percent dicetyl phosphoric acid shows
the stronger intermolecular bonding of this acid component.
Microscopy showed two forms of aggregates in these systems with a
large amount of amorphous material which melted at a temperature
near 70°. An interpretation of these mixed systems is that two
forms of mixed aggregates arise. In one the presence of small
amounts of dicetyl phosphoric acid affects the transition of
dipalmitoyl lecithin so that it is in the liquid-crystal form
over a wide temperature range. The other aggregate contains
lecithin incorporated into the dicetyl phosphoric acid structure
which has a transition temperature a few degrees lower than
dicetyl phosphoric acid alone.

The diagrams shown in Curves IV and V, Figure 5 and Table I
permit a comparison of closely related analogs of dipalmitoyl
lecithin and dipalmitoyl phosphatidyl ethanolamine whose thermal
diagrams are shown in Curves I and III. The compounds which

contain the ether linkage to glycerine in place of the ester
linkage exhibit transitions that are much like those of the normal
lipid type but at a small increase in temperature. This may be
the result of closer packing of the hydrocarbon chains with increased Van der Waals forces in the ether bonded compounds.

In the series of mixtures of dipalmitoyl lecithin with
L-2,3-dihexadecyl-glycerine-1-phosphoryl ethanolamine not only are
the compounds different in their polar moieties but also in the
linkage of the hydrocarbon chain to glycerine. This series of
thermal diagrams (Fig. 6) resembles the ones depicted in Figure 3
with some differences. The transition for dipalmitoyl lecithin is
present and sharp in the equi-molar mixture with the ether linked
ethanolamine lipid but is broadened and at a lower temperature in
the mixture with phosphatidyl ethanolamine. It is conceivable
that the dipalmitoyl lecithin becomes intermixed with the more
closely related phosphatidyl ethanolamine more readily than it
does with the somewhat different ether linked ethanolamine analog.
The lecithin then loses its structural characteristics and its
sharp transition in the equi-molar mixture with phosphatidyl
ethanolamine.

Chapman (1966) reviewed some of the biologic significance
of the liquid crystalline state of membrane lipids as a result of
the studies of the physical properties of lipid systems. It is
interesting, although still hazardous, to draw inferences concerning the nature of biomembranes from the results described here
for lipid-water systems. In the mixture of lipids found in most
biomembranes, it is possible that some of the lipid types could
influence the transition from the gel to lamellar form more
strongly than other types. This effect would be related to the
nature of the polar groups and their interactions with the polar
groups of other molecules. In this connection, the mesomorphic
transitions of bi-ionic lipids could be strongly influenced by
the incorporation of small amounts of anionic lipid. Furthermore,
the lipids in biomembranes may not be distributed uniformly
through the bilayer but may exist in regions or domains with
different compositions and physical characteristics. These
associations of lipids would in large measure be determined by the
interactions between the polar groups of like and unlike molecules
with each other and with water. As seen in some of the thermal
diagrams given here, these lipid associations or domains may
undergo liquid crystalline transitions at different temperatures.
Conceivably, changes in the environment may also influence these
regions differently. It is then possible that some microscopic
regions of the membrane may be in the fluid smectic state while
others remain in the gel form. This could strongly affect the
permeabilities of these lipid structures as shown by de Gier et al.

(1969) in studies of the effect of temperature on the permeability of liposomes.

The author is grateful to Dr. W.T. Norton for reviewing the manuscript and Mr. C. Gryte for technical assistance.

REFERENCES

Bangham, A.D., Standish, M.M., and Watkins, J.C. (1965). J. Mol. Biol., 13:238.

Chapman, D., (1966). Ann. N.Y. Acad. Sciences, 137:745.

Chapman, D., Byrne, P., and Shipley, G.G., (1966). Proc. Roy. Soc. London A 290:115.

Chapman, D., Williams, R.M., and Ladbrooke, B.D., (1967). Chem. Phys. Lipids, 1:445.

de Gier, J., Mandersloot, J.G., and Van Deenen, L.L.M., (1969). Bioch. Biophy. Acta, 173:143.

Ladbrooke, B.D., Jenkinson, T.J., Kamat, V.B., and Chapman, D., (1968A). Bioch. Biophy. Acta, 164:101.

Ladbrooke, B.D., Williams, R.M., and Chapman, D., (1968B). Bioch. Biophy. Acta, 150:333.

Ladbrooke, B.D., Williams, R.M., and Chapman, D., (1968C). Bioch. Biophys. Acta, 150:333.

Steim, J.M., Tourtellotte, M.E., Reinert, S.C., McElhaney, R.N., and Rader, R.L., (1969). Proc. Nat. Acad. Sci. U.S., 63:104.

Tomasz, A., (1968). Proc. Nat. Acad. Sci. U.S., 59:86.

Zull, J.E., and Hopfinger, A.J., (1969). Science, 165:512.

THE PHYSICAL STATE OF LIPIDS OF BIOLOGICAL IMPORTANCE:

CHOLESTERYL ESTERS, CHOLESTEROL, TRIGLYCERIDE

Donald M. Small[+]

Biophysics Unit, Dept. of Medicine, Boston University

School of Medicine, Boston, Massachusetts

Introduction

Alpha and beta serum lipoproteins and the lipids of the lesions of atherosclerosis contain a large fraction of cholesterol esters. Further, the lipid droplets of the fatty streak of the atherosclerotic lesions are probably nearly pure cholesteryl ester. Although there is a vast literature concerning the metabolism (1,2) and transport of cholesterol and its esters, very little is known of the physical state of the biologically important cholesteryl esters and even less about the physical state of these esters in mixtures of other naturally occurring lipids. Even more important, the state of these compounds in vivo has not even been studied. While it is true that pure cholesteryl esters of saturated fatty acids have been carefully examined (3-7), there has been no report of the state of monounsaturated and polyunsaturated cholesteryl esters as a function of temperature. Yet, it is these most important but neglected unsaturated and polyunsaturated esters which make up the bulk of cholesteryl esters of biologic or pathologic origin. The purpose of this report is to discuss briefly the physical state of cholesteryl esters of saturated, cis monounsaturated and cis polyunsaturated long chain fatty acids. Next, since about 70-80% of the cholesteryl esters in α and β lipoprotein (8) and atheromatous lesions (9-12) are cholesteryl oleate and cholesteryl linoleate a condensed binary phase diagram will be presented to show the physical state of all mixtures of these two esters over a temperature range (0°C to 90°C) encompassing most temperatures encountered by living organisms. Finally, because free cholesterol and triglycerides are also present in very high

+ Markle Scholar in Academic Medicine

concentration in certain lipoprotein fractions (2,8) and possibly
in atheromatous plaques (9-12), the interactions of these substances with each other and with cholesteryl esters will be discussed. The specific interrelations will be illustrated using the
binary phase diagrams, cholesterol-cholesteryl oleate, cholesterol-
triolein and triolein-cholesteryl oleate and the ternary system
(3 components) cholesteryl oleate-cholesterol-triolein.

While cholesterol and triglycerides melt from a crystalline
form directly to an isotropic liquid, cholesteryl esters behave
differently. They form thermotropic liquid crystals (13,14) or
mesophases (4). It is well known that many organic substances do
not melt simply from a crystalline solid to an isotropic liquid,
but melt through a series of mesomorphic phases before finally
melting to an isotropic liquid (4,13,14). These mesomorphic phases
have been called "liquid crystalline" phases since they have
characteristics of both liquids and crystals. The long chained
fatty acid esters of cholesterol characteristically show two mesophases, a cholesteric liquid crystalline phase and a smectic liquid
crystalline phase. The cholesteric liquid crystal phase forms from
an isotropic melt. It has a conic focal texture with a negative
sign of birefringence and often exhibits a wide range of reflected
colors. The smectic liquid crystalline phase forms from the
cholesteric phase. It is more grossly opaque, colorless, and
transmits polarized light more intensely than the cholesteric mesophase. The smectic mesophase has a very obvious conic focal texture
but the sign of birefringence is positive.

Certain esters (e.g. cholesteryl myristate) have inantiatropic
phase transformations (5,6,15), that is, the transitions are reversible and occur either on heating or cooling. However, many of
the liquid crystalline phases of other cholesteryl esters are monotropic with respect to a crystalline phase, that is, they have a
lower transition temperature than the melting point of the crystal
and do not appear when the crystal is melted. They only appear when
the melt is cooled below the melting point of the crystal. These
monotropic mesophases may be more or less stable. Nearly all of
the liquid crystal transitions (Isotropic melt ⟷ cholesteric
mesophase; cholesteric mesophase ⟷ smectic mesophase) are
perfectly reversible, if they can be supercooled and reheated before crystallization of a true crystal occurs. If however,
crystallization occurs to a crystal of higher melting point, no
liquid crystalline transformation will form on reheating; the
crystal will simply melt to an isotropic liquid. Some of the monotropic mesophases are very stable and remain unchanged for several
days or even weeks and months, whereas others are highly unstable
and true crystals grow from the liquid crystals within seconds after
the phase appears. Although in some cases the monotropic liquid
crystals seem to be stable, it would be expected that, with proper
nucleating agents and temperature or with long periods of time,

crystals would finally grow from these phases since the crystalline phases with melting points higher than the cholesteric to isotropic transitions are, from the thermodynamic point of view, the most stable state.

Materials and Methods

Materials

Cholesteryl esters of saturated fatty acids of the purest grade were obtained from Eastman Kodak and from Hormel Institute, Austin, Minn. Esters of unsaturated fatty acids were obtained from the Hormel Institute and were judged to be 99% pure with less than 5% trans double bonds. The esters were stored at -4°C in small vials under nitrogen.

Methods

All pure compounds or mixtures of pure compounds were observed grossly, by polarizing microscope, and by differential scanning calorimetry (DSC).

Preparation or Mixtures

Mixtures were prepared by two techniques. 1) Solutions of pure components were made in appropriate organic solvents mixed to give a desired weight proportion and the solvent evaporated under N_2. 2) Pure compounds were weighed directly onto the microscopic slide, melted and mixed by stirring with a small platinum wire. For some of the gross studies compounds were weighed into thick-walled glass tubes melted under a stream of N_2 and sealed. For DSC runs compounds were weighed directly into the DSC pan. A melting curve was obtained and the mixture was allowed to remain well above the melting point for 1 hour to assure mixing. A cooling curve was then obtained. The rest of the DSC procedure is as outlined below.

Microscopic and gross examination of sample.

The sample was placed on a clean glass slide and a coverslip was placed over the sample so that one edge of the coverslip came in contact with the slide and the other with a thin (0.16 mm) piece of glass. This arrangement produces a wedge-shaped sample varying in thickness from 0 to 0.16 mm. The temperature of the sample was then raised on the stage of a polarizing microscope fitted with a special heating chamber described previously (16) at about 1-2°/min until the sample melted to the isotropic state. The sample was then cooled at about 1-2°/min and both the gross and microscopic changes were observed. After each phase change the temperature was reversed at a rate of 1°/min to verify the reversibility of the phase change. When all liquid crystalline phase transformations were observed, the sample was allowed to cool slowly to 20°C until crystallization occurred. The sample was then reheated slowly to note the transition temperature of the crystalline phase. Rapid quenching of the sample from the isotropic melt was then performed by placing the sample on an

aluminum block at -20°C or on a piece of dry ice. The sample was
reheated and the phase changes noted. Finally, the sample was
melted to an isotropic liquid and the entire procedure repeated.
Each sample was studied in duplicate.

The changes in gross appearance of the sample were followed by
observing the wedge of sample against a black background on a
heated stage under a bright light over a wide range of angles of
the incident beam. The first evidence of phase change noted in
going from isotropic liquid to the cholesteric liquid crystalline
phase was often a sudden appearance of a purple or blue color in
the thicker part of the sample. Under crossed nicols this phase
appeared to be isotropic unless the sample was perturbed by pressing
on or moving the coverslip. When perturbed the sample became bi-
refringent. In some cases there appeared to be a change from this
apparent isotropic condition (homeotropic) to a fine conic focal
texture with an obvious negative sign of birefringence somewhat
below the original change from colorless to purple. Grossly a
slight turbidity is noted coincident with this change. Gray (5)
has previously described these changes for some saturated cholesteryl
esters. As the sample is cooled and the cholesteric phase transforms
to the smectic phase, an obvious turbidity occurs in the gross
specimen and all color is lost. A marked sudden increase in the
birefringence occurs to a conic focal texture with an obvious
positive birefringence. The accuracy of determining transition
temperatures by the gross microscopic methods is limited to $\pm$ 0.2°C.

Differential Scanning Calorimetry. (see Fig. 1) Two to four mg of
sample were weighed into an aluminum sample pan on a Kahn Microbalance
This sample was sealed and placed in a differential scanning
calorimeter fitted to a DuPont 900 Differential Thermal Analyzer.
An empty sealed aluminum sample pan served as the reference standard.
A heating curve was obtained at 2.5, 5 or 10°/min until the sample
was entirely melted. The sample was then cooled at the same rate
until the liquid crystalline phase transformations were observed.
The sample was then reheated at the same rate to observe the reversibility of the phase transformations. The sample was then
cooled to -50°C and allowed to remain at that temperature for 30
min. A new heating curve was then obtained at 2.5, 5 or 10°C/min
until the sample melted. The entire procedure was then repeated.
The transition temperature was taken from the intersection of the
flat part of the curve with line drawn through the steepest part of
the melting curve. Repeated studies on the same sample and different
sized samples of the same material showed that the transition
temperature of pure materials were accurate to $\pm$ 0.5°C. The transition temperatures in the liquid crystalline state always varied
by less than a degree for both heating and cooling, showing the
almost perfect reversibility of the liquid-cholesteric and
cholesteric-smectic transitions. These temperatures agreed ($\pm$ 0.5°C)
with the transitions observed grossly or by microscope.

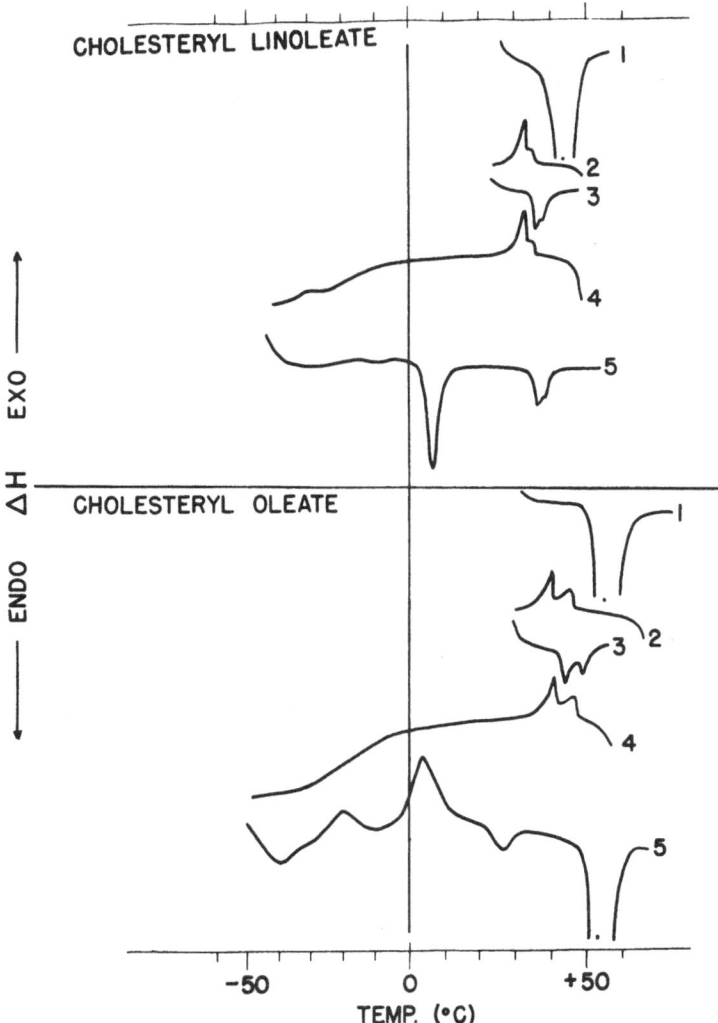

Figure 1. <u>Differential scanning calorimetry curves of cholesteryl</u>
<u>linoleate and cholesteryl oleate</u>. Above-cholesteryl linoleate.
1) First heating curve from room temperature; C_1 melts at 42°C.
2) First cooling curve from melt. Exothermic liquid crystalline
transformations at 36.5 and 34°C are present. 3) Reheating from a
liquid crystalline state. The liquid crystalline transformations
are reversible. 4) Recooling to -50°C. 5) Reheating after 30 min
at -50°C. The endothermic peak at 3°C represents the melting of
crystalline phase C_2 to the smectic liquid crystalline phase.
Below-cholesteryl oleate. The numbers 1-5 refer to heating and
cooling curves as stated above. Note that C_1 melts at 51°C in both
1 and 5. The liquid crystalline transitions are at 47.5 and 42°C.

The heat of transitions can be calculated from triangulation or planography of the areas under the curves. These heats of transition have previously been reported by Barrall, Porter and Johnson (17) for the liquid crystalline transitions of some cholesteryl esters of saturated fatty acids. Both isotropic to cholesteric and cholesteric to smectic transitions are small, being of the order of 0.5 Cal/gm. These changes are illustrated in Fig. 1. The heats of these two transitions for the unsaturated cholesteryl esters are likewise small, being of the same order as those reported for saturated esters. The cholesteric to isotropic heat of transition is about half that of the smectic to cholesteric.

Results

The results will be presented in diagrammatic fashion utilized for instance by Gray (10). The symbols used are the following:

C_1 = crystal having a melting point higher than the highest
      liquid crystal transition
C_2 = crystals having melting point lower than the highest
      liquid crystal transition
Cholesteric = cholesteric mesophase
Smectic   = smectic mesophase
Isotropic   = isotropic liquid melt
(   )   refer to monotropic phases

The temperature above or beside the arrow indicates the temperature transition from one state to another. A dotted arrow indicates a transition occurring with supercooling.

Saturated Fatty Acid Esters of Cholesterol

Cholesteryl decanoate

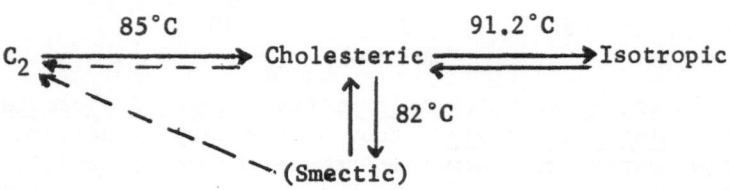

The crystalline form of cholesteryl decanoate (C_2) melts sharply to the cholesteric mesophases at 85°C. The cholesteric mesophase is enantiotropic and melts quite sharply to an isotropic liquid at 91.2°C. On cooling the liquid to cholesteric phase is reversible. Cooling further there is a cholesteric to smectic transformation at 82°C which is also reversible. This monotropic smectic mesophase is quite stable and only after several hours at lower temperatures does C_2 grow from the smectic mesophase.

Cholesteryl laurate (cholesteryl dodecanate)

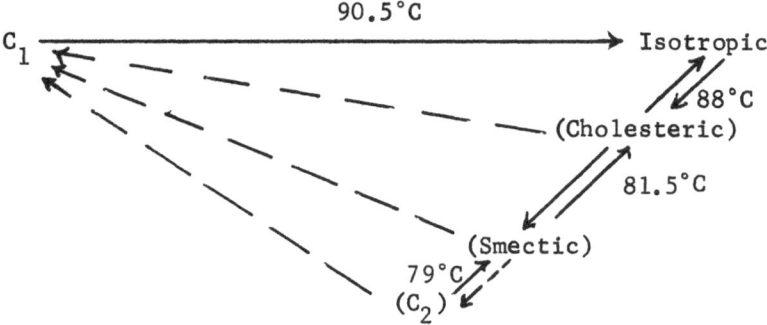

C_1 melts at 90.5° to an isotropic liquid. On cooling there are transformations to the cholesteric phase at 88° and to a smectic phase at 81.5°. If the sample is allowed to cool quickly a second crystalline form C_2 forms from the smectic mesophase. This crystalline form melts to the smectic mesophase at 79°C. The two monotropic mesophases formed in this system are very unstable and within minutes crystals of C_1 will grow from both the cholesteric and the smectic mesophase. The crystals thus formed (C_1) melt at 90.5°C.

Cholesteryl myristate (cholesteryl tetradecanoate)

$$C_2 \xrightleftharpoons{72°C} \text{Smectic} \xrightleftharpoons{79.8°C} \text{Cholesteric} \xrightleftharpoons{85.5°C} \text{Isotropic}$$

Cholesteryl myristate shows two enantiotropic mesomorphic phase transformations. C_2 melts to the smectic phase at 72°C and the smectic phase melts to the cholesteric phase at 79.8°C. The cholesteric phase melts to an isotropic liquid at 85.5°C. The liquid-cholesteric transformation and the cholesteric-smectic transformation show no supercooling effects. However, the smectic phase may be cooled well below 72° before recrystallization of C_2 occurs. Thus the transitions of cholesteryl myristate are similar to those of cholesteryl laurate except no C_1 crystal form is found in cholesteryl myristate.

Cholesteryl palmitate (cholesteryl hexadecanoate)

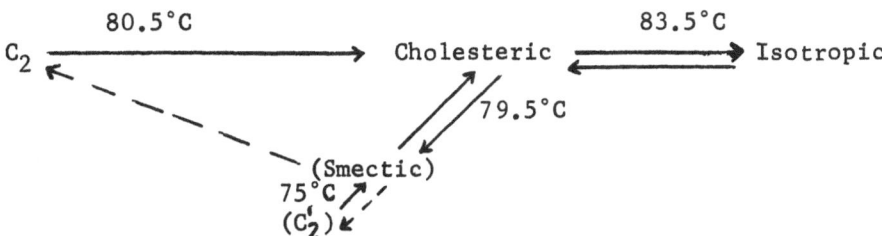

Cholesteryl palmitate (CP) shows 1 enantiotropic mesophase transition (CH⟷I). The behavior of CP is similar to that of
cholesteryl dodecanoate, although the phase transformations are
lower. A second unstable monotropic crystalline phase (C_2') forms
when the smectic phase is supercooled rapidly. C_2' melts to SM at
75°C.

Cholesteryl stearate (cholesteryl octadecanoate)

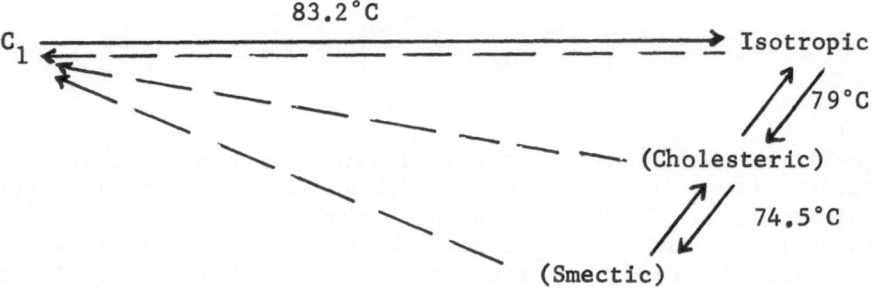

Cholesteryl stearate shows two monotropic mesophase transformations
at 79 and 74.5°C. These phases are highly unstable and crystals
of C_1 grow from both liquid crystalline phases very quickly after
their formation. Further, the liquid phase cooled slightly below
83.2° will grow crystals of C_1.

These phase transformations are very similar to those described by Gray (5) and by Barrall, Porter and Johnson (6). There
are slight differences in the transition temperatures of some of
the esters which may reflect purity or variations in technique.

Esters of Monounsaturated and Polyunsaturated Fatty Acids

Cholesteryl palmitoleate (cholesteryl cis 7-8 hexadecenoate)

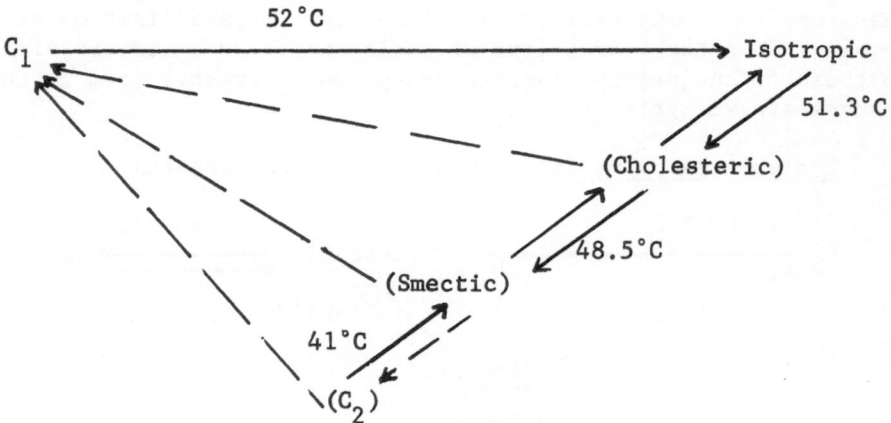

Crystals of cholesteryl palmitoleate (C_1) melt to an isotropic liquid at 52°C. On cooling, the isotropic liquid forms a cholesteric phase at 51.3° and the smectic mesophase at 48.5°C. These phases are reversible without supercooling effects. When the smectic mesophase is cooled to room temperature a second crystalline phase (C_2) forms. These crystals melt to the smectic mesophase at 41°C. Samples left at room temperature for several days undergo a polymorphic crystal transformation from C_2 to C_1. The monotropic cholesteric and smectic mesophases of this cholesterol ester are fairly stable.

Cholesteryl oleate (cholesteryl cis 9-10 octadecenoate)

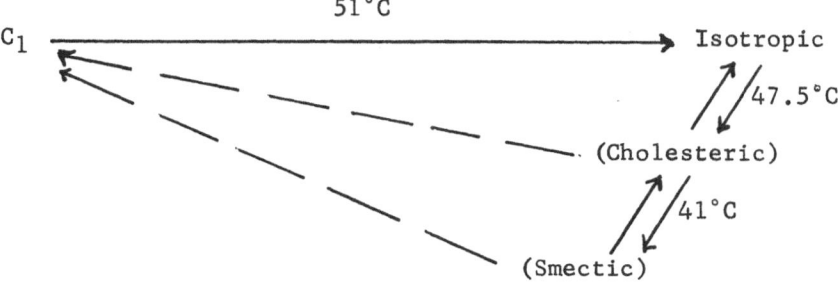

Crystals of cholesteryl oleate (C_1) melt sharply to an isotropic melt at 51°C. Monotropic cholesteric and smectic phases are formed at 47.5°C and 42°C, respectively. The monotropic cholesteric mesophase is fairly stable, but C_1 crystals will grow from the smectic mesophase cooled to room temperature after several minutes.

Polyunsaturated Fatty Acid Esters of Cholesterol

Cholesteryl linoleate (cholesteryl cis 9-10,12-13 octadecadienoate)

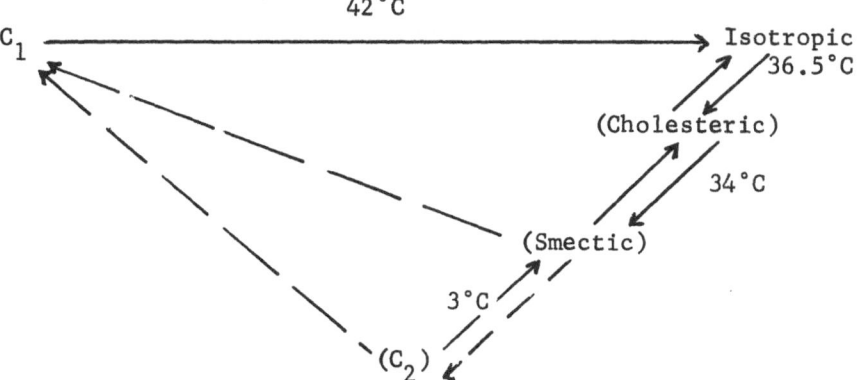

Crystalline cholesteryl linoleate (C_1) melts to the isotropic

liquid at 42°C. Two reversible monotropic phase transformations
to the cholesteric phase and smectic phase occur at 36.5°C and at
34 °C. On cooling to room temperature, recrystallization from the
smectic phase to C_1 will occur very slowly (only after days). If
the sample is cooled rapidly to -40°, a second crystalline form,
C_2, will be formed which melts at 3°C to the smectic mesophase.

Cholesteryl linolenate (cholesteryl cis 9-10, 12-13, 15-16
octadecatrienoate)

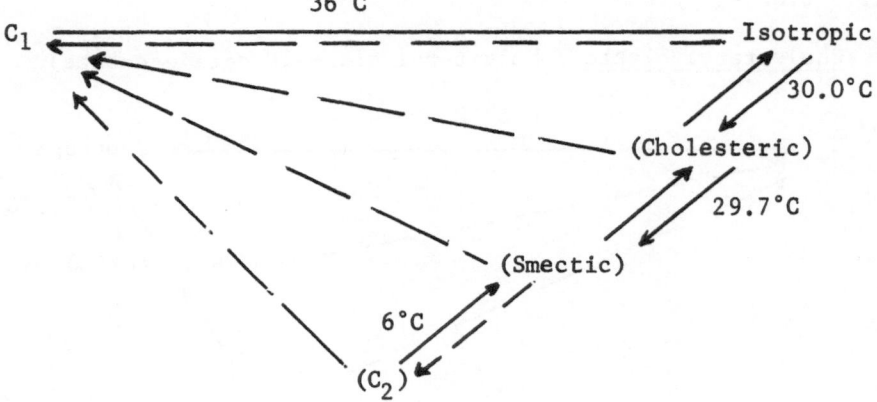

On the initial heating, crystals of cholesteryl linoleate melt to
an isotropic liquid at 36°C.   On cooling two monotropic mesophases are formed. The cholesteric phase forms abruptly at 30.0°C
and at   29.7° the cholesteric phase transforms to the smectic
phase. Note that there is only 0.3° between the phase transformations. This was difficult to detect on the DSC curve as the
cholesteric-liquid endotherm appeared as a shoulder on the smectic-
cholesteric endotherm. However, these transformations were quite
evident from gross and microscopic examination of the sample. The
mesophases of cholesteryl linoleate are unstable and C_1 grows from
a supercooled isotropic phase, the cholesteric phase and the
smectic mesophase. If the sample is cooled rapidly to -40°C the
smectic phase transforms to a second crystalline phase (C_2) which
melts to the smectic phase at 6°C.

Cholesteryl cis 11-12, 14-15, 17-18 eicoso trienoate

$$C_2 \xrightleftharpoons[\quad]{27°C} \text{Smectic} \xrightleftharpoons[\quad]{40°C} \text{?Cholesteric?} \xrightarrow{\quad} \text{Isotropic}$$

This ester of the trienoic C20 fatty acid belongs to the linolenic
fatty acid series (that is, the first double bond counting from the
end methyl group of the fatty acid occurs at the 3 position). Acids
of the linoleic series and the oleic series have their first
double bond placed at the 6 and 9 positions respectively. The
crystals of this cholesteryl ester melt at 27° to the smectic

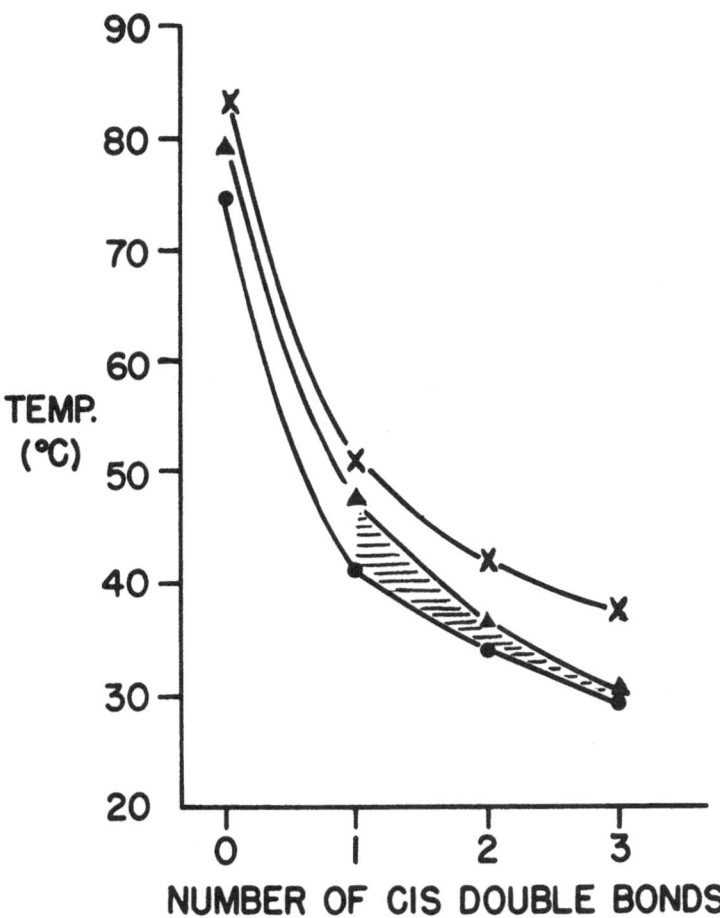

Figure 2. <u>Transition temperature vs the number of cis double
bonds in C-18 fatty acids esterified with cholesterol</u>. X, the
crystal $(C_1) \longrightarrow$ isotropic transition; ▲ , the cholesteric ⟷
isotropic transition; ● , the smectic ⟷ cholesteric transition.
Shaded area refers to the temperature domain of the cholesteric
mesophase. For further explanation see text.

mesophase. No cholesteric mesophase was observed. However, when
the smectic mesophase melts to the isotropic phase a slight flash
of color seems to occur coincident with the transition which
suggests that there may be a cholesteric phase which exists over
such a small temperature range that it is undetected by microscope or DSC.

Cholesteryl arachidonate (cholesteryl cis 5-6,8-9,11-12,14-15
eisoco-tetraenoate)

19°C
C ⇌ ────────── ────────── ────────── ──────────→ Isotropic

Cholesteryl arachidonate shows no mesophases. Crystals (C) of this
ester melt sharply at 19.0° to the isotropic liquid. On cooling
recrystallization to this crystalline phase occurs. In some
slightly less pure samples a second polymorphic form was present
at lower temperatures and on heating melted at 15° and immediately
recrystallized to C which melted again at 19°C.

Cholesteryl cis 4-5,7-8,10-11,13-14,16-17,19-20 **docoso
hexenoate**
        This C_{22} hexenoic acid ester of the linolenate series was a
liquid at 0°C. It did solidify at -72°, but no liquid crystalline
phases were noted on melting.

Results show that both the melting points of the crystalline
material and the liquid crystalline transitions of the unsaturated
and polyunsaturated cholesteryl esters are considerably lower than
those of their saturated homologues. To illustrate this graphically
the crystalline to isotropic transitions ($C_1 \longrightarrow$ isotropic), the
isotropic to cholesteric transitions and cholesteric to smectic
transitions are plotted as a function of the number of cis double
bonds in C18 fatty acids in Figure 2. There is a fall in all 3
transition temperatures as the number of double bonds increases.
Further, considering only the unsaturated cholesteryl esters the
temperature domain of the cholesteric mesophase (shaded area in
Fig. 2) decreases sharply with the number of double bonds. Thus,
for cholesterol oleate the cholesteric phase extends from 42 to
47.5° or covers a domain of 5.5°. The dieonic ester, cholesteryl
linoleate, has a domain of 2.5°C, but cholesteryl linolenate has a
domain of only 0.3°C. In the C20 homologue of cholesteryl linolenate no cholesteric phase could be clearly detected. Further, in
those polyunsaturated cholesteryl esters having 4-6 cis double
bonds, the melting points of the crystals were considerably lower
and no mesophases were observed.

While the specific position of the cis double bonds is un-
doubtedly important in determining the phases present and the
transition temperatures of the phase it can be generally stated

that the more double bonds the lower the transition temperatures. Cholesteryl esters having four or more cis double bonds do not form mesophases.

Two Component Systems

Cholesteryl oleate-cholesteryl linoleate condensed binary phase diagram. Because the 2 most abundant esters of cholesterol in biological systems are cholesteryl oleate and cholesteryl linoleate I have studied the binary mixtures of these 2 cholesterol esters to determine the physical state of mixtures of these compounds as a function of temperature. In Fig. 3 an attempt has been made to construct the condensed binary phase diagram of cholesteryl oleate and cholesteryl linoleate. The temperature is plotted on the vertical axis the percent cholesteryl linoleate or cholesteryl oleate on the horizontal axis. This diagram has been constructed from the DSC heating curves and from gross and microscopic examination. At the lower temperatures it appears that the cholesteryl oleate and cholesteryl linoleate from a compound (X) which contains about 70% cholesteryl oleate and 30% cholesteryl linoleate. Below 3°C in concentrations of cholesteryl linoleate greater than about 65% a crystalline solid solution (SS) of the 2 esters is present (lower right side of Fig. 3). The crystalline form of SS is probably identical to the C_2 form of cholesteryl linoleate which melts to the smectic phase at 3°C (point D). Mixtures containing a high proportion of cholesteryl linoleate form a smectic liquid crystalline phase (SM) at temperatures above 3°C. Depending on the temperature this phase (SM) can contain up to about 50% cholesteryl oleate (point E). The compound X of cholesteryl oleate and cholesteryl linoleate has a melting point about 30°C and melts to form a smectic mesophase and crystals of cholesteryl oleate. Mixtures containing less than 30% cholesteryl oleate form a cholesteric mesophase (CH) before melting to an isotropic solution (I).

For the sake of explanation let us examine several mixtures of different composition of see what happens as the temperature is raised. For example, observe a mixture containing 20% cholesteryl linoleate and 80% cholesteryl oleate as the temperature is increased along line abcdef. Between 0°C and 27°C (point a) this mixture is made of two phases, a polymorphic crystalline type CR' of cholesteryl oleate and crystals of the compound. As the temperature is raised CR' undergoes a polymorphic change to CR at about 27°C. CR is identical to the stable crystalline form (C_1) of cholesteryl oleate. At 30°C (point b) the behavior is similar to a peritectic system and theoretically 3 phases can coexist, CR, X and SM. However, as X melts it forms more CR and a second phase SM such that at a temperature just above 30°C only CR and SM are present (point c). On further heating SM transforms to the cholesteric mesophase (CH) at point d. In the process more CR is

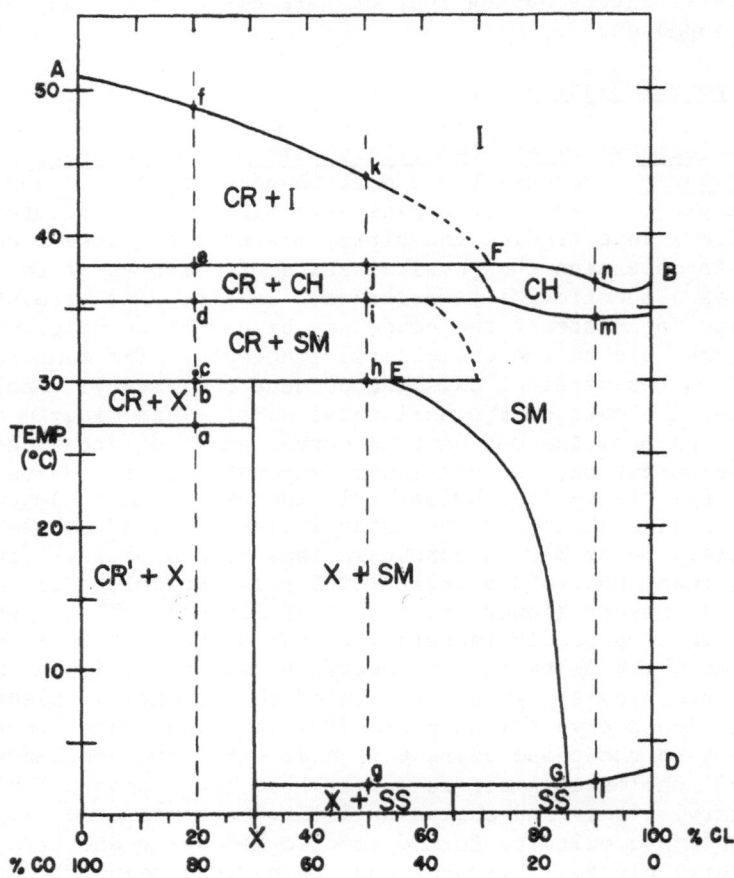

Figure 3. <u>The cholesteryl oleate-cholesteryl linoleate condensed</u>
<u>phase diagram</u>. Vertical axis, temperature °C; horizontal axis
weight % cholesteryl linoleate (CL) or cholesteryl oleate (CO).
CR, crystals of cholesteryl oleate (C_1). CR' polymorphic form of
cholesteryl oleate; X, the crystalline compound of cholesteryl
oleate and cholesteryl linoleate; SS, solid solution of cholesteryl
linoleate containing limited quantities of cholesteryl oleate; SM,
smectic liquid crystalline phase; CH, cholesteric liquid crystalline phase; I, isotropic liquid phase. Points A, B, D, E, F, G
are points explained in the text. Lines abcdef, ghijk and lmn are
explained in the text. Phase boundaries represented by broken
lines are only approximate.

formed. Between d and c gradual melting of CR occurs and the
composition of the melt (I) becomes richer in cholesteryl oleate
(following line fF). At point f the last of CR melts producing an
isotropic solution (I) of composition identical to the original
mixture.

Now consider the behavior of a mixture having a composition
of 50% of each component as it is heated along line ghijk. At
$0^{\circ}C$ this mixture is made up of two solids, compound X and the
solid solution (SS) containing about 35% cholesterol oleate. At
about $2^{\circ}C$ (point g) the solid solution melts to form a mixture of
X and smectic liquid crystalline phase (SM) having a composition
of about 85% cholesteryl linoleate (point g). As the temperature
is increased along gh more X melts to SM having a composition given
by GE. At h theoretically 3 phases can be present (as with point
b, see above), CR, X and SM. At this temperature the reaction
$X \longrightarrow CR + SM$ occurs and at a slightly higher temperature only CR
and SM remain. Points i, j and k are analogous to points d, e and
f respectively. The behavior of this mixture is similar to the
previous mixture (80 cholesteryl oleate/ 20 cholesteryl linoleate)
above $30^{\circ}C$, except that the final melting point of CR (point k) is
somewhat lower because the mixture contains more cholesteryl
linoleate.

Finally consider a mixture containing 90 cholesteryl linoleate,
10 cholesteryl oleate as it is melted along lmn. At $0^{\circ}C$ this mixture is a single phase of crystalline solid solution (SS). At
3° (point l) SS melts directly into another single phase SM. At
$34.5^{\circ}C$ (point m) there is a phase change to a second liquid
crystalline phase, the cholesteric phase (CH). At $37^{\circ}C$ (point n)
CH melts to an isotropic liquid.

One important point should be made about this phase diagram.
As was pointed out in the earlier section on pure cholesteryl
esters, both the mesophases of cholesteryl oleate and cholesteryl
linoleate are monotropic with respect to crystalling phase C_1.
However, the mesophases of the oleate ester are unstable and cholesteryl oleate crystals (C_1) can be readily made to form from the
mesophases by cooling or by being allowed to stay at room temperature for several hours. However, the mesophases of cholesteryl
linoleate are quite stable under certain conditions. The specific
conditions under which this diagram was produced involve the cooling of the mixture from isotropic melt to $-50^{\circ}C$ at a rate of
5°/min. The mixture is then allowed to stand for 30 min at $-50^{\circ}C$
and then reheated at a rate of 5°/min. While crystallization of
C_1 invariably takes place (labelled CR on fig 3) in mixtures having 30% or more cholesteryl oleate no recrystallization of C_1
takes place with less cholesteryl oleate. AF is the final melting
of the oleate crystals while FB is the melting of the cholesteric

phase. Further, if each mixture is allowed to cool slowly to 25°C
and is then left for 48-72 hrs those mixtures containing less than
30% cholesteryl oleate remained in the smectic mesophase (SM) at
25°C. On heating, these mixtures form the cholesteric phase (CH)
at 34.5 to 35.5°C and melt to an isotropic solution along FB. Many
of the mixtures were stable for 6 months. Nevertheless, since SM
and CH are probably monotropic phases one would expect that under
the appropriate conditions of time, nucleation and temperature a
more stable crystalline form (C₁) would be present. Thus, the
true phase (equilibrium) diagram would contain no monotropic phases
and line FB would be replaced with a line representing the melting
of C_1. It is possible that a eutectic exists between the stable
crystalline form (C_1) of cholesteryl oleate and cholesteryl
linoleate. Whether this possible eutectic would have a lower melting point than the mesophases at that composition has not yet been
determined. If the eutectic melting point was lower then a stable
mesophase would be present at the eutectic composition. Experiments are now in progress to resolve this problem. Nevertheless,
from a biologic point of view it seems likely that the phase diagram presented, although probably not truly representing equilibrium conditions, has importance because of the great stability of
the smectic and cholesteric mesophases containing large amounts of
cholesterol linoleate.

   Cholesteryl oleate-cholesterol condensed binary phase diagram.
The condensed phase diagram of cholesterol and its ester of oleic
acid is presented in Fig. 4. The melting point of crystals of
cholesterol is depressed continuously along line AC by increasing
amounts of cholesterol oleate. The melting point of cholesterol
oleate is also depressed by small quantities of cholesterol along
line BC. Point C represents a eutectic composition of 95% cholesteryl oleate, 5% cholesterol (8.4 mole% cholesterol). This
composition melts quite sharply at 46°C to an isotropic fluid. To
illustrate the phase diagram, let us observe a mixture containing
96% cholesterol oleate - 4% cholesterol. At 25°C crystals having
a sphereolyte configuration are present. At 46°C, melting is not
complete until the temperature is raised to 47.5°C. Now, examine
a mixture containing 92% cholesteryl oleate and 8% cholesterol.
Again, one recognizes at 25°C a sphereolyte type of configuration
of the crystals (Fig. 5a). These crystals melt between 45.5 and
46°C. However, there remains a radial network of thin crystals of
cholesterol which now form a ghost-like outline of the sphereolyte
(Fig 5b). These crystals gradually melt as the temperature is
raised, melting completely only at 65°C to an isotropic liquid.
The condensed phase diagram may be divided into the following
regions. Below line DCE-pure cholesterol crystals and crystals
of cholesterol oleate exist. There is a eutectic composition at
point C which has the lowest melting point (46°C). Above ACB a
single isotropic melt is present. In the region ADC crystals of

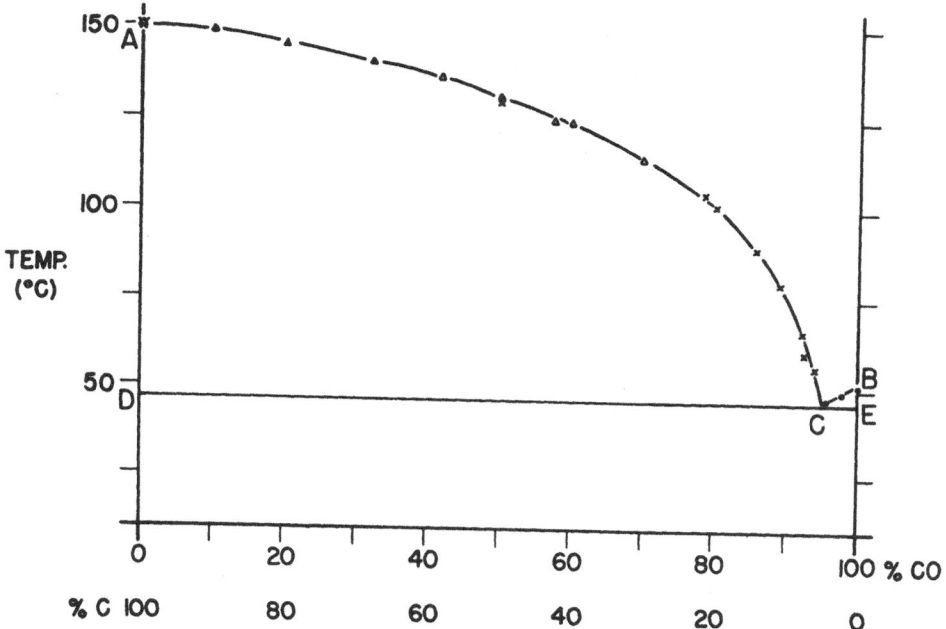

Figure 4. <u>Cholesterol-cholesteryl oleate condensed binary phase</u>
<u>diagram</u>. Vertical axis, temperature ($^{\circ}C$); horizontal axis weight
% cholesteryl oleate (CO) or cholesterol (C). The triangles represent peak of melting curve of cholesterol by DSC, X the final
microscopic melting point of crystals of cholesterol in each mixture and the solid circles the final melting points of cholesterol
oleate. Point A is the melting point of pure cholesterol, Point B
the melting point of the crystalline (C_1) form of cholesteryl
oleate and point C the melting point of the eutectic composition.

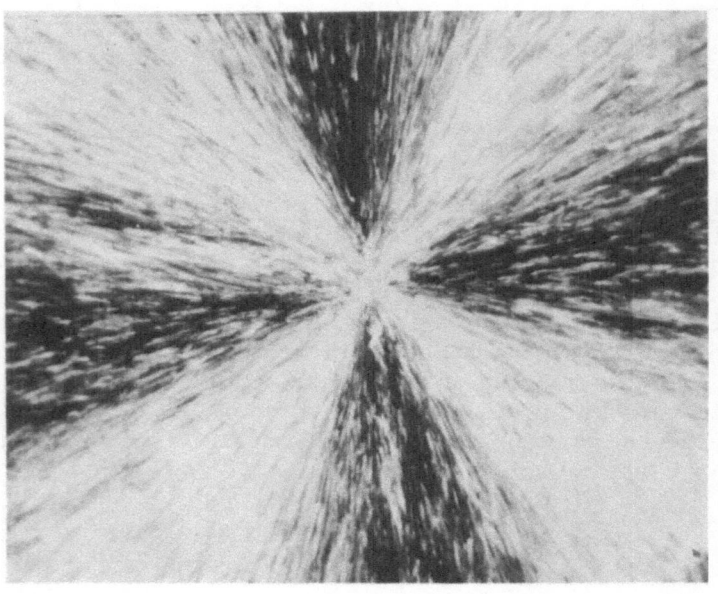

Figure 5. Mixture containing 92% cholesteryl oleate-8% cholesterol. Crossed nicols, 50X. Left, mixture at 35°C showing a sphereolitic crystalline configuration, consisting primarily of crystalline (C_l) cholesteryl oleate. Right, same mixture at 56°C after the eutectic composition had melted. Only a fine network of cholesterol crystals remain. Bright spot in center is dust particle, which served as a nidus for crystallization. For further explanation see text.

cholesterol are in equilibrium with a liquid having a composition
lying along AC. In the area marked BEC crystals of cholesteryl
oleate are in equilibrium with a liquid lying along line BC. It
should be noted that those mixtures containing a high proportion
of cholesteryl oleate (greater than 80%) can be supercooled readily
to produce liquid crystalline transformations similar to those
formed with cholesteryl oleate alone. However, since these transformations are monotropic and rather unstable, they will not be
discussed further.

 Cholesteryl oleate-triolein condensed binary phase diagram.
The binary phase diagram formed by the mixture of triolein and
cholesteryl oleate is given in Fig. 6. The melting point of the
crystalline form (C_1) of cholesteryl oleate is depressed by increasing concentrations of triolein. There appears to be a eutectic at about 4% cholesteryl oleate--96% triolein, although the
melting point of the eutectic mixture was only half a degree lower
than that of pure triolein. On heating a mixture containing 50%
triolein and 50% cholesteryl oleate from 0°C, one notes that the
eutectic mixture (nearly pure triolein) melts at about 4°C. Cholesteryl oleate crystals remain and gradually melt until only an
isotropic solution is present (46.5°C). Mixtures containing less
than 5% cholesteryl oleate could not be nucleated or made to grow
cholesteryl oleate crystals. The phase diagram is thus composed
of the following regions. Above line ACB a single isotropic liquid
phase is present. Below DCE crystals of triolein and of cholesteryl
oleate are present. In the very small zone labeled ADC, there is
a mixture of pure triolein and crystals of cholesteryl oleate
which melt along line AC. In zone BCE crystals of cholesteryl
oleate are present which melt along line CB.

 The cholesterol-triolein condensed phase diagram. In Fig. 7
the phase diagram of cholesterol-triolein system is presented.
Triolein, like cholesteryl oleate (Fig. 4), also depresses the melting point of cholesterol. The line AC, in fact, gives the solubility of cholesterol in triolein as a function of temperature.
The shape of AC suggests that a eutectic is present at C but this
could not be demonstrated clearly by microscope or DSC curves.
Partington (18) produced similar diagrams for long chain fatty
acids and concluded, as must also be concluded from this diagram,
that no compounds are formed between either long chain fatty acids
or triolein and cholesterol.

Three Component Systems (Ternary systems)

 Cholesterol-cholesteryl oleate-triolein. In order to get an
idea of the interrelations of these three lipids as a function of
temperature, a large series of mixtures of the three lipids were
made up in thick-walled glass tubes and the mixtures sealed under

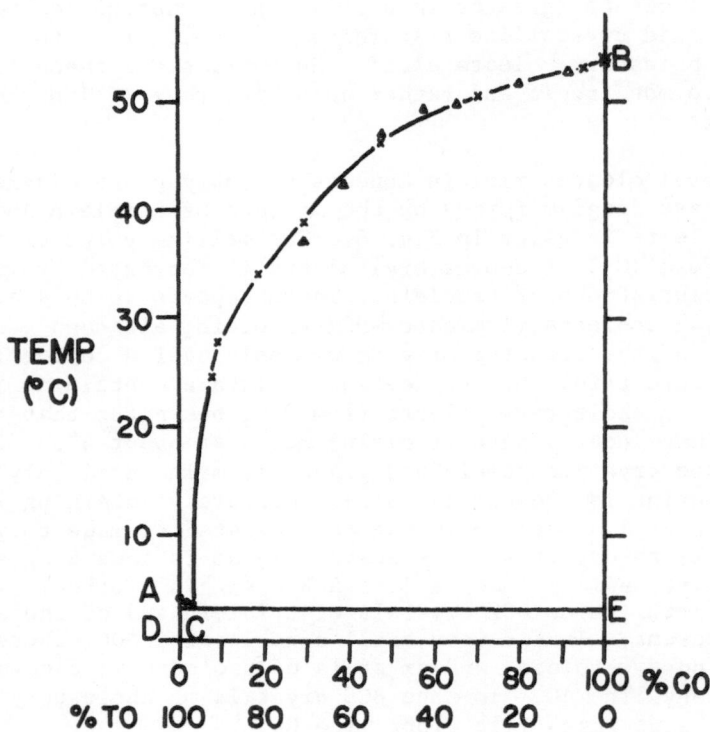

Figure 6. <u>Triolein-cholesteryl oleate condensed binary phase</u>
<u>diagram</u>. Vertical axis, temperature, horizontal axis percent cho-
lesteryl oleate (CO) or triolein (TO). Triangles represent the
peak of the cholesteryl oleate melting DSC curve, X the final
melting point of cholesteryl oleate in the mixture and the solid
circles represent the final melting point of triolein. Point A
represents the melting point of pure triolein, B the melting point
of crystals (C_1) of pure cholesteryl oleate and C the melting
point of the eutectic composition.

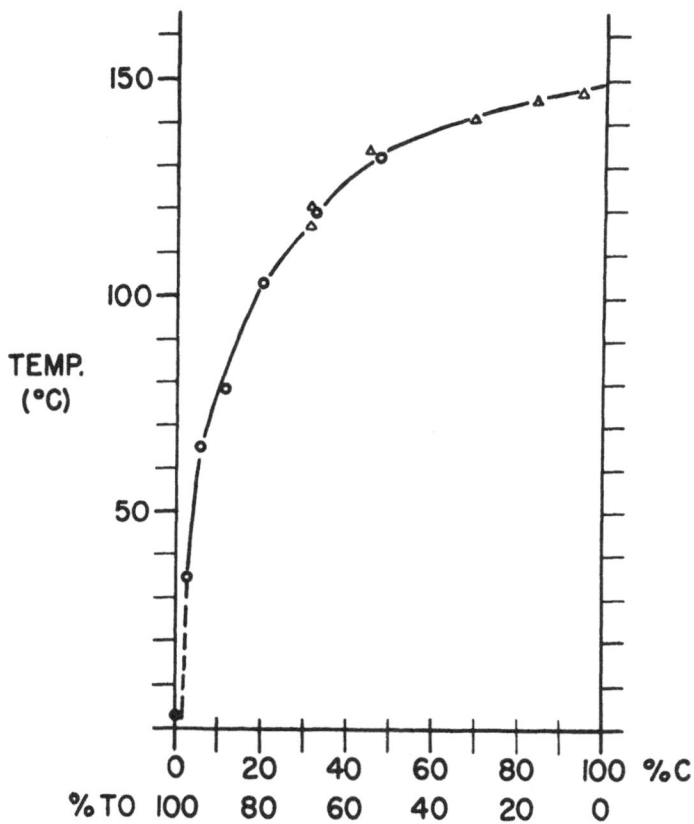

Figure 7. Triolein-cholesterol condensed binary phase diagram.
Vertical axis, temperature °C, horizontal axis percent triolein
(TO) or cholesterol (C). Triangles represent peak of the DSC
melting curve of cholesterol, open circles represent the final
melting point of cholesterol crystals in triolein and the solid
circle represents the melting point of pure triolein. Although
a eutectic may be present at about 1% cholesterol it could not
be clearly defined.

nitrogen.   Each mixture contained approximately 1 gram.   The mixtures were then heated until all solids had melted, shaken and
were then cooled to 0°C.   After one week the temperature of the
mixtures was raised at approximately 5° levels and allowed to stay
at each temperature level for a period of 7-14 days.   The mixtures
were agitated intermittently and observed for the amount of solid
present.   When equilibrium was reached (no further decrease in the
amount of solid present) the temperature was raised to the next
level.   When the mixture was completely melted the temperature was
noted.   The phase diagram given in Figure 8 shows only the composition range of the mixtures which were completely liquid at the
temperature of the experiment.

At 25°C only mixtures very close to 100% triolein are liquid.
This could have been predicted from the binary phase diagrams
cholesterol-triolein and cholesteryl oleate-triolein.   As the temperature is increased, the cholesteryl oleate melts into the triolein but only a small amount of cholesterol is melted so that at
38° the zone of one liquid phase has increased appreciably only in
relation to cholesteryl oleate.   At 52°C cholesteryl oleate and
triolein are mutually soluble in all proportions.   A total of
about 12% cholesterol can be dissolved in a mixture fairly high in
cholesteryl oleate.   As the temperature is increased the amount of
cholesterol that can be dissolved in mixtures of cholesteryl oleate
and triolein gradually increases.   Of course, at 150° the entire
system is one single oily phase.

Discussion

Generally speaking, from a biologic point of view, lipids may
be lumped into two large classes.   The first class are those lipids
which are involved in the structure of plasma membranes, the membranes of cellular organelles and the circulating lipoproteins.
The second class are those lipids which are primarily stored to be
either used later as sources of energy or as building blocks for
structural lipids.   The former group includes primarily the complex
lipids such as phospholipids, glycolipids and sulfolipids, which
for the most part are insoluble, but swell in water to form lyo-
tropic crystals (16,19,20).   Many individual species and mixtures
of these lipids are capable of forming an endless series of bimolecular leaflets (the lamellar liquid crystalline phase) which
bears analogy to the plasma and organelle membranes found within
the cell.   Cholesterol, while not capable of swelling alone in
water, is often found in association with plasma membranes and
indeed has been shown to interdigitate between lecithin molecules
in the bilayers of the lecithin liquid crystalline lattice (21,22,
23) and the lipids of the erythrocyte membrane (24).   Thus, while
pure cholesterol has a melting point greatly above normal body
temperature, it can be solubilized within a lamellar liquid
crystalline lattice in ratios up to 1 molecule of cholesterol

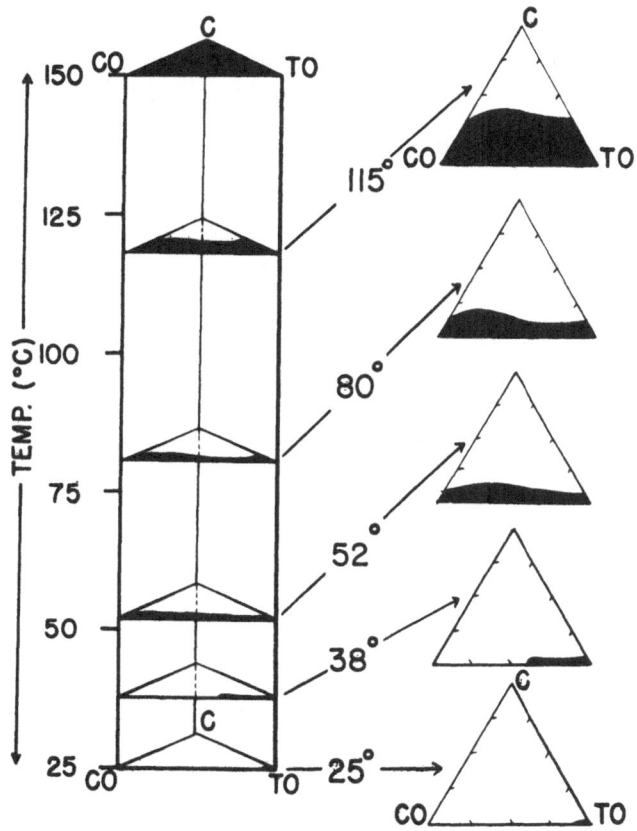

Figure 8. <u>The cholesteryl oleate-cholesterol-triolein ternary</u>
<u>phase diagrams as a function of temperature</u>. On the left the
ternary phase diagram is expressed as a regular prism with each
edge representing one of the three components. Representative
ternary (3 component) systems at specific temperatures are shown
on the right. CO, cholesteryl oleate; C, cholesterol; TO,
triolein. The darkened area of the triangle represents those
compositions forming one single isotropic liquid. The rest of
the phase diagram is made up of mixtures of two or three phases.
As the temperature is increased the domain of compositions which
are in the completely melted state increases. For further explanation see text.

to 1 molecule of phospholipid (21 22 23). The storage lipids, for
the most part, are less polar lipids and while some, such as triglycerides, fatty acids and free cholesterol are capable of spreading at an oil-water interface (19), others such as the long chain
fatty acid esters of cholesterol are almost completely nonpolar
and will not spread. The major storage lipids include triglyceride,
cholesteryl esters of long chain fatty acids and to some extent
free cholesterol and fatty acids. Because living organisms require
mobility, it would seem important that the storage lipids be maintained in a liquid state. It is hard to imagine how storage lipids
in a crystalline state could readily diffuse to be metabolized by
enzymes or even to be mobilized for transport. Thus, I have begun
a series of studies designed to define the physical state of a
number of chemical classes of those lipids which can be roughly
defined as storage lipids. Biologically, these lipids are found in
adipose tissue, liver, adrenal glands, the gonads and circulating
lipoproteins in small drops which, on superficial examination,
appear to be composed of a single liquid (oil) phase. In this preliminary paper the physical state of cholesteryl esters, free
cholesterol and triglyceride alone and in mixtures is reported.
These studies give a general idea of the physical state of individual components and solubility relationships of mixtures of
these lipids as a function of temperature.

The present studies show that the individual cholesteryl esters
of biological origin show a marked variability in their transition
temperatures. The esters of saturated long chain fatty acids have
melting points some 40° above body temperature. This makes them
unsuitable to function as storage lipids, at least in the pure
state, since they would be solid and quite insoluble. On the other
hand, the esters of the unsaturated fatty acids have transition
temperatures and melting points fairly close to body temperature.
It would, however, appear that the monounsaturated esters,
cholesteryl oleate and cholesteryl palmitoleate, have transition
temperatures far enough above body temperature and unstable enough
monotropic liquid crystalline phases that they would probably also
be unsuitable as a storage lipid in pure states. On the other
hand, cholesteryl linoleate, which has 2 cis double bonds, could
exist, at least in part, in a liquid crystalline state at body
temperature, and therefore might be considered suitable as a
storage lipid. The binary system cholesteryl oleate and cholesteryl
linoleate show that there are a range of compositions high in
cholesteryl linoleate which form very stable liquid crystalline
phases in the temperature range present in warm-blooded animals.
While the liquid crystalline phase consists of molecules in a
partly organized state which gives the mesomorphic phases a viscosity considerably higher than that of the liquid phase (13,14),
these states are rather fluid and therefore may represent a suitable
structure for storage lipid. However, because of the viscosity and
the partial orientation of the molecules within these phases, it is

a non-liquid state. There are, obviously, other cholesteryl esters present in both beta lipoproteins and in the cholesteryl ester fraction isolated from lipids of the plaque. These include small amounts of saturated esters (mainly cholesteryl palmitate) and some polyunsaturated fatty acid esters. It would be premature to try to predict the physical state of the lipids in the plaque without 1) examining the effect of the other esters on the physical state of the total mixture and 2) without further careful study of both the chemical composition of the lipid and the distribution of lipid within the plaque. However, droplets of lipid present a single phase and the composition of that phase in the long run must be determined by the physical characteristics of the molecules within the phase.

In the past it has been shown that large amounts of cholesterol can be solubilized in phospholipid bilayers (21,22,23). What about the solubility of cholesterol in the major storage lipids, such as triglycerides and cholesterol ester? The present work clearly shows that the solubility of free cholesterol in either cholesteryl oleate (Fig. 4) or triolein (Fig. 7) is very small at body temperature. Further, the three component system shows that cholesterol is only sparingly soluble in mixtures of cholesteryl oleate and triolein and only when temperatures well above body temperatures are large amounts of cholesterol solubilized. Reports on the amount of free cholesterol in atherosclerotic plaques vary. Luddy et al (12) and Dayton et al (11) found quite large amounts of free cholesterol in plaques taken from autopsy specimens. However, these high values could be due to postportem hydrolysis of cholesterol ester. Smith et al (9,10) find somewhat less free cholesterol. Theoretically the free cholesterol could partition between an oil phase which consists primarily of cholesteryl ester and into the phospholipid bilayers of cell membranes and organelles associated with the plaque. It is also possible that cholesterol is solubilized by other substances (protein, nucleotides, glycoproteins, etc.), but this has not yet been shown to be quantitatively important. Is the total amount of cholesterol present greater than the total solubilizing capacity of these two systems? This cannot be answered at the present time, but it might appear that in the plaques containing a high amount of "amorphous material" (9,10) that there is an excess of free cholesterol beyond that which can be solubilized readily by 1) an oil phase containing small amounts of glyceride and large amounts of cholesteryl ester and 2) the phospholipids present in that plaque. If this is the case, one might expect to find crystals of true cholesterol occurring in these lesions.

Triglycerides are storage lipids found in adipose tissue, liver and lipoproteins of low density. In higher animals dietary fat is hydrolyzed in the intestinal lumen, absorbed and then re-synthesized to triglyceride in the intestinal mucosal cell. It is

quite possible that the process of transport out of the phase
either of molecules in their intact chemical state or through
biochemical reactions such as hydrolysis might proceed at a much
slower rate and thus allow the accumulation of lipid in this state.
In the atherosclerotic plaque, it has been shown that a large
proportion of the lipid is cholesterol ester (9-12). Although no
vigorous studies on the physical state of the lipids in the
atherosclerotic plaque have been carried out to date, Stewart (25)
has shown that some substance (presumably lipid) present in both
betalipoprotein and in atherosclerotic plaques was anisotropic and
had a negative sign of birefringence. Unfortunately, the substance observed was not analyzed and careful temperature studies
were not carried out. He, nevertheless, suggested that perhaps
the lipids in arterial walls and perhaps in beta lipoprotein
could be in the liquid crystalline state. Considering the sign
of birefringence, the liquid crystalline state would have to be
the cholesteric liquid crystalline phase. Since about 60-80% of
cholesteryl esters of the beta lipoproteins (2,8) and of the
cholesteryl esters found in atherosclerotic lesions (9-12) are
cholesteryl oleate and cholesteryl linoleate, the condensed phase
diagram of these systems may be important. Again, the condensed
phase diagram shows that at body temperature mixtures of cholesteryl
oleate and cholesteryl linoleate in which there is less than about
30% cholesteryl oleate present are in a fairly stable cholesteric
liquid crystalline phase. Thus, it is possible that the anisotropic
lipid droplet which Stewart observed with negative birefringence
represented the cholesteric liquid crystalline phase. If that were
the case, one would predict that these droplets would be made up
of almost pure cholesterol esters with very little triolein or free
cholesterol dissolved in them. We have, in fact, recently found
that isolated lipid droplets of the fatty streaks of human atherosclerotic plaques are nearly pure cholesteryl ester and are present
in a liquid crystalline phase which has a transition point from
cholesteric to isotropic (varying on the source of the lipid) from
37° to 42°. Smith et al (9,10) have recently described histo-
logically two types of oil droplets occurring in the atherosclerotic
plaque, the first type, perifibrinous lipid, is extracellular and
has a cholesteryl oleate to cholesteryl linoleate ratio very
similar to beta lipoprotein in that it is relatively high in
cholesteryl linoleate. On the other hand, the second type designated, intracellular fat droplets, have a different distribution
of fatty acids in their cholesteryl esters, being very high in
cholesteryl oleate. Although no attempt was made to examine the
physical state of these deposits in the natural state, one might
predict that if these droplets are in the liquid state that they
would be unstable and that precipitation of cholesteryl oleate
could readily occur within these drops. Once the excess cholesteryl
oleate was precipitated, it would not be normally removed and could
act as a nidus for accumulation of further esters. It could well
be that a portion   of the cholesteryl oleate in the plaque is in

released into the lymph in large dietary (exogenous) chylomicrons. These particles are transported in the plasma to the liver and other parts of the body by the systemic circulation. Smaller endogenous chylomicrons are probably synthesized primarily in the liver and released into the circulation. While these particles have been called lipoproteins they are really small emulsion droplets to which a very small amount of protein is attached. The oil phase is largely triglyceride and the emulsifiers are primarily phospholipids, free cholesterol and perhaps to some minor degree fatty acid and protein. The composition of the oil phase of exogenous chylomicrons like that of adipose tissue is analogous to the oil phase shown in Fig. 8 for the cholesteryl oleate-cholesterol-triolein diagram at 25°C (i.e. very little cholesteryl ester or free cholesterol is present in the oil phase). The monolayer of emulsifier at the surface of the oil droplet has a composition which obeys the cholesterol-lecithin-water phase diagrams (21,22, 23) in that less than one cholesterol molecule per phospholipid molecule is present in this layer (26).

This discussion has been largely speculative in nature principally for two reasons: 1) data concerning the composition and physical state of specific lipid phases occurring naturally or pathologically is not readily available and 2) the application of the data obtained in the present studies is limited because the number and types of lipid species studied has been limited. Nevertheless, the broad outline of the interrelations between cholesterol, its esters and triglycerides described here may be applicable to the interrelations of these lipids in biological systems and in pathological states. Only time and more data will tell.

Acknowledgements

The author wishes to thank Mr. John Steiner for excellent technical assistance and Mrs. C. Castles for secretarial help.

Supported by NIH Public Health Research Grant AM 11453.

REFERENCES

1.   Bloch, K.   Science, 150:3692, 19, 1965.

2.   Frederickson, D.S., R.I. Levy and R.S. Lees. New Eng. J.
      Med. 276:1, 34; 276:3, 148; 276:4, 215 and 276:5, 273, 1967.

3.   Jaeger, Rec. Trav. chim. 25:334, 1906.

4.   Friedel, G. Ann. Physique, 17:273, 1922.

5.   Gray, G.W. J. Chem. Soc. pt. 3, 3733, 1956.

6.   Barrall, E.M., II., R.S. Porter and J.F. Johnson. J.
      Physical Chem. 70:385, 1966.

7.   Chistyakov, I.G. Soviet Physics - Crystall. 8:1, 57, 1963.

8.   Freeman, N.K., F.T. Lindgren and A.V. Nichols. Progress in
      Chem. of Fats and other Lipids. 6:215, 1963.

9.   Smith, E.B., P.H. Evans and M.D. Downham. J. Atheroscler.
      Res. 7:171, 1967.

10.   Smith, E.B., R.S. Slater and P.K. Chu. J. Atheroscler. Res.
      8:399, 1968.

11.   Dayton, S., S. Hashimoto and M.L. Pearce. Circulation, 32:
      911, 1965.

12.   Luddy, F.E., R.A. Barford and R.W. Riemenschneider. J. Biol.
      Chem. 232:843, 1958.

13.   Brown, G.H. and W.G. Shaw. Chemical Rev. 56:6, 1049, 1957.

14.   Gray, G.W. Molecular Structure and the Properties of Liquid
      Crystals. Academic Press Inc., New York, N.Y., 1962, p. 114.

15.   Barrall, E.M., R.S. Porter and J.F. Johnson. Mol. Crystals,
      3:103, 1967.

16.   Small, D. M. J. Lipid Res., 8:551, 1967.

17.   Barrall, E.M., II, R.S. Porter and J.F. Johnson. J. Physiol.
      Chem. 71:1224, 1967.

18.   Partington, J.R. J. Chem. Soc. 99:313, 1911.

19.   Small, D.M. J. Am. Oil Chem. Soc. 45:108, 1968.

20.   Dervichian, D.G. in Progress in Biophysics and Molecular
      Biology, 14, 263. (Butler, J. and H. Huxley, Eds. Academic
      Press, New York, 1964).

21. Small, D.M., M. Bourges and D.G. Dervichian. Nature, 211: 816, 1966.

22. Small, D.M. and M. Bourges. Mol. Crystals, 1:541, 1966.

23. Bourges, M., D.M. Small and D.G. Dervichian. Biochim. et Biophys. Acta 137:157, 1967.

24. Rand,R.P. and V. Luzzati. Biophys. J. 8:1, 125, 1968.

25. Stewart, G.T. in Ordered Fluids and Liquid Crystals. Adv. in Chem. Series 63, p. 141, 1967.

26. Zilversmit, D.B. J. Lipid Res. 9:180, 1968.

THE EFFECT OF HYDROCARBON CONFIGURATION AND CHOLESTEROL ON

INTERACTIONS OF CHOLINE PHOSPHOLIPIDS WITH SULFATIDE*

Morris B. Abramson and Robert Katzman

The Saul R. Korey Department of Neurology, Albert

Einstein College of Medicine, Bronx, New York

ABSTRACT

Mixtures of sphingomyelin and sulfatide or lecithin and sulfatide showed a titration capacity in the acid range that was greater than that of any of these lipids titrated alone. This is explained as the results of the ionic interaction of the sulfate and quaternary amine of neighboring molecules thus making the phosphate of the choline lipid available for titration.

This effect was greatest for sulfatide with sphingomyelin or with dipalmitoyl lecithin and decreased with egg lecithin and soybean lecithin. This order follows the increase in molecular cross-sectional area of the choline lipid and the increased separation of the charged groups of the mixed lipids. Incorporation of the neutral lipid, cerebroside, into sphingomyelin-sulfatide had little effect. Cholesterol, however, decreased the interaction of sulfatide with sphingomyelin or lecithin. This probably resulted from the separation of the charged lipids by the molecules of cholesterol. These data point to an intermolecular ionic linkage of choline lipids with anionic compounds in lipid membrane systems. The absence of any effect of cholesterol on the titration of choline lipids alone supports the view that lecithin and sphingomyelin without acidic components exist as dipolar ions with intramolecular bonding.

*This work was supported by Multiple Sclerosis Grant 503A, U.S. Department of the Interior, Office of Saline Water Grant 14-01-0001-1277, and U.S. Public Health Service Grants NB 03356 and NB 01450.

The addition of salts to systems of sphingomyelin-sulfatide or lecithin-sulfatide at pH 4.0 gave a cation-proton exchange with $K^+ = Na^+ >$ choline$^+$.

Sodium dodecyl sulfate mixed with sphingomyelin or lecithin gave reactions similar to but less than those of sulfatide with these lipids.

Titrations were also performed on mixed dispersions of dipalmitoyl lecithin and sodium dicetyl phosphate. These systems showed that all the phosphate groups of the latter were available for titration and lends weight to the assumptions made in the interpretations of the systems containing sulfatide.
- - - -

Despite the widespread occurrence and involvement of the choline phospholipids, lecithin and sphingomyelin, in biologic membranes (O'Brien, 1967), our knowledge of the molecular structures of these compounds in membrane systems and their inter-action with other lipid components is fragmentary. In a report from this laboratory (Abramson and Katzman, 1968), we described the change of the dipolar structure of sphingomyelin or lecithin systems upon the addition of sulfatide. In these mixed lipid aggregates, the interaction between the fully ionized acid groups of sulfatide with the positive region of the choline of a neigh-boring lipid molecule leads to the freeing of the phosphate of that molecule. The change of the phosphate group is easily recognized by the difference in the titration characteristics of the single lipids and the mixed lipids between pH 3.5 and 10.

In this paper we describe further studies of the ionic inter-actions of choline lipids. The effects of different hydrocarbon structures and the inclusion of cholesterol on the interaction with sulfatide is given.

EXPERIMENTAL

Materials

Sphingomyelin obtained from freshly slaughtered beef brain was purified by the method described earlier (Abramson, et al., 1965). Sulfatide, also from beef brain, was extracted and converted to the sodium salts as in earlier work (Abramson, et al., 1967). Egg lecithin (Sylvana) was purified by silicic acid chromatography. The L-α-dipalmitoyl lecithin was from Mann Research Laboratories, New York. The cerebrosides were a gift from Dr. W.T. Norton. Cholesterol was obtained from Eastman Kodak. Sodium dodecyl sulfate and dicetyl phosphoric acid were

obtained from K and K Laboratories, Plainview, New York. The
former was recrystallized 3 times. Dicetyl phosphoric acid was
converted to the sodium salt by dissolving in chloroform-methanol
(2:1, v/v) and partitioning with 0.1M NaCl maintained at pH 8.5.
The sodium salt was separated and repartitioned. All lipids used
gave single bands on thin-layer chromatograms performed as
described below.

Dispersed Systems

Lipid systems were prepared in a 20 ml glass tube. Egg
lecithin which was stored at -10° in solution in chloroform-
methanol (2:1, v/v) was deposited in a weighed tube by vacuum
evaporation of the solvent. The other lipids used were in solid
form and were weighed into the glass tube. For single and mixed
lipids, the weighed solids were first dissolved in chloroform-
methanol (2:1, v/v). To dissolve the sodium sulfatide required
gentle warming. On evaporation of the solvent by nitrogen and
then in vacuum, a deposit of lipids was formed on the bottom of
the tube. After adding 5 ml water, the system cooled in an ice-
water bath, was dispersed by a 1 minute exposure of ultrasonic
radiation using an MSE ultrasonic generator. All systems containing anionic lipids dispersed readily and showed low turbidity
with no tendency to settle. Addition of cholesterol produced
more turbid systems. Representative systems produced this way
were studied by polarized light and electron microscopy. They
showed structures similar to those reported by Bangham, et al.,
1965.

Titrations

Titrations were performed at $24 \pm 1°$ C in the 20 ml tube
covered by a plastic cap. A stirrer, combined glass and reference
electrode, 0.1 ml microburet (Manostat), and a tube supplying CO_2-
free moist nitrogen, fit tightly into the plastic cover. Additions
of 1 to 3 μl of standard 0.100 N NaOH or HCl were made by means of
the microburet, and the system after mixing was stirred at a very
slow constant rate for 10 to 30 minutes until a stable pH value
was measured on a pH meter (Corning Model 12). Most titrations
were performed with duplicate samples. Blank titrations were made
under identical conditions except for the absence of lipids in the
system. By subtracting the blank from the titration curve, a
corrected curve was obtained.

Thin-layer Chromatography

To assess the extent of degradation of the compounds during the course of the experiments, 0.1 ml samples were taken before and after the titration. In some instances, additional samples were taken at selected points during the titration. These samples were dried by a stream of nitrogen and dissolved in chloroform-methanol (2:1, v/v). Thin-layer chromatograms were obtained in silica gel G (Merck). Small quantities of the original lipids were run on each chromatogram as standards. Chromatograms were developed with chloroform-methanol-water-ammonia (29%) (70:30:4:1, by volume). They were stained with iodine vapor and charred with sulfuric acid. Sphingomyelin, sulfatide and dipalmitoyl lecithin showed very slight or no degradation products. Egg lecithin underwent a small amount (<5%) of degradation after titration in the basic range. Soybean lecithin, however, degraded to a much greater degree.

RESULTS

Effect of Area of Choline Molecule

Titrations were carried out on aqueous systems of sodium sulfatide, sphingomyelin, and three forms of lecithin. Comparable systems were prepared using sodium sulfatide mixed with sphingomyelin or lecithin. The titrations of these mixed systems under similar conditions showed an increased acid capacity in the pH range studied compared with the titration of sodium sulfatide alone or of any choline lipid alone. For these mixed systems, we then envisage analogous reactions between the anionic sulfate with the positive nitrogen of the choline moiety (Fig. 1). In comparing the extent of interaction of the 2 lipids, we measured the number of μmole H^+ bound per μmole of 1:1 lipid complex between pH 7 and 3.5. Titrations to lower pH levels were avoided because of flocculation and the greater possibility of degradation of the lipids. Since the titration curve is nearly vertical at pH 7 (Fig. 2), measurements can be made from this point.

Quantitative differences are shown by the different choline compounds (Table 1). Whereas the system of either sphingomyelin or dipalmitoyl lecithin with sulfatide showed extremely large acid capacity, reacting with 0.42 - 0.44 μmole of H^+/μmole complex between pH 7.0 and 3.5, the other two systems with egg or soybean lecithin and sulfatide showed much smaller effects.

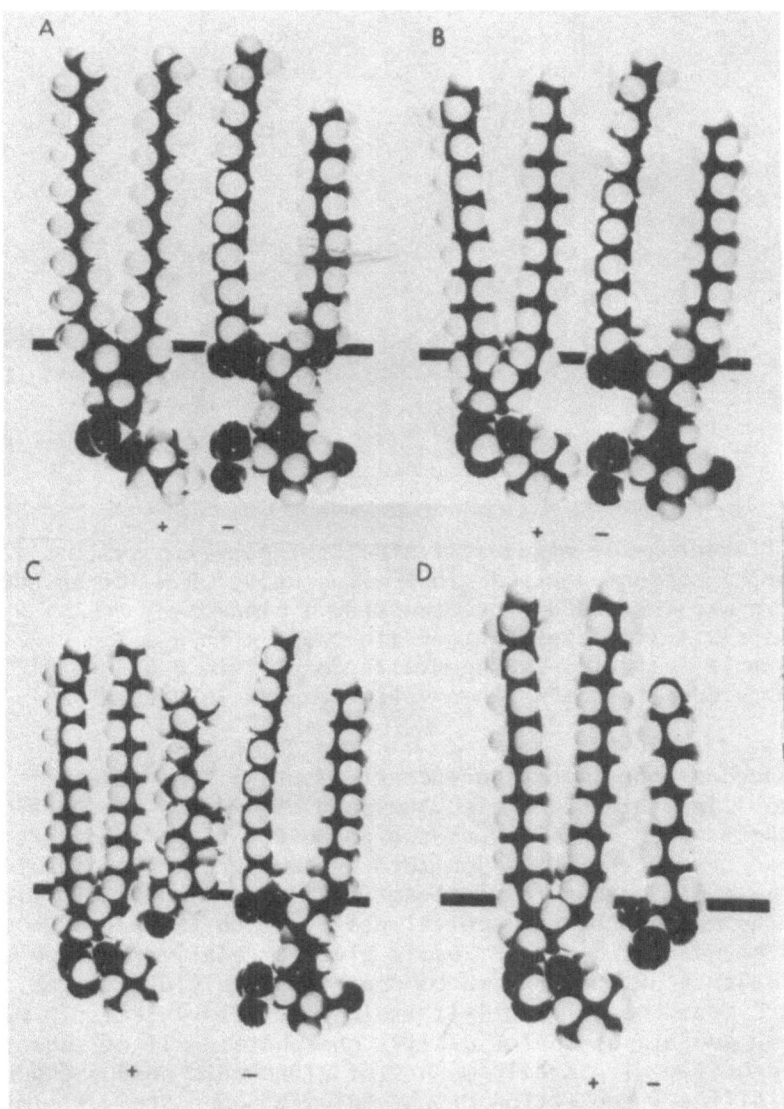

Fig. 1. Corey-Pauling Models of interacting lipid molecules. The
continuous aqueous phase is below the horizontal line. The
$-N(CH_3)_3^+$ of sphingomyelin or lecithin and $-SO_3^-$ groups of sulfatide
are indicated by their charges. The metallic cations are not
shown.
    A. sphingomyelin - sulfatide
    B. dipalmitoyl lecithin-sulfatide
    C. dipalmitoyl lecithin-cholesterol-sulfatide
    D. dipalmitoyl lecithin-dodecyl sulfate

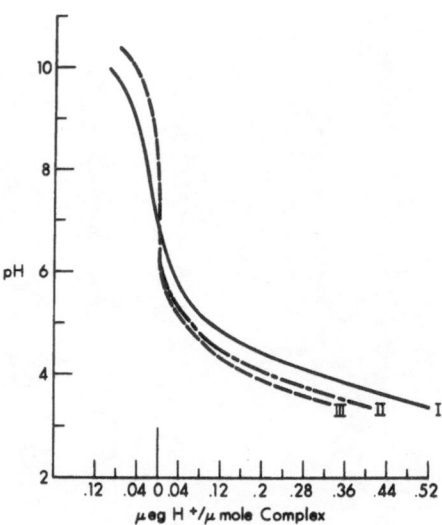

Fig. 2. Titrations of mixed sulfatide-sphingomyelin systems (1:1 molar ratio) performed at 24° in 5 ml using 0.100 N HCl or NaOH. Each curve was corrected by subtracting a blank.
I   7.9 μmole sulfatide-sphingomyelin complex in H₂O.
II   7.7 μmole sulfatide-sphingomyelin complex in 0.01 M NaCl.
III 7.5 μmole sulfatide-sphingomyelin complex in 0.05 M NaCl.

Analogous experiments were carried out with mixtures of dipalmitoyl lecithin and the sodium salt of dicetyl phosphoric acid. These experiments served two purposes: first to demonstrate the degree of dispersion of lecithin systems and the availability of all the ionic groups for interactions; secondly to determine whether the measure of the equivalents of H⁺ that react with these mixtures between two set pH levels gives a relative measure of the equivalents of the ionized component in the lipid systems. We prepared systems containing 9-11 μmole dipalmitoyl lecithin with increasing amounts of sodium dicetyl phosphate. All of these dispersed easily with small amounts of ultrasonic energy and had low turbidities. The action of the anionic phosphate was similar to that of sulfatide in increasing the dispersibility of lecithin.

Typical titration curves of such mixed systems in water or 0.06M NaCl are shown in Fig. 3. Identifiable equivalence points in salt-free systems were seen at pH 4.25 and 9.3. In this range, the contribution of the dipalmitoyl lecithin to the titration is small. In Fig. 4 is shown the reasonably good agreement of the μequivalents H⁺ that react in this range with the μmoles sodium dicetyl phosphate present in the systems. This gives support to the view that the anionic component is completely codispersed with

TABLE I: Effect of Dimensions of Choline Lipid on its Interaction with Sodium Sulfatide. Titrations were performed at 24°.

The acid capacity of the mixed lipid systems can be compared with the single lipids given below.

Systems (1:1 molar ratio)	Acid Capacity (μmole H⁺/μmole complex)		Monolayer Radius (Å) Choline Lipid[a]
	pH 7.0-4.0	pH 7.0-3.5	
Sphingomyelin-sulfatide	0.30	0.44	3.7
Dipalmitoyl lecithin-sulfatide	0.29	0.42	3.7
Egg lecithin-sulfatide	0.19	0.30	4.3
Soybean lecithin-sulfatide	0.10	0.21	4.5
Single Lipids	(μeq H⁺/μmole lipid)		
Sphingomyelin	0.083	0.14	
Egg lecithin	0.044	0.058	
Dipalmitoyl lecithin	0.078	0.12	
Sodium sulfatide	0.085	0.16	
Phosphatidylinositol	0.28	0.37[b]	

a The radius was calculated from the monolayer molecular area at moderate surface pressure (20 dyne/cm), from data of Shah and Schulman (1967 a,b,c).
b Abramson, et al., (1968)

the lecithin and that all the ionic groups are exposed to the aqueous medium. To carry further the parallel between these systems and those with lecithin and sulfatide, we assume that the former could not be titrated below some arbitrarily set pH level such as pH 6. A linear relation is seen in Fig. 4 for the μequivalents H⁺ that react from the equivalence points of pH 9.3 to this cut-off point of pH 6. In this way these μequivalents of H⁺ can give a measure of the relative amounts of anionic compound present in the mixture.

Effect of Cholesterol or Cerebroside on the Ionic Interaction of Lipids

We titrated 1:1 mixtures of sulfatide with choline lipid admixed with varying amounts of cholesterol. The presence of cholesterol in all instances decreased the interaction of the other two lipids (Table 2). The presence of 0.5 μmole cholesterol per μmole complex reduced the acid capacity of sphingomyelin-

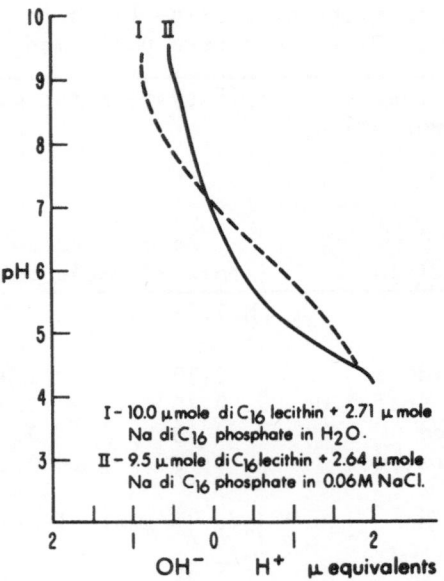

Fig. 3. Titrations of dipalmitoyl lecithin codispersed with sodium
salt of dicetyl phosphoric acid at T = 24°. The minor contribution
of the lecithin was subtracted. All systems were brought to pH
10 with NaOH and then titrated with HCl.
 I 10.0 μmole dipalmitoyl lecithin + 2.7 μmole Na dicetyl
   phosphate in H_2O.
 II 9.5 μmole dipalmitoyl lecithin + 2.6 μmole Na dicetyl
   phosphate in 0.06M NaCl.

sulfatide systems to that found for the egg lecithin-sulfatide
system alone. Furthermore, a 1 μmole mixture of cholesterol with
1 μmole sphingomyelin-sulfatide complex had an acid capacity
comparable with that of egg lecithin-sulfatide containing a 0.5
mole ratio of cholesterol. In systems with dipalmitoyl lecithin
replacing sphingomyelin and also mixed with sulfatide and cholesterol, the effect of the cholesterol was essentially the same as
with the sphingomyelin systems.

  Since myelin contains a relatively large percentage of cerebrosides (kerasin and phrenosin) which are structurally related
to both sphingomyelin and sulfatide, it was desirable to study
the effect of the uncharged cerebrosides on the other two.
Sphingomyelin admixed with phrenosin or phrenosin and kerasin
showed minor differences in titration behavior compared with
sphingomyelin alone. This was also true for sulfatide with
phrenosin. Systems of sphingomyelin and sulfatide complex

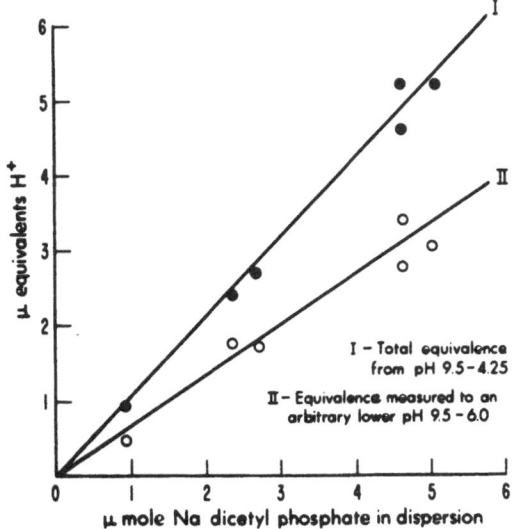

Fig. 4. Relation between the μmole of Na dicetyl phosphate (co-dispersed with 0.5 - 11.0 μmole dipalmitoyl lecithin) and the μequivalents of H⁺ reacted in titrations in aqueous systems. This relation shows the complete dispersion of the lipids with availability of the acid groups for reactions.

   I Total number of μequivalents H⁺ that react between equivalence points indicated by the titration curves (pH 9.3 - 4.25)

  II Number of μequivalents H⁺ that react from the equivalence point at pH 9.3 to an arbitrary lower value at pH 6.0.

containing phrenosin showed almost the same titration capacity as observed in the absence of phrenosin (Table 2).

Reactions with Sodium Dodecyl Sulfate

The question arises whether an ionic interaction takes place between a choline lipid and an insoluble long-chain anion with a structure simpler than that of sulfatide. To study such an interaction, sodium dodecyl sulfate was dissolved in chloroform-methanol with choline lipids, and the dried solid was dispersed in water as in the previous experiments. The titration characteristics of these systems are given in Table 3.

An interaction between the long-chain alkyl sulfate with the choline is evident from the titration data. The acid capacity of the phosphate, however, is somewhat less than in the comparable system containing sulfatide. Other similarities, however, do persist; the effect with egg lecithin is less than with sphingo-

TABLE II: Effect of Neutral Lipids on Titration Characteristics of Ionic Lipids in Aqueous Systems at 24°.

Choline Lipid (A)		Acidic Lipid (B)		Neutral Lipid		Acid Capacity pH 7.0-3.5 (μmole H$^+$/μmole Lipid(A))
Sph*	1.0	-		Chol	2.3	0.091
Egg PC	1.0	-		Chol	1.2	0.063
						(μmole H$^+$/μmole A-B complex 1:1)
Sph	1.0	Sulf	1.0	-		0.44
Sph	1.0	Sulf	1.0	Chol	0.5	0.31
Sph	1.0	Sulf	1.0	Chol	1.0	0.24
Dipalm PC	1.0	Sulf	1.0	-		0.42
Dipalm PC	1.0	Sulf	1.0	Chol	0.5	0.35
Dipalm PC	1.0	Sulf	1.0	Chol	1.0	0.23
Egg PC	1.0	Sulf	1.0	-		0.30
Egg PC	1.0	Sulf	1.0	Chol	0.5	0.24
Sph	1.0	Sulf	1.0	Phren	1.0	0.42
Sph	1.0	Sulf	1.0	Phren	2.2	0.38
Sph	1.0	Sulf	0.5	Phren	2.2	0.42

*Sph - sphingomyelin   Dipalm - dipalmitoyl
 PC - lecithin   Chol - cholesterol
 Sulf - sulfatide   Phren - phrenosin

myelin, and the inclusion of cholesterol reduces the effect.

Cation-Proton Exchange

When solutions of calcium chloride or alkali metal chloride were added to aqueous systems of the single lipid, sulfatide, lecithin or sphingomyelin, very small and non-reproducible changes in pH resulted. When a mixed system of sulfatide and sphingomyelin or sulfatide and egg lecithin was brought to pH 4.00 by the addition of hydrochloric acid, and solutions of these chloride salts, also brought to pH 4.00, were added, a sharp drop in pH resulted. A definite relationship was found for the H$^+$ ion released and the type and concentration of cation added to the system. To study this relationship the experimental procedure

TABLE III: Titration Characteristics of Aqueous Dispersions of Systems Containing Sodium Dodecyl Sulfate (NaDS) at 24°.

	Molar Ratio		Acid Capacity μmole H^+/μmole pH 7.0-3.5
NaDS			0.055
			μmole H^+/μmole Complex
Sphingomyelin+NaDS	1:1		0.29
Egg Lecithin+NaDS	1:1		0.18
Sphingomyelin+NaDS	1:1	Cholesterol 0.5	0.16
Egg Lecithin+NaDS	1:1	Cholesterol 1.0	0.065

followed was the addition of a small amount of a concentrated solution of the salt, the system was then permitted to come to a stable pH. The volume of standard NaOH required to return the system to pH 4.00 gave a measure of the cation-proton interaction. With increasing concentrations of salt the release of protons diminished so that little further change resulted at concentrations greater than 0.05 M univalent cation or 1 mM $CaCl_2$ (Fig. 5). For a 1:1 sphingomyelin-sulfatide system at pH 4.00 and 24°, the effects of KCl and NaCl were roughly alike. At a concentration of 0.05 M salt, 0.24 - 0.26 μmole H^+ was exchanged per μmole of complex, with a small further increase at 0.100 M salt. Calcium chloride solutions produced an equal effect at a concentration of 1 mM. Solutions of choline chloride produced a somewhat smaller effect, exchanging 0.10 μmole H^+/μmole complex at a concentration of 0.05 M choline. A noteworthy observation is that the total effect of the calcium ion was not greater than that of the alkali metals, as it is in other acidic lipid systems (phosphatidic acid or phosphatidylinositol). This may be the result of the reduced ability of the bivalent cation to bridge two negative charge sites in the mixed lipid systems studied here.

In mixed systems consisting of egg lecithin and sulfatide at pH 4.0 and 24°, the exchange of cations for H^+ was less than in comparable sphingomyelin-sulfatide systems. At 0.05 M NaCl or KCl, 0.09 μmole of H^+ per μmole complex was exchanged. In these systems, the effectiveness of choline was the same as NaCl or KCl.

The interaction of cations with sphingomyelin-sulfatide systems is also shown by their titrations in 0.01 and 0.05 M salt as compared with salt free systems (Fig. 2). Here again, a greater effect of NaCl or KCl as compared with choline chloride was shown.

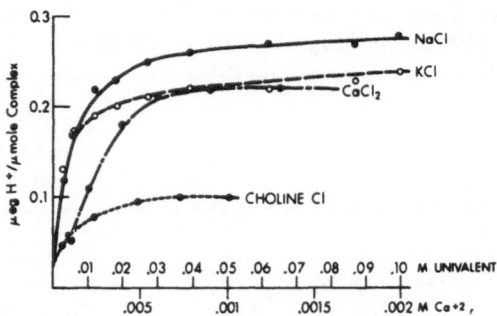

Fig. 5. Cation-proton exchange with mixed sulfatide-sphingomyelin systems (1:1 molar ratio) maintained at pH 4.0 and 24°. Each system contained 7.5 to 7.8 μmoles of 1:1 complex in 5 ml. After each addition of concentrated salt solution at pH 4.0 the μeq of NaOH required to maintain constant pH gave a measure of the μeq H+ exchanged.

DISCUSSION

In explaining the differences in the extent of the inter-action of sulfatide with the four choline lipids studied, we compare the hydrocarbon chains in these compounds. Unlike dipalmitoyl lecithin in which the acyl chains are fully saturated, egg lecithin contains approximately 50% unsaturated fatty acids, while in soybean lecithin 75% of the hydrocarbon chains are un-saturated with many polyunsaturations (Shah and Schulman, 1967a). Sphingomyelin contains a small percent of unsaturated chains. It has been shown (Shah and Schulman, 1967b; Demel et al., 1967) that the area of choline lipids as measured by monolayer experiments increased with the unsaturation of the hydrocarbon chain. Assum-ing that the lipid aggregates we studied are at moderate and roughly equal surface pressures, then using the radius calculated from the molecular area at 20 dyne/cm (Shah and Schulman, 1967a, b,c), the distance between a negative charge site on the sulfatide molecule and the positive nitrogen of the neighboring choline lipid would probably increase as the area of the fatty acid chains increase. If such factors as polarization and dielectric constants are presumed to be unchanged, the extent of the ionic interaction of the two lipids would decrease with the square of the charge separation. Although the actual distance between charges cannot be given precisely, the increased dimensions of the choline lipid give some indication of the increasing separation as the lipid becomes more unsaturated. The resulting decrease in acid capacity in Table 1 may be compared with the changes in radius of the choline lipid.

When cholesterol, sulfatide, and choline lipid are dissolved in warm chloroform-methanol (2:1, v/v) and stirred, the 3 components become intermixed, and on evaporation of the solvent, the solid residue is this mixture of lipids. We assume that when this solid is dispersed in an aqueous medium, each particle contains a mixture of the lipids. The decreased interaction between sulfatide and the choline lipids (Table 2) can be interpreted as the result of the interpenetration of the cholesterol between the other lipid molecules.

It is interesting to relate these changes in the titration characteristics of mixed lipid systems containing cholesterol with some of the data available from studies of mixed monolayers. Shah and Schulman (1967b) find that lecithin-cholesterol monolayers in the absence of Ca^{+2} ions remain in the liquid state except at high surface pressures. If this condition persists with the additional presence of sulfatide, the mobility of the molecules in the lamellar aggregates should permit the arrangement leading to minimum free energy. Since the ionic interaction between the sulfatide and choline lipid is reduced by increasing the concentration of cholesterol in the lipid systems, this leads us to the view that the coulombic forces are not able to "squeeze out" the cholesterol molecules. This may be either the result of ion-dipole interaction between cholesterol and the choline lipid or due to the geometric structures of these molecules which permit a favorable packing arrangement. From Table 3, we can see that cholesterol produces a similar decrease in the interaction between sodium dodecyl sulfate and choline lipids such as occurs with sulfatide and choline lipids. If we assume an association of cholesterol with the choline lipid persisting in the presence of a third component (sulfatide), we may then use data obtained from mixed monolayers for the average molecular areas in lecithin-cholesterol systems (Table 4). A 1:1 molar mixture of dipalmitoyl lecithin-cholesterol monolayer at 20 dyne/cm surface pressure has an area of 80 $\overset{o}{A}{}^2$ for the two molecules, whereas a 1:0.5 mixture of these lipids has an average molecular area of 45 $\overset{o}{A}{}^2$, giving an apparent area of 68 $\overset{o}{A}{}^2$ for the lecithin-cholesterol association. At an equal surface pressure, a mixed egg lecithin monolayer in 1:0.5 mole ratio with cholesterol has an area of 84 $\overset{o}{A}{}^2$ for the mixture and for egg lecithin alone, a molecular area of 71 $\overset{o}{A}{}^2$. The reasonable agreement between the changes in the combined molecular area for the lecithin cholesterol association and the decreased interaction between sulfatide and lecithin adds weight to a relatively simple explanation based upon the combined area of the cholesterol with the choline lipid as the predominant factor in determining the extent of interaction with the sulfatide in these mixtures. This is further supported by the almost identical results obtained with sphingomyelin and dipalmitoyl lecithin which, despite their different nonpolar structures, are alike in molecular area and in their phosphoryl choline portions.

TABLE IV: Relation Between I. Molecular Area of Lecithin-Cholesterol Association and II. Acid Capacity of 1:1 Complex of Lecithin-Sulfatide with Added Cholesterol

Molar ratio	Average Molecular area[a] $\overset{o}{A}{}^2$	I Combined molecular area lecithin-cholesterol association $\overset{o}{A}{}^2$	II Acid Capacity 1:1 lecithin-sulfatide complex pH 7 - 3.5 μmole H^+/μmole complex
Dipalmitoyl lecithin-cholesterol 1:0.5	45	68	0.31
Dipalmitoyl lecithin-cholesterol 1:1	40	80	0.23
Egg lecithin	71	-	0.30
Egg lecithin-cholesterol 1:0.5	56	84	0.24
Soybean lecithin	78	-	0.21

[a]Adapted from monolayer measurements Shah and Schulman (1967 b).

As shown by Table 2, the addition of phrenosin to the sphingo-myelin-sulfatide complex does not have the same effect as choles-terol. From the structural similarity of phrenosin and the other two lipids as well as their occurrence together in biologic systems, it would appear that the cerebroside (phrenosin) enters the lipid lamellar structure with its molecules paralled to the other sphingolipids, the polar galactose moiety of the cerebroside extending into the aqueous medium. However, the positions taken by the cerebroside molecules are between the units of bimolecular complexes and not between the molecules of the complex, as though the cerebroside is unable to break the ionic bond. In this way, the ionic characteristics of the complex remain unchanged.

To explain the smaller effect produced by sodium dodecyl sulfate than by sulfatide in the reaction with the choline lipids, the structural features of these two sulfates are compared. In sodium dodecyl sulfate the negative charge site is not far removed from the hydrocarbon chain. At the lipid-aqueous interface, the anionic site is close to the phase boundary. Sulfatide is

somewhat more hydrophilic due to the galactose present so that
the sulfate will be further below the interface than in the case
of the alkyl sulfate. As seen in Fig. 1, Corey-Pauling models of
the choline lipids and sulfatide show the choline-positive and
sulfatide-negative groups on a plane parallel to the lipoid-water
interface, but this arrangement is not seen when sulfatide is
replaced by an alkyl sulfate.

It is well-established that the acid and base groups of
choline lipids are not available for titration (except at very low
and very high pH levels) and are presumably bound in some manner.
This may be an intermolecular salt linkage between the quaternary
amine of one molecule and the phosphate of another as suggested by
Pethica (1965). Alternatively, it may involve an intramolecular
bond between the ionic groups, or as suggested by Sundaralingam
(1968) a bond between the nitrogen and the phosphate-ester oxygen.
Intermolecular ion binding between choline lipid molecules could
add to membrane cohesion. Our titrations of sphingomyelin-choles-
terol reveal no increase in titrability of the acid group as compared with sphingomyelin alone. Similar results were found with
egg lecithin-cholesterol systems. This may be evidence of the
absence of intermolecular ion bonds between lecithin or sphingomyelin molecules. However, in the presence of an amphipathic
anion (sulfatide or dodecyl sulfate), an intermolecular ionic
linkage is established.

The release of H^+ ions that takes place on adding salt
solutions to our choline lipid-sulfatide systems can be interpreted
along either of two alternative paths: 1: The effect may be the
result of a change in the phosphate that was made available by the
interaction of the sulfatide with the choline lipid. This reaction
may resemble the changes that take place when cations are added to
incompletely ionized phosphatidic acid or phosphatidylinositol
(Abramson et al., 1968). Another possible explanation that would
be unique for the inter-lipid ionic bond is: II: The increased
ion concentrations in the system may lead to a rupture of the bond
between the sulfate and the positive site of the choline molecule.
This then permits the phosphate to re-establish its normal intramolecular phosphoryl-choline bonding. Since the phosphate has
been shown to be ionic in lecithin and sphingomyelin (Abramson
et al., 1965), the change from an incompletely ionized to a fully
ionized phosphate leads to a release of protons. If the latter
mechanism is the actual one, it signifies the breaking of an ionic
intermolecular linkage between lipids, with attending structural
changes, produced by cations in solution. Earlier studies of
cation-proton exchange with phosphatidylinositol systems in water
at pH 3.5 showed that on bringing the system to 0.05 M Na^+ or K^+,
roughly 0.16 μmole H^+/μmole lipid was exchanged, whereas titrations
showed the maximum protons available as determined by the fraction

of unionized phosphatidylinositol at pH 3.5 is 0.30. In the
system of sphingomyelin-sulfatide, at pH 4.0 the H^+ ions available
for exchange are 0.30 μmole/μmole complex, while the exchange with
alkali metal ions is 0.24 - 0.26 μmole/μmole complex. A much
greater proportionate change takes place than with phosphatidylinositol. This may be an indication that the two reactions are
not alike, and that the H^+ ion release detected in our mixed lipid
systems results from the rupture of the intermolecular bond as
described in mechanism II.

In a theoretical treatment of lecithin in water systems,
Parsegian considers the positive choline regions of the lecithin
as constituting a diffuse layer of counterions bordering the
surface of lipid molecules given negative charges by the ionic
phosphate (Parsegian, 1967). The positive charges are free to
diffuse within a distance of 5Å from the surface at the ends of the
flexible $-CH_2-CH_2-$ linkage. In our studies of mixed lipids we
can visualize the positive choline as shifting its position from
a counterion for phosphate to the sulfate. The limitation of the
5Å length of the CH_2 linkage then explains the decreased effect
when the intermolecular distances increase with increased
unsaturation of the lecithin molecule.

REFERENCES

Abramson, M.B., and Katzman, R. (1968), Science, 161, 576.
Abramson, M.B., Colacicco, G., Curci, R., and Rapport, M.M. (1968),
     Biochemistry, 7, 1692.
Abramson, M.B., Katzman, R., Curci, R., and Wilson, C.E. (1967),
     Biochemistry, 6, 295.
Abramson, M.B., Norton, W.T., and Katzman, R. (1965), J. Biol.
     Chem., 240, 2389.
Bangham, A.D., Standish, M.M. and Watkins, J.C. (1965), J. Mol.
     Biol., 13, 238.
O'Brien, J.S. (1967), J. Theoret. Biol., 15, 307.
Parsegian, V.A. (1967), Science, 156, 939.
Pethica, B.A. (1965) In Surface Activity and the Microbiol Cell,
     Soc. Chem. Ind. (London) 19, p. 85.
Shah, D.O., and Schulman, J.H. (1967a), J. Colloid Interface
     Sci., 25, 107.
Shah, D.O., and Schulman, J.H. (1967b), Jour. Lipid Res., 8, 215.
Shah, D.O. and Schulman, J.H. (1967c), Biochim. Biophys. Acta,
     135, 184.
Sundaralingam, M. (1968), Nature, 217, 35.

LIPID-POLYMER INTERACTION IN MONOLAYERS: EFFECT OF

CONFORMATION OF POLY-L-LYSINE ON STEARIC ACID MONOLAYERS

Dinesh O. Shah

Surface Chemistry Laboratory, Marine Biology Division
Lamont-Doherty Geological Observatory of Columbia
University, Palisades, New York

ABSTRACT

Surface pressures and surface potentials of stearic acid
monolayers were measured at various pH values in the presence
and absence of poly-L-lysine in the subsolution. The presence of
poly-L-lysine strikingly influences the state of stearic acid
monolayers. Surface potential measurements indicated that the
maximum interaction between poly-L-lysine and stearic acid
monolayers occurred between pH 10 and 11. Poly-L-lysine solutions
exhibited surface activity in the same pH range. Air bubbles
covered with poly-L-lysine films showed maximum stability at pH 11.
These results indicate that in the pH range 10-11, where coil-to-
helix transition occurs in solution, poly-L-lysine has partial or
complete helical conformation which causes the slowest rate of
drainage of water in bubble lamellae, and which also exhibits
surface activity and hence, increases the interaction of poly-L-
lysine with stearic acid monolayers. The implications of these
findings for lipid-protein associations in biomembranes are
discussed.

INTRODUCTION

Lipid-protein interactions are of great interest to
understand the structure and function of biological membranes.
Various approaches have been taken to elucidate the interaction
of lipids and proteins in biological membranes. In recent
years, nuclear magnetic resonance and electron spin resonance
spectroscopy have been used profitably to investigate these
interactions (1-4). Phospholipid bilayers and monolayers have
served as useful models for these studies (5-6). The monolayer

*Lamont-Doherty Geological Observatory Contribution No. 1404.

approach has been found very useful to understand molecular
mechanisms presumably occurring at the cell surface (7-10).

Earlier studies on lipid-protein interaction in monolayers
were reported by Schulman and his co-workers (11-12) who
investigated the interaction of albumin, globulin and haemoglobin
with cholesterol, cephalin, cardiolipin, alkyl sulfate and alkyl
trimethylammonium monolayers. Eley and Hedge (13,14) studied
protein-protein and protein-lipid interactions in fibrinogen,
thrombin, albumin, lecithin and cephalin monolayers. The
interaction of synthetic dihydroceramide lactoside monolayers
with globulin, albumin, and ribonuclease was investigated by
Colacicco, Rapport and Shapiro (15). These workers (16) have
also shown from their studies on the interaction of apoprotein
with various lipid monolayers, that the unusual surface activity
of apoprotein may be intimately related to the mechanism of
formation of the lipo-protein. Recently, Arnold and Pak (17)
have investigated protein-protein interaction in monolayers.

To investigate the interaction of water with films,
Trapeznikov (18) and Garrett (19) have studied the stability of
bubbles covered with a monolayer of surface-active materials. In
general, the stability of such bubbles is related to the rate of
drainage of water in the bubble lamellae. The interaction of
polar groups with water (i.e. hydration of polar groups) impedes
the drainage of water in the lamellae, and, hence, increases the
time required to reach a critical thickness where bubble lamellae
break. Therefore, more strongly hydrated molecules increase the
bubble stability. This method was used in the present study to
investigate the hydration of stearic acid and poly-L-lysine films.

It has been recognized that both ionic and hydrophobic
interactions play a role in the lipid-protein association. A
simple model system of stearic acid and poly-L-lysine was
selected to investigate various aspects of the ionic interaction
in the present studies, since the ionic properties of stearic acid
monolayers (20,23) and of poly-L-lysine solutions (24) have been
established. The objective of the present studies was to
investigate how the ionization of carboxyl groups in the monolayer
and the conformation of poly-L-lysine in the subsolution influence
interactions at the interface.

EXPERIMENTAL

Materials: Poly-L-lysine hydrochloride (molecular weight
100,000 - 200,000) was bought from Mann Research Laboratories
Inc. (New York, N.Y. 10006). Highly purified (>99%) stearic
acid was purchased from Applied Science Laboratories, Inc.
(State College, Pa., 16801). Inorganic chemicals of reagent
grade and distilled-deionized water were used in all experiments.

For pH close to 2, the solutions of 0.05 M HCl were used; for
pH 3 to 6, 0.05 M buffer solutions of citric acid-sodium citrate
were used; for pH 7 to 9, 0.05 M buffer solutions of tris-HCl were
used; for pH 10 to 11, 0.05 M buffer solutions of glycine-NaOH
were used; for pH 12 to 13, 0.05 M and 0.1 M solutions of NaOH
were used. The buffer solutions were prepared according to
Biochemists' handbook (25). A stock solution of 5 mg poly-L-lysine
per ml of distilled water was prepared. 2.4 ml (containing 12 mg
of poly-L-lysine) of this solution was added to 100 ml of the
subsolution for surface measurements. The stearic acid was
dissolved in chloroform-methanol-hexane (1:1:3 v/v/v) in a
concentration of about 0.8 mg/ml.

Methods: The surface pressure was measured by a modified
Wilhelmy plate method, and the surface potential by a radioactive
electrode, as described previously (26). The state of the
monolayers was determined qualitatively by the talc method (27).
The monolayers of stearic acid were spread on buffered subsolutions
in the presence and absence of poly-L-lysine (12 mg/100 ml
subsolution).

Bubble stability: The survival time (i.e., the time
interval between the formation and collapse) of bubbles was
measured with a stopwatch after producing a small air bubble by
a dropper (tip diameter 1 mm)under monolayers and subsolutions
in the following manner. When a monolayer was compressed to its
limiting area ($\simeq$20 A^2/molecule), a bubble was produced on each
side of the compression glass barrier. For subsolutions
containing poly-L-lysine, the monolayer side of the compression
barrier showed surface properties of stearic acid + poly-L-lysine,
whereas the other side of the barrier showed those of adsorbed
film of poly-L-lysine alone. At least ten measurements were made
for bubble stability. It should be pointed out that since the
collapse of a bubble produces considerable structural reorganization
and rearrangement of molecules in the monolayer, a second bubble
should not be produced in the same region of the monolayer.
Therefore, all ten bubbles were produced in different parts of
the monolayer and their average survival time was calculated.

The surface tension of buffered solutions of poly-L-lysine
was measured with a Roller-Smith surface tensiometer. The
surface pressure (π) of poly-L-lysine solutions is defined as
$\pi = \gamma_o - \gamma_p$, where γ_o is the interfacial tension without
poly-L-lysine and γ_p is that with poly-L-lysine in the
subsolution; hence, π represents the lowering of the surface
tension of buffer solutions by the presence of poly-L-lysine.

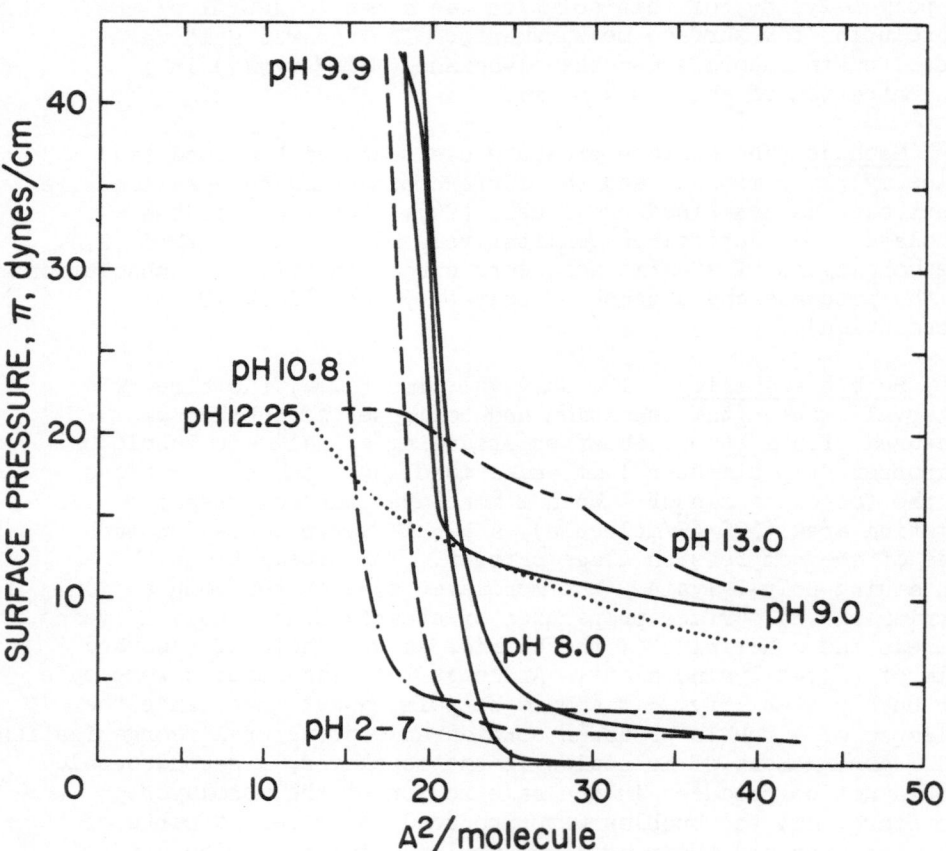

<u>Fig. 1.</u>   Surface pressure-area curves of stearic acid monolayers
          on buffered subsolutions at various pH values at 22°C.

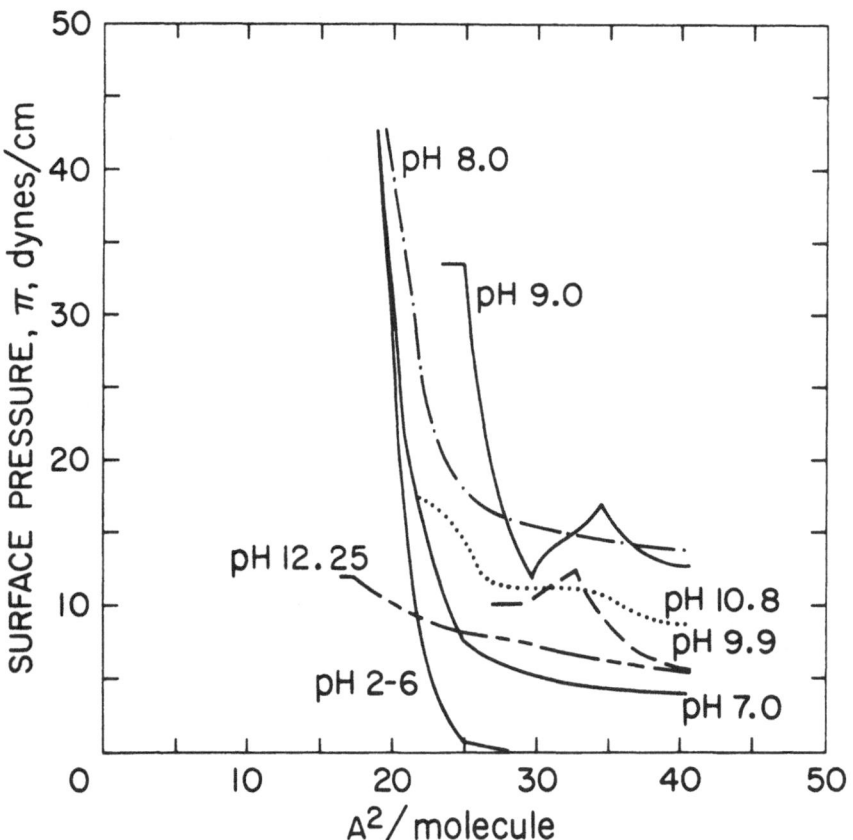

<u>Fig. 2.</u> Surface pressure-area curves of stearic acid monolayers
on buffered subsolutions containing 0.02 mg/ml of poly-
L-lysine at various pH values at 22°C.

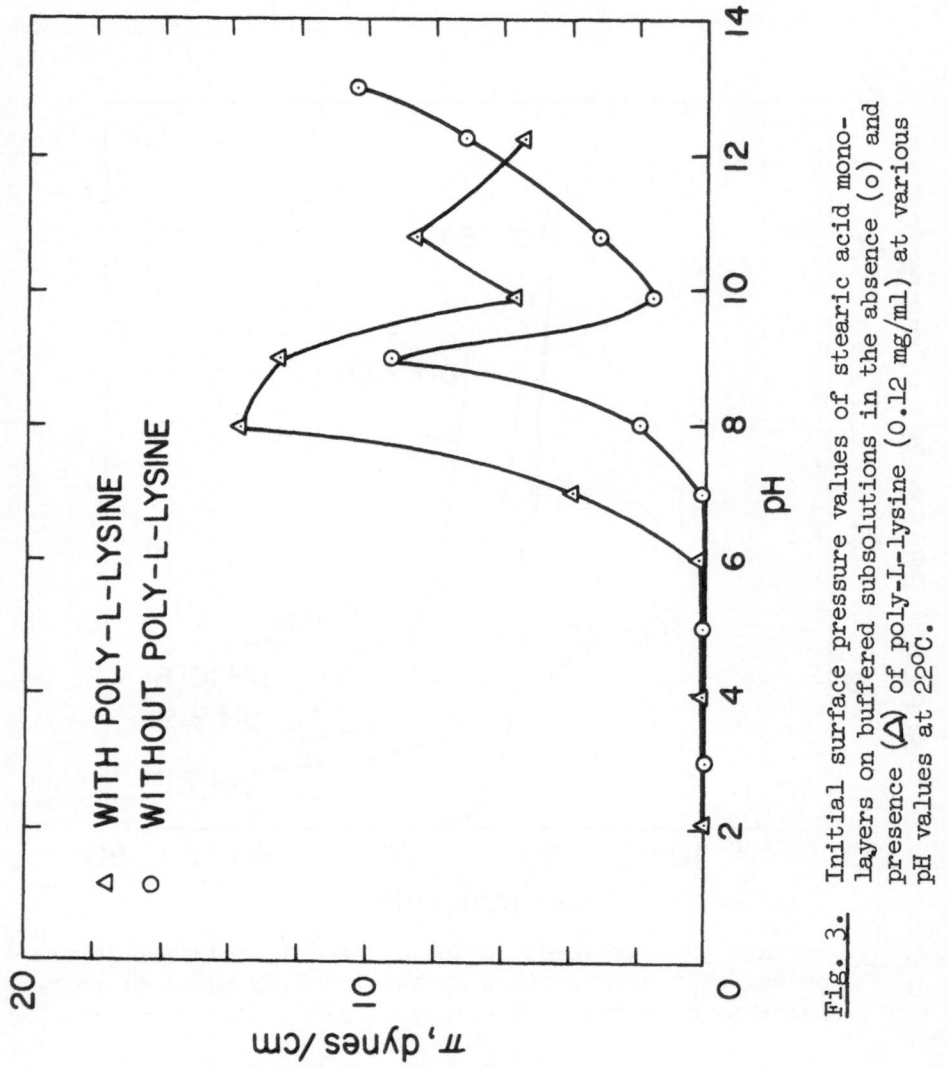

<u>Fig. 3.</u> Initial surface pressure values of stearic acid mono-
layers on buffered subsolutions in the absence (o) and
presence (△) of poly-L-lysine (0.12 mg/ml) at various
pH values at 22°C.

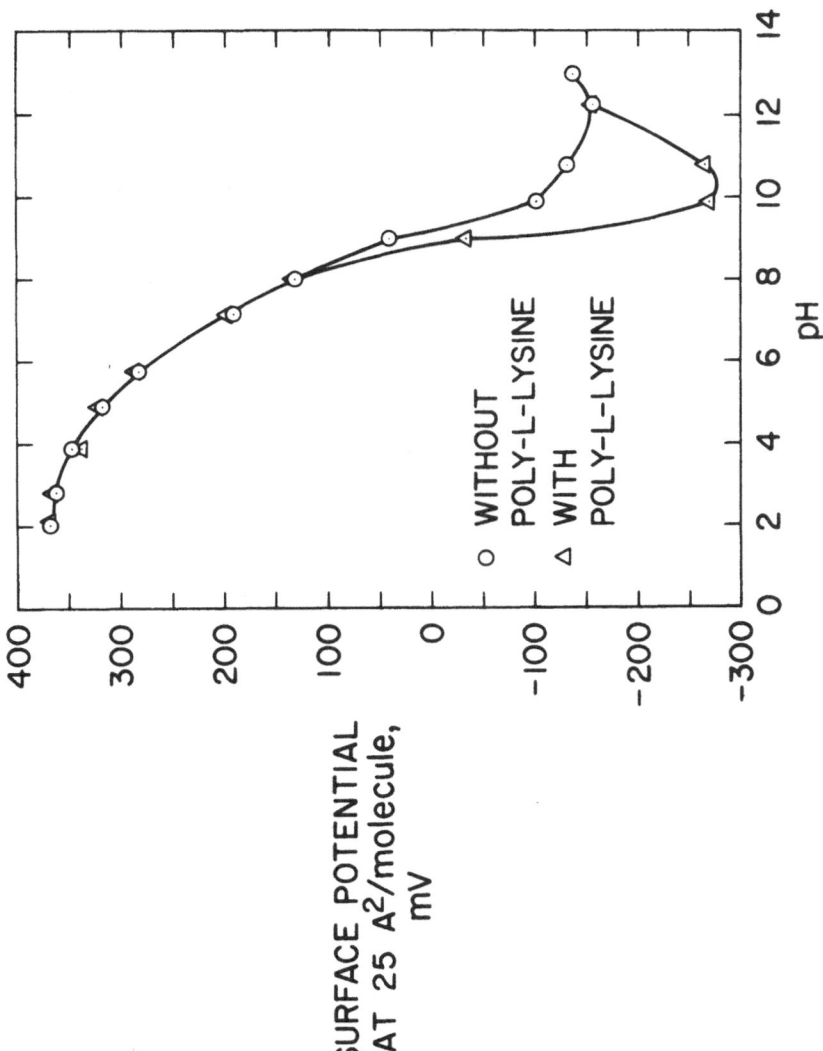

<u>Fig. 4.</u> Surface potentials of stearic acid monolayers at 25 Å^2/ molecule on buffered subsolutions in the absence (o) and presence (△) of poly-L-lysine (0.12 mg/ml) at various pH values at 22°C.

RESULTS

Figures 1 and 2 show surface pressure-area curves of stearic acid monolayers in the absence and presence of poly-L-lysine in subsolutions of different pH values. It has been shown (28) that the plateau region in the surface pressure-area curves of stearic acid at large area per molecule is directly related to repulsion in the monolayers because of ionization of carboxyl groups. Figure 3 shows the surface pressure values where the plateau region begins for different pH values. In general, the presence of poly-L-lysine increases these initial surface pressure values and shifts the curve to the left (or acid side).

Figure 4 shows the surface potentials of stearic acid monolayers at 25 A^2/molecule in the presence and absence of poly-L-lysine at various pH values. It is evident that the interaction of poly-L-lysine with stearic acid monolayers lowers the surface potential and that the maximum interaction occurs in the pH range 10 to 11.

Table I summarizes the state of stearic acid monolayers near the collapse pressure in the presence and absence of poly-L-lysine in the subsolution at various pH values. It shows that the interaction between stearic acid monolayers and poly-L-lysine in the pH range 9 to 11 solidifies the monolayers.

The upper part of figure 5 shows the data of Applequist and Doty (24) on poly-L-lysine solutions. The lower part of figure 5 shows our data on the bubble stability of stearic acid monolayers in the presence of poly-L-lysine in the subsolution. The bubble stability for stearic acid monolayers alone, which is not shown in figure 5, did not exceed 10-15 seconds over the whole pH range. The surface activity (or surface pressure) of poly-L-lysine solutions and bubble stability at various pH values are also shown in figure 5. It is evident from figure 5 that at pH 11, the conformation of poly-L-lysine molecules, which is nearly helical and surface-active, affords maximum stability to bubble lamellae.

DISCUSSION

Figure 1 shows that for pH values from 2 to 9, the limiting area of stearic acid is approximately the same ($\approx$20 A^2/molecule), implying that the monolayers are insoluble in this pH range. At pH 9.9 and 10.8, the limiting areas are respectively 16 and 18 A^2/molecule, which may be due to slight solubility of ionized stearic acid molecules in the subsolution, or to rearrangement of molecules in the monolayers. The initial surface pressure values

TABLE I

pH	Subsolution	Subsolutions without Poly-L-lysine — The state of monolayers	Subsolutions with Poly-L-lysine — The state of monolayers
2.0	HCL solution	liquid	liquid
2.8	citric acid-sodium citrate buffer	liquid	liquid
3.9	citric acid-sodium citrate buffer	liquid	liquid
4.9	citric acid-sodium citrate buffer	liquid	liquid
5.75	citric acid-sodium citrate buffer	liquid	liquid
7.15	tris - HCl	solid	solid
8.0	tris - HCl	solid	solid
9.0	tris - HCl	gel (+)*	solid
9.9	glycine-NaOH buffer	gel (+)	solid
10.8	glycine-NaOH buffer	gel (++)	solid
12.25	NaOH solution	gel (+++)	gel

*The number of + signs indicates qualitatively the increasing surface viscosity of the monolayers.

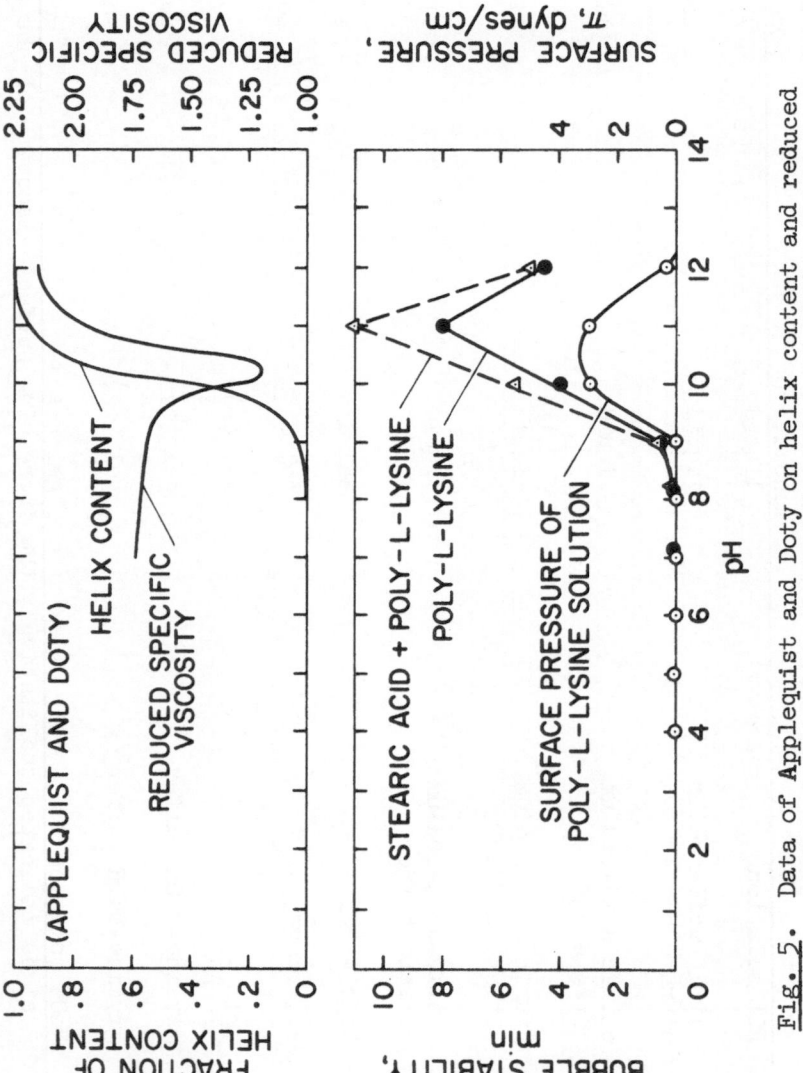

<u>Fig. 5.</u>   Data of Applequist and Doty on helix content and reduced
specific viscosity of poly-L-lysine solutions (upper
part) surface pressure (or surface activity), and bubble
stability of poly-L-lysine solutions (lower part). The
bubble stability of stearic acid monolayers in the
presence of poly-L-lysine is shown by a broken line,
whereas that of stearic acid alone was 10-15 seconds in
the whole pH range.

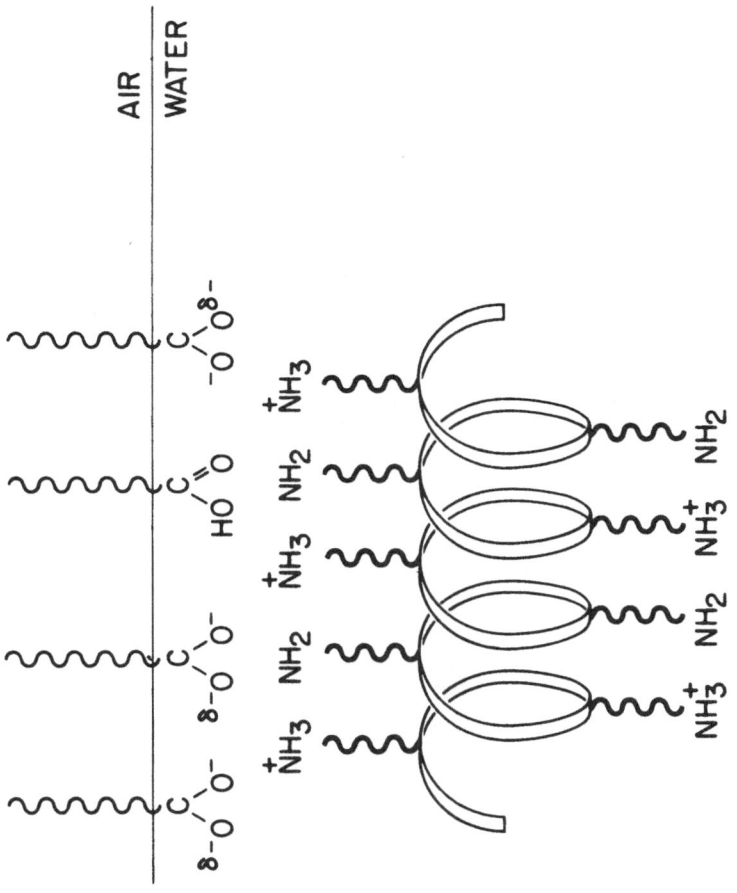

<u>Fig. 6.</u> Schematic representation of interaction between stearic acid and poly-l-lysine in monolayers at pH 11. δ- represents a partial ionic charge on oxygen atoms.

of the plateau region indicate the extent of ionic repulsion in monolayers (fig. 1). As shown in figure 2, the presence of poly-L-lysine strikingly alters the surface pressure-area curves of stearic acid monolayers above, but not below pH 6, indicating that poly-L-lysine does not interact with stearic acid monolayers in the pH range 2 to 6.

It is interesting to compare the effect of cations such as calcium with poly-L-lysine on stearic acid monolayers. It has been shown (29-31) that the binding of calcium ions to stearic acid monolayers begins to occur at about pH 5; this causes condensation of the monolayers. In contrast, the interaction between poly-L-lysine and stearic acid expands the monolayers presumably due to penetration of side chains of poly-L-lysine in the monolayers. The kinks in the surface pressure area curves at pH 9.0 and 9.9 at about 35 A^2/molecule indicate the areas at which presumably some of the penetrated side chains of poly-L-lysine are squeezed out of the monolayers (fig. 2).

Figure 3 shows the initial surface pressure values of the plateau region in the presence and absence of poly-L-lysine in subsolutions. Stearic acid monolayers without poly-L-lysine show a maximum at pH 9, whereas, the maximum occurs at pH 8.0 in the presence of poly-L-lysine in the subsolution.

It has been shown (28, 32) that at pH 9, where 50% of the molecules are ionized (i.e., pK = 9), there is maximum separation between the molecules in stearic acid monolayers. It is clear that in the presence of poly-L-lysine in subsolutions, the maximum separation occurs at pH 8.0, which suggests that the pK value has shifted by one pH unit. A similar decrease in the pK of oleic acid by the presence of calcium ions was reported by Benzonana and Desnuelle (33). A second maximum observed at pH 11 may be due to the penetration of poly-L-lysine into the monolayers. In general, the initial surface pressure values above pH 6 are higher in the presence of poly-L-lysine in the subsolution than in the absence of it. This can be explained as follows: because of coulombic attraction the cationic side chains of poly-L-lysine may penetrate the negatively charged stearic acid monolayers. This will increase the surface concentration of molecules and, hence, the surface pressure in monolayers.

Figure 4 shows the surface potentials of stearic acid monolayers in the presence and absence of poly-L-lysine in subsolutions. It shows that the maximum interaction occurs in the pH range 10-11 where the surface potential decreases by about 175-185 mv. It is interesting to note that the presence of calcium ions in the subsolution also decreases the surface potential of stearic acid monolayers by about 200 mv (31). Hence, the

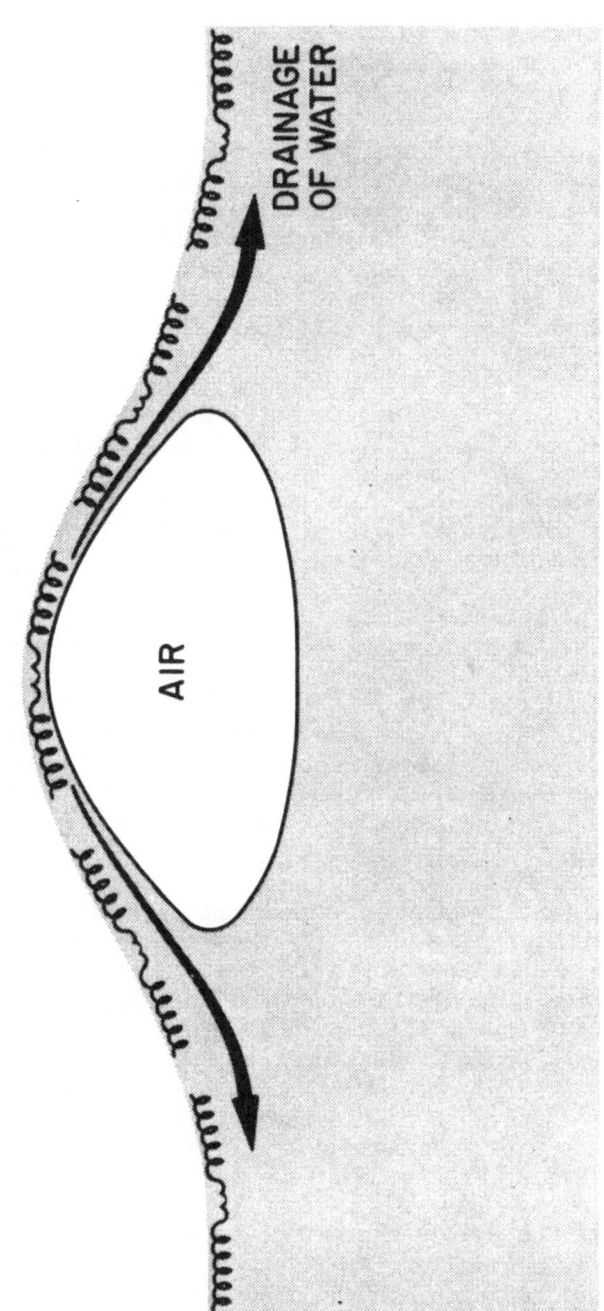

Fig. 7. Schematic representation of the mechanism governing the stability of an air bubble covered with a film. A faster rate of drainage of water in the lamellae decreases the bubble stability.

mechanism of action of poly-L-lysine on stearic acid monolayers is presumably the same as that of calcium ions. In contrast, the interaction of calcium ions with lecithin, sphingomyelin, cardio-lipin and dicetyl phosphate monolayers increases the surface potentials (26, 34, 35).

Applequist and Doty (24), using birefringence, viscosity, sedimentation and optical rotatory dispersion measurements, have shown that random coil-to-helix transition in poly-L-lysine solu-tions occurs in the pH range 10 to 11. The dissociation of NH_3^+ groups also takes place in this pH range. Figure 6 schematically shows the interaction of poly-L-lysine with stearic acid monolayers at pH 11 where poly-L-lysine molecules have nearly helical conformation.

Figure 5 shows that poly-L-lysine solutions exhibit surface activity (or surface pressure) in the pH 10 to 11. Although the surface pressure of poly-L-lysine is very low (3 dynes/cm), it strikingly influences the bubble stability. Below pH 9, the bubble stability of poly-L-lysine solution is about 10 seconds, whereas at pH 11, its bubble stability is 8 minutes.

The bubble stability of stearic acid monolayers in the absence of poly-L-lysine is about 10-15 seconds over the whole pH range, whereas that of stearic acid in the presence of poly-L-lysine is about 11 minutes at pH 11. It is evident that this enhancement of bubble stability is due to the presence of poly-L-lysine since it also shows maximum bubble stability (8 mins.) without stearic acid monolayers. The upper part of figure 5 shows that at pH 11, poly-L-lysine has almost helical conformation in solution.

Figure 7 schematically shows the mechanism of bubble stability in the presence of a film. If the rate of drainage is rapid, the bubble has a shorter survival time. If the molecules in the film impede the drainage of water due to film-water interaction, the stability of the bubble increases. The data presented in this paper suggest that helical conformation of poly-L-lysine decreases the rate of drainage of water in the bubble lamellae and, hence, increases the bubble stability. The interaction between stearic acid and poly-L-lysine causes further increase in the bubble stability.

The question may be raised whether poly-L-lysine molecules at pH 11 retain their helical conformation at the interface or not. In this author's opinion, the helical conformation is preserved at the interface. If poly-L-lysine molecules are denatured at the interface, one would not observe the striking properties in the pH range 10 to 11. Moreover, using deuterium exchange, infra-red spectroscopy and electron diffraction methods to study skimmed monolayers, Malcolm (36, 37) has shown that helical conformation of poly-peptides is retained in the monolayer at the air-water interface.

It is difficult to extrapolate the results of monolayer studies to interactions in biological membranes. However, the correlation of these studies with those of others on biological membranes should be mentioned. It has been established by various workers with different techniques that membrane proteins have partially helical conformation (38-40). From the results reported in this paper it appears that partially helical conformation of proteins in biological membranes may cause maximal interaction with water (i.e., maximum hydration) and enhance the interaction with lipids presumably because of the surface activity of the proteins in such conformation.

In summary, the results presented in this paper indicate that the lipid-polymer interaction in monolayers is strikingly influenced by the conformation of the polymer. Extensive studies on the interaction between various lipids and poly-peptides in monolayers are in progress in this author's laboratory.

ACKNOWLEDGEMENTS

The author wishes to express his sincere gratitude to the National Research Council for providing an NRC-NASA Resident Research Associateship, and to Dr. L.P. Zill and Dr. M.R. Heinrich of Ames Research Center, and Dr. O.A. Roels of Lamont-Doherty Geological Observatory, for providing helpful suggestions and encouragement at various stages of this work. Part of this work was supported by Sea Grant GH-16, and was completed at the Lamont-Doherty Geological Observatory of Columbia University. Special thanks are also extended to Mr. E.J. Murphy (Senior Research Scientist) for help in the preparation and to Dr. O.A. Roels for critical review of this manuscript.

REFERENCES

1. D. Chapman, V.B. Kamat, J. De Gier, and S.A. Penkett, J. Mol. Biol., 31 (1968) 101.
2. D. Chapman, V.B. Kamat, J. De Gier, and S.A. Penkett, Nature 213 (1967) 74.
3. M.D. Barratt, D.K. Green, and D. Chapman, Biochim. Biophys. Acta, 152 (1968) 20.
4. J.M. Steim, O.J. Edner and F.G. Bargoot, Science, 162 (1968) 909.

5. T. Hanai, D.A. Haydon and J. Taylor, J. Theoret. Biol., 9 (1965) 422.

6. L.M. Tsofina, E.A. Liberman, and A.V.Babakov, Nature, 212 (1966) 681.

7. D.O. Shah and J.H. Schulman, J. Colloid Interface Sci. 25 (1967) 107.

8. D.O. Shah and J.H. Schulman, Advances Chem. Ser., 84 (1968) 189.

9. A.D. Bangham, B.A. Pethica and G.V.F. Seaman, Biochem. J. 69 (1958) 12.

10. B.A. Pethica and J.H. Schulman, Biochem. J., 53 (1953) 177.

11. P. Doty and J.H. Schulman, Faraday Soc. Disc., London 6 (1949) 27.

12. R. Matalon and J.H. Schulman, Faraday Soc. Disc., 6 (1949) 27.

13. D.D. Eley and D.G. Hedge, Proc. Roy. Soc., Ser. B, 145 (1956) 554.

14. D.D. Eley and D.G. Hedge, J. Colloid Sci., 11 (1956) 445.

15. G. Collacicco, M.M. Rapport, and D. Shapiro, J. Colloid Sci., 25 (1967) 5.

16. G. Camejo, G. Colacicco, and M.M. Rapport, J. Lipid Res., 9 (1968) 562.

17. J.D. Arnold and C.Y. Pak, J. Am. Oil Chemists' Soc., 45 (1968) 1.

18. A.A. Trapeznikov, Acta Physicochemica (U.S.S.R.), 13, (1940) 265.

19. W.D. Garrett, Deep-Sea Res., 14 (1967) 661.

20. D.O. Shah, J. Colloid Interface Sci., in press

21. J. Bagg, M.D. Haber and H.P. Gregor. J. Colloid Interface Sci., 22 (1966) 138.

22. A.P. Christodoulou and H.L. Rosano, Advances Chem. Series, 84 (1968) 210.

23. E.D. Goddard and J.A. Ackilli, J. Colloid Sci., 18 (1963) 585.

24. J. Applequist and P. Doty, In Polyamino Acids, Polypeptides, and Proteins, Ed. M.A. Stahmann. The University of Wisconsin Press, Madison, (1962) 161.

25. C. Long, ed. "Biochemists' Handbook", Van Nostrand, Princeton (1961) p. 30.

26. D.O. Shah and J.H. Schulman, J. Lipid Res. 6 (1965) 341.

27. D.O. Shah and J.H. Schulman, Lipids, 2 (1967) 1.

28. D.O. Shah, J. Colloid Interface Sci., in press, (1969).

29. J. Bagg, M.B. Abramson, M. Fichman, M.D. Haber and H.P. Gregor, J. Am. Chem. Soc., 86 (1964) 2759.

30. D.W. Deamer, D.W. Meek and D.G. Cornwell, J. Liipid Res., 8 (1967) 255.

31. J.V. Sanders and J. A. Spink, Nature, 175 (1955) 644.

32. J. Bagg, M.D. Haber, and H.P. Gregor, J. Colloid Interface Sci., 22 (1966) 138.

33. G. Benzonana and P. Desnuelle, Biochim. Biophys. Acta, 164 (1968) 47.

34. D.O. Shah and J.H. Schulman, J. Lipid Res., 8 (1967) 227.

35. D.O. Shah and J.H. Schulman, Biochim. Biophys. Acta, 135 (1967) 184.

36. B.R. Malcolm, Proc. Roy. Soc. (London) Ser A. 305 (1968) 363.

37. B.R. Malcolm, Nature, 219 (1968) 929.

38. J.M. Steim, Advan. Chem. Ser., 84 (1968) 259.

39. J. Lenard and S.J. Singer, Proc. Natl. Acad. Sci., 56 (1966) 1828.

40. D.W. Urry, M. Mednieks, and E. Bejnarowicz, Proc. Natl. Acad. Sci., 57 (1967) 1043.

INTERACTION OF DNA WITH POSITIVELY CHARGED MONOLAYERS

M. A. Frommer*, I.R. Miller and A. Khaïat

Polymer Department, Weizmann Institute of Science

Rehovoth, Israel

ABSTRACT

Surface pressures and potentials of monolayer complexes between DNA and cetyltrimethylammonium or positively charged polypeptides (copolymers of L-lysine and L-phenylalanine) were measured. The areas per molecule of the different components in the surface were determined by direct measurement of their surface concentrations, using tritium labelled substances.

It was found that the surface pressures of the condensed monolayers were only slightly affected by the presence of the DNA in the surface. The effect of the DNA is more pronounced on the surface potential and on the surface pressure of the expanded monolayers. The effect of DNA on these properties is analogous to that of high salt concentrations.

A model for the structure of the DNA containing surface complex is proposed.

INTRODUCTION

The surface properties of DNA adsorbed at the air-water (1,2) mercury-water (3,4) and solid-water (5,6) interfaces have been studied quite extensively. However, due to the high solubility of DNA there is no information about its properties when

---

*Present address: Hydronautics Israel Ltd., Kiryat Weizmann, Rehovoth, Israel

spread as a monolayer at the A/W interface. The claimed stability
of monolayers of the pure sodium salt of DNA on 2 N NaCl substrate
reported by James and Mazia 15 years ago (7) was found to be
incorrect. This is presumably due to high protein content in the
DNA used by these authors. It was shown by Kleinschmidt and
co-workers (8) and by Cheesman (9) that binding DNA to proteins
stabilizes it at the A/W interface. Stable monolayers of DNA
can also be obtained by substituting some of its counter-ions
with surface active ions. In this article we shall describe
observations on the properties of long chain quarternary ammonium
salts of DNA as well as on the interaction of DNA with monolayers
of a basic polypeptide.

EXPERIMENTAL:

Materials

    Calf thymus DNA was purchased from Worthington Corp. Its
tritiation by the Wilzbach method is described in our previous
publication (2).

    Cetyltrimethyl ammonium bromide (CTABr) was a gift from
"Fluka AG". Elementary analysis of this product yields practically
the theoretical formula $C_{19}H_{42}NBr$. To prepare tritiated CTA hexa-
decylamine (a Fluka AG product) was allowed to react for 24 h. at
room temperature with an equivalent amount of tritiated methyl
iodide (obtained from the Radiochemical Center, Amersham, England)
in nitromethane over sodium bicarbonate powder. A large
excess of methyl iodide was then added and the reaction was
permitted to continue at 60°C for two days. The solution concentration was determined on the basis of nitrogen content.

    Solutions of long chain quarternary ammonium salts of DNA
were prepared by mixing solutions of 0.5-0.7 mg/cc of DNA in
$1:1 \times 10^{-3}$ N NaCl: isopropyl alcohol, and 0.6-0.8 mg/cc of cetyltrimethylammonium bromide in the same mixed solvent (12,13). The
concentrations of the DNA were determined by optical density
measurements of the aqueous solution prior to addition of isopropyl
alcohol. In the case of CTABr solution, its concentration was
calculated from the dry weight of the long chain salt.

    Copolymers of L-lysine hydrobromide and L-phenylalanine were
supplied by "Yeda" Research and Development Co., Rehovoth, Israel.
Their exact composition as well as their concentration in the
spreading solutions were determined by amino acid analyzer manu-
factured by Beckman-Spinco. All other salts and liquids were
commercial products of highest purity grade and in many cases were
further purified by distillation (i-PrOH) crystallization ($MgCl_2$) or

TABLE 1

Influence of: a) amount of CTA-DNA spread, b) long chain salt content in the spreading solution and c) salt concentration in the substrate on: 1) the ratio of the measured DNA surface concentration to that expected from the amount spread, and on 2) the surface concentration of the DNA

CTA Mole DNA Nucleotides	Conc. of DNA in spread solu. micropole "p" ml	0.001 N				0.01 N				0.1 N				1 N			
		2×10^{-4}		2×10^{-3}		2×10^{-4}		2×10^{-3}		2×10^{-4}		2×10^{-3}		2×10^{-4}		2×10^{-3}	
		ratio	$\frac{mg}{m^2}$	ratio	$\frac{mg}{m^2}$	ratio	$\frac{mg}{m^2}$	ratio	$\frac{mg}{m^2}$	ratio	$\frac{mg}{m^2}$	ratio	$\frac{mg}{m^2}$	ratio	$\frac{mg}{m^2}$	ratio	$\frac{mg}{m^2}$
1.34	0.69	0.9	0.41	0.8	3.7	0.9	0.41	0.5	2.3	0.8	0.37	0.4	1.8	0.6	0.28	0.09	0.41
0.89	0.87	1.0	0.57	0.8	4.6	1.0	0.57	0.5	2.8	0.9	0.51	0.3	1.7	0.6	0.34	0.1	0.57
0.85	0.93	0.8	0.49	0.6	3.6	0.9	0.55	0.3	1.8	0.9	0.55	0.3	1.8	0.8	0.49	0.1	0.61
0.45	1.15	0.8	0.61	0.5	3.8	1.0	0.77	0.4	3.1	1.0	0.77	0.3	2.3	0.7	0.53	0.1	0.77
0.22	1.38	0.5	0.46	0.3	2.7	0.7	0.64	0.3	2.7	0.8	0.73	0.2	1.8	0.6	0.55	0.1	0.92
0.09	1.57	0.3	0.31	0.1	1.0	0.5	0.52	0.2	2.1	0.7	0.73	0.3	3.1	0.6	0.62	0.2	2.1
0.04	1.77	0.2	0.23	0.08	0.94	0.4	0.47	0.1	1.2	0.5	0.59	0.3	3.5	0.4	0.47	0.3	3.5

heating to high temperatures (NaCl).

Techniques

Measurements of surface concentration, surface pressure, and surface potential were performed as described in previous communications (10,11). Monolayers of 1.15:1 L-lysine:L-phenyl alanine were spread from a 1 mg per cc solution of 1:2 dichloroacetic acid: chloroform. Spreading was performed by the Trurnit method (18) on a glass rod sticking out above the liquid surface.

RESULTS

Surface Concentrations of Monolayers of Cetyltrimethylammonium Salts of DNA (CTA-DNA)

The surface concentrations of the components of spread monolayers of CTA/DNA were determined by utilizing tritiated DNA or tritiated CTA and measuring the surface radioactivity. It was found that if the CTA content in the CTA/DNA complex was above 0.1 mole CTA per mole nucleotides, and the area per CTA ion was higher than about 90 $Å^2$, the spread CTA remained quantitatively in the surface. Table 1 summarizes the influences of: a) the amount of CTA/DNA spread, b) the long chain salt content in the spreading solution, and c) salt concentration in the substrate, on the ratio of the measured surface concentration of H^3-DNA to that expected from the amount spread. It is obvious from the results of Table 1 that increasing the volume spread, from 0.001 ml to 0.01 ml, causes dissolution of the H^3-DNA from monolayer, and the ratio of the measured surface concentration to the "theoretical" one decreases. However, when the salt concentration in the substrate is low and the CTA content high enough, practically all the DNA remains at the interface. This is in full agreement with the reported non-solubility of long chain quarternary ammonium salts of DNA (12,13).

Considering now the changes in the stability of the monolayer with the salt concentration, one finds two contradictory influences. When the CTA to DNA ratio is high, an increase in the salt concentration of the substrate causes increased solubility of the DNA from the interfacial film. On the other hand, when the CTA content in the monolayer is low, increasing NaCl concentration enhances the persistence of the DNA in the monolayer. In intermediate compositions when about half of the counter-ions of the DNA are sodium ions, maximal DNA retention in the monolayer is found in medium salt concentrations. These observations demonstrate that increasing salt concentration causes, on the one hand, an increased exchange of the long chain counter-ions with sodium ions, and thus to an increased dissolution of the DNA from the monolayer, and on the other hand, to an enhanced adsorption of the DNA at the air-water interface. When the CTA content is high, the influence of the exchange reaction predominates and increasing salt concentration enhances desorption of DNA, while, in sodium rich monolayers, the result is the contrary. The highest surface concentration obtained was about 4.6 mg per m^2, corresponding to 12 $\overset{\circ}{A}^2$ per nucleotide. This area of 12 $\overset{\circ}{A}^2$ per nucleotide is significantly smaller than the value of 35 $\overset{\circ}{A}^2$ per nucleotide reported for native DNA adsorbed at the mercury-water interface (3,4) and it seems therefore that the excess of the insoluble CTA-DNA complex remains in the surface forming a thick interfacial film.

Surface Pressure and Potential of CTA-DNA Monolayers

It is obvious from Table 1 that monolayers of CTA-DNA remain quantitatively on the surface of a low salt concentration substrate, provided the CTA to DNA ratio is higher than about 0.5 mole CTA per nucleotide. It was possible therefore to measure the dependence of the surface pressure (π) and surface potential (ΔV) on the area occupied by such monolayers. The experimental results are presented in Figures 1 and 2.

Figure 1 describes the π-A curves of monolayers of cetyltrimethyl-ammonium chloride (CTACl) on 1 N NaCl solution and of CTA-DNA salts of varying CTA content on 10^{-3} N NaCl solutions. It is obvious that the surface pressure of all CTA/DNA salts at high surface areas is very small, and is even smaller than that of CTACl occupying the same area on 1 N NaCl. Moreover, Figure 1 also demonstrates the unexpected observation that the "limiting area" occupied by a CTA residue is almost unaffected by its being bound to DNA and is similar to that of CTACl on high salt concentration substrate. The slight shift to higher "limiting area" for CTA to DNA ratio of 0.64 and 0.75 mole per equivalent can account for steric effects exerted by the DNA in the sublayer on the CTA monolayer. It seems that the maximal areas per CTA at CTA/DNA ratios around 0.7 is related to the dificulty of binding CTA ions anchored at the inter-face to the negative groups on the DNA double helix which point toward the solution. It is thus assumed that about one third of the charges on the DNA in the surface area is inaccessible to the CTA ions. Figure 3 describes a possible surface configuration conforming with these steric requirements. The area occupied by a CTA residue when the CTA to DNA ratio is about 0.7 is approximately 110 $\overset{\circ}{A}^2$. It is obvious from Fig.3 that when increasing the CTA:DNA ratio above 0.7, the area per CTA ion decreases since the additional CTACl molecules may arrange themselves between two CTA/DNA molecules. Eventually the average area per CTA, at CTA/DNA ratios the DNA becomes too soluble to penetrate the CTA monolayer, and it tends to

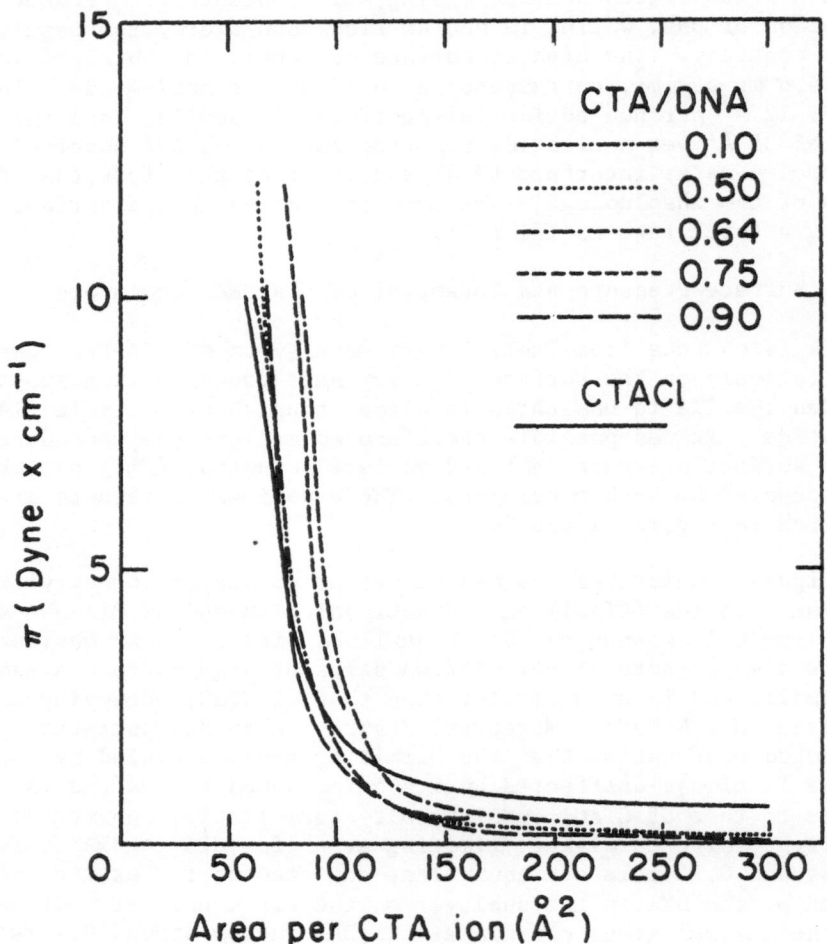

Fig. 1. The dependence of surface pressure-π on the area-(A) occupied by a monolayer of CTACl on 1 N NaCl substrate (full curve), and by monolayers of CTA/DNA of varying CTA content on 10^{-3} N NaCl substrate. The composition of the CTA/DNA monolayers are expressed in unit of moles CTA per nucleotide residue.

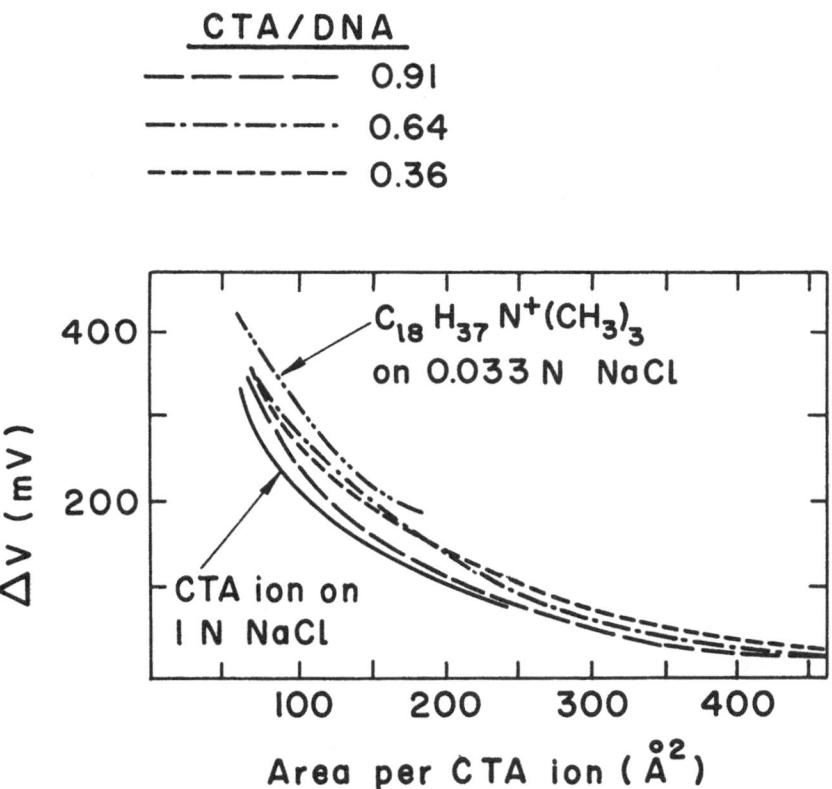

Fig. 2. The dependence of surface potential-ΔV, on the area-(A) occupied by monolayer of CTACl on NaCl substrate, by monolayer of of $C_{18}H_{37}N^+(CH_2)_3Cl^-$ on 0.033 N NaCl substrate, and by monolayers of CTA/DNA of varying CTA content on 10^{-3} N NaCl substrate. The compositions of the CTA/DNA monolayers are expressed in units of moles CTA per nucleotide residue. The data on the behaviour of the $C_{18}H_{37}N^+(CH_3)_3Cl^-$ monolayer is reproduced from Davies' measurements (14).

form a sublayer without any effect on the area per CTA.

The assumption inherent in this model, that the DNA retains its native structure, is based on the weak interaction forces evident from the small effect on the surface tension. Moreover, it will be shown in the discussion of the surface potential data, that the distorting vector of these forces is probably very small and the interacting monolayer charges can conform with the DNA structure. Supporting evidence for this assumption is the native structure of the NaDNA regenerated from the CTA/DNA precipitate (15). The presence of the DNA at the interface (proved by our surface concentration measurements presented above) can easily be confirmed by its influence on the shape of the π-A isotherms. As shown in Figure 4, the "character" of π-A curves of CTACl changes gradually from an "expanded" isotherm at low salt concentrations to a condensed one at high salt concentrations. The CTA/DNA monolayers which were spread on a 10^{-3} NaCl substrate are of a "condensed" type, with a low surface pressure at high areas.

The surface potential-area (ΔV-A) curves of CTACl on 1 N NaCl solution of $C_{18}H_{37}$ N $(CH_3)_3Cl^-$ on 0.033 N NaCl substrate, and of CTA/DNA salts of varying CTA content, on 10^{-3} N NaCl substrate are presented in Figure 2. We see again that the surface potential properties of the CTA/DNA monolayers are determined quite exclusively by the long chain hydrophobic counter-ion. Again, the practical identity of the ΔV-A curves of CTACl on 1 N NaCl solution and the various CTA/DNA monolayers on 10^{-3} N NaCl cannot be explained by dissolution of the DNA. As shown by Davies (14) a tenfold increase in the salt concentration of the substrate causes a decrease of 55 mV in the ΔV in the ΔV-A curves of monolayers of $C_{18}H_{37}$ N $^+(CH_3)_3$. In Figure 2 Davies's results dealing with $C_{18}H_{37}N^+(CH_3)_3$ monolayers on 0.033 N NaCl concentration are compared with our results and indicate that the DNA attached to the CTA monolayer acts as if it would contribute a local ionic strength of ≈ 1. This is much larger than our experimental error of ± 15 mV. The small shift to higher ΔV values in the CTA/DNA monolayers cannot be interpreted more quantitatively due to lack of knowledge on the contribution to the surface potentials of the small fraction of DNA anchored at the interface, and the non-avoidable configurational changes in the long chain salt.

Monolayers of Lysine-Phenylalanine Copolymers and Their Interaction With DNA

Figure 5 describes the dependence of π-A curves of monolayers of copolymers of L-lysine:L-phenylalanine on the composition of the substrate. Let us consider first the influence of the salt concentration and of the composition of the copolymer on the π-A relation.

The limiting area occupied by a lysine residue increases considerably with increasing phenylalanine content in the copolymer. This is expected of course, since the hydrophobic phenylalanine groups are those which anchor the polymeric molecule at the interface. Consequently increasing phenylalanine content in the copolymer by a factor of about ten should cause an increase in the limiting area per lysine residue by a similar factor and this is indeed the case. However, whereas the π-A relation of the 1:9.6 L-lysine: L-phenylalanine copolymer is independent of salt concentration in the range 5×10^{-3} M to 1 M [Na^+] (and is practically independent of pH), the surface pressure-area relation of 1.15:1 L-lysine:L-phenylalanine copolypeptide depends considerably on the ionic strength. The limiting area per <u>amino acid</u> residue of the 1:9.6 copolymer is $21.5A^2$, practically independent of salt concentration and pH. The limiting area per amino acid residue of the 1.15:1 copolypeptide is however about $9A^2$ on 10^{-3} N NaCl substrate, and increases to $17A^2$ on 0.1 - 1 N NaCl solution. Limiting areas of 15-17 A^2 per amino acid residue are typical of many proteins and polypeptides at the air-water interface (19,20,22) and are usually attributed to the β-keratin conformation. The increased limiting area of the 1:9.6 copolymer is typical of phenylalanine polypeptides (21).

The extraordinarily low limiting area of 1.15:1 L-lysine:L-phenylalanine on 10^{-3} N NaCl solution seems to result from a partial dissolution of the lysine residues in the substrate. Indeed it is well known (22,23) that polylysine monolayers can only be spread on very concentrated salt solutions. The penetration of the lysine groups into the surface at high ionic strength is therefore confirmed by our measurements. We do not believe that the abnormal low limiting area of 9 A^2 per amino acid residue might be due to helical conformation, since such conformation of charged polypeptide is expected to be more stable on a high, rather than on a low, ionic strength substrate.

In a preliminary qualitative study of the adsorption of DNA onto a monolayer of copolypeptide of 1:1 L-lysine: L-phenylalanine, reported in part in our previous communication (11), it was found a) that DNA adsorbs from a low salt concentration substrate [e.g. 0.004 M $(NH_4)_2SO_4$] onto such monolayers at a rate corresponding approximately to a diffusion controlled mechanism, b) that the amount of DNA adsorbed on a fairly compressed copolypeptide monolayers is smaller than that expected from a stochiometric ratio between the free amino groups and the phosphate anions. Thus, when the area occupied by an amino acid residue was 19.5 A^2 only 0.65 nucleotides were adsorbed per each lysine residue at the interface. The amount adsorbed was decreased to 0.4 nucleotides per lysine residue when the monolayer was further compressed and the area available per amino acid residue was only 10 A^2. Since in the bulk phase the interaction between polylysine and DNA was found to be a stochiometric one (15,16) it seems that steric hindrance at the interface is responsibl

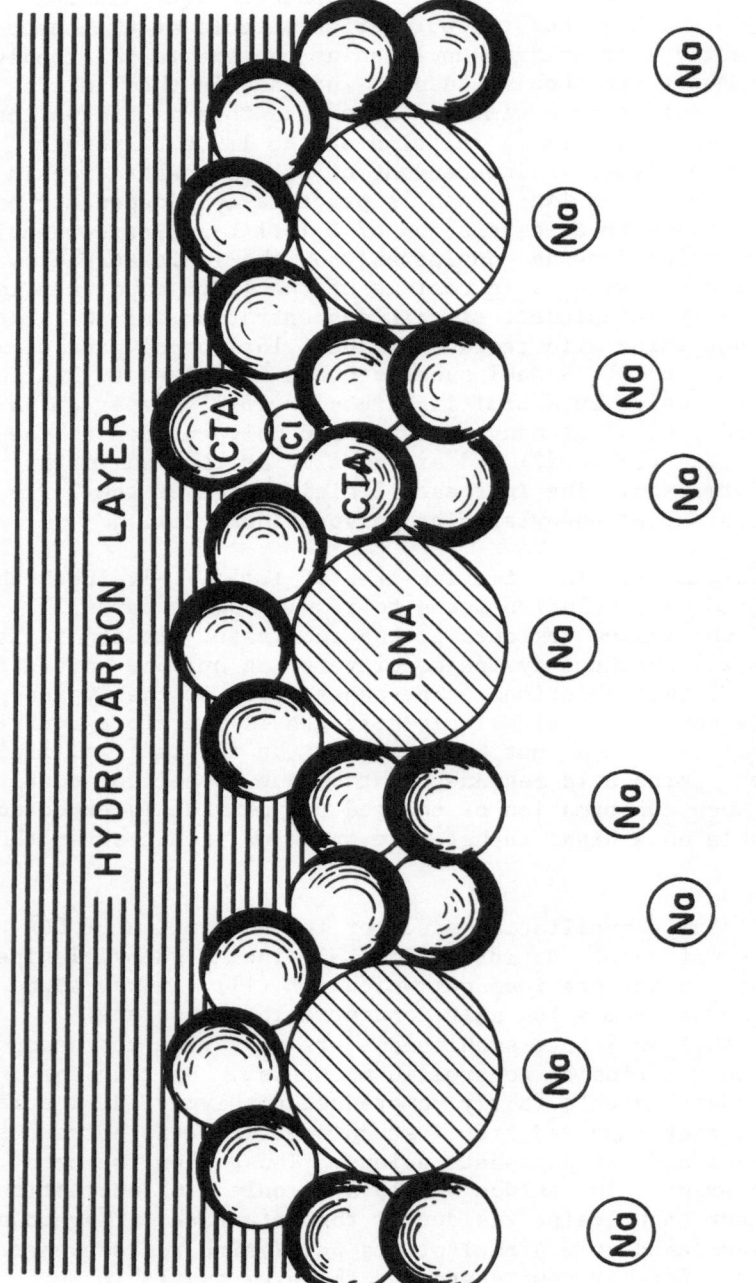

Fig. 3. Schematic representation of the arrangement of CTA ions around the DNA bihelix at the air-water interface.

for the non-stochiometric ratio. This restriction indicates that
the interacting DNA tends to preserve its stiff double helical
structure. It might be recalled that in the adsorption of polyacids
on positively charged mercury it was found that only a small
fraction of the polyelectrolyte is anchored at the interface (17)
and neutralized by surface charges. In the present case it is also
possible that not all the adsorbed DNA charges are neutralized by
the charges of the polypeptide monolayers, even though the interaction between the polymeric charges in the surface may be facili-
tated by protrusion of the positively charged lysine residues into
the sublayer.

The influence of substrate DNA on the surface pressure of a
monolayer of 1.15:1 L-lysine:L-phenylalanine is shown also in
Figure 5. As in the case of monolayers of CTA/DNA, the interaction
of the DNA with the basic polypeptide changes the shape of its
π-A curve from that typical to a "gaseous" monolayer to that
characteristic of condensed ones. However, the interaction between
the DNA and the monolayer does not cause an increase in the limiting
area.

DISCUSSION

The most striking phenomenon displayed either by monolayers of
CTA/DNA or by monolayers of basic polypeptides containing DNA is
the fact that despite its unquestionable existence at the interface
(proved by radioactive and surface pressure and potential measurements) DNA penetrates the compressed positively charged monolayer
only weakly, and has little effect on the limiting areas. This
resembles the "surface potential and tension paradox" described in
a previous communication on "the adsorption of DNA at the air-water
interface" (1). Adsorption of DNA at the A/W interface is not
followed by any detectable changes in surface tension or potential.
It was shown that this can happen only when a very small fraction of
the adsorbed molecule is anchored at the surface, or when the
adsorption forces are too weak to cause an appreciable reorientation
of the residues near the surface. Similarly, it seems that when
the DNA is held at the surface by interaction with hydrophobic
counter-ions, only a small fraction of the molecule is really anchored
at the air boundary, and the electrostatic binding takes place in
the monolayer.

What is the conformation of the DNA at the interface? To know
this we have to determine the exact composition of the DNA at the
surface. Let us consider the implications of a 1:1 interaction of
DNA with CTA. If as in the bulk phase all sodium ions of the DNA
are substituted by CTA ions to yield pure CTA/DNA (12), then if the
DNA retains its double helical conformation it must affect the
structure of the long chain salt at the surface. This is because the

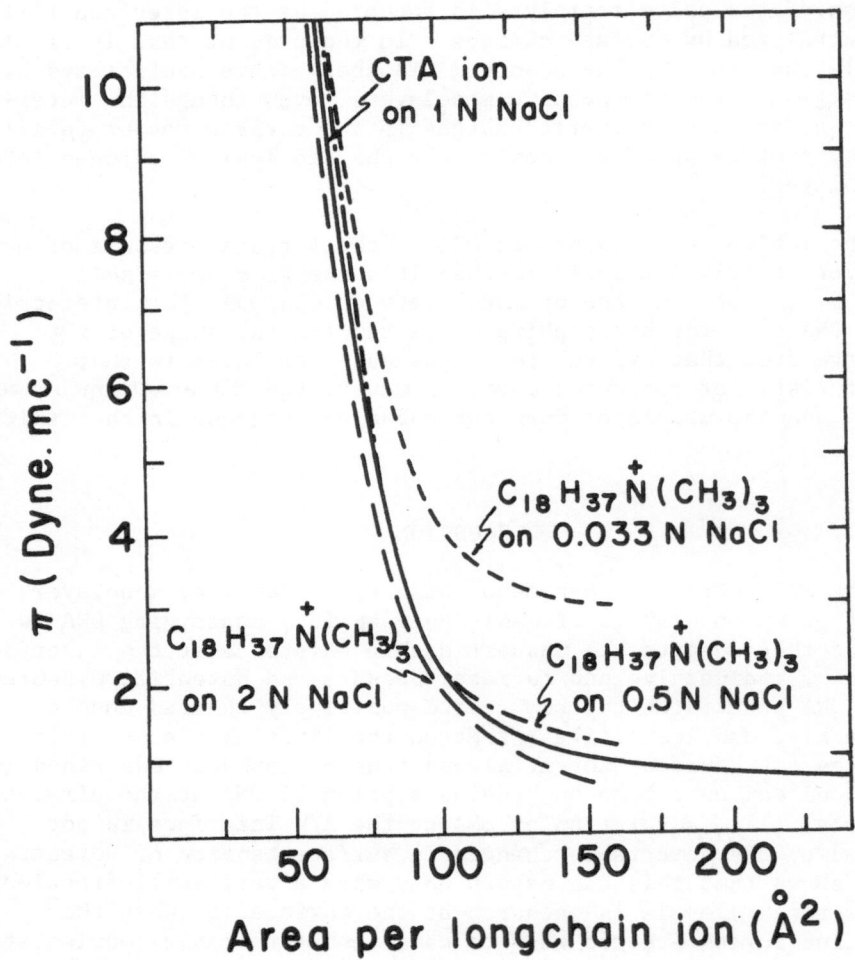

Fig. 4. The dependence of surface pressure-π on the area-(A) occupied by a monolayer of CTACl on 1 N NaCl substrate, and by monolayers of $C_{18}H_{37}N^+(CH_3)_3Cl^-$ on 2 N, 0.5 N, and 0.033 N NaCl substrate. The data on the behaviour of the $C_{18}H_{37}N^+(CH_3)_3Cl^-$ monolayer is reproduced from Davies' measurements (14).

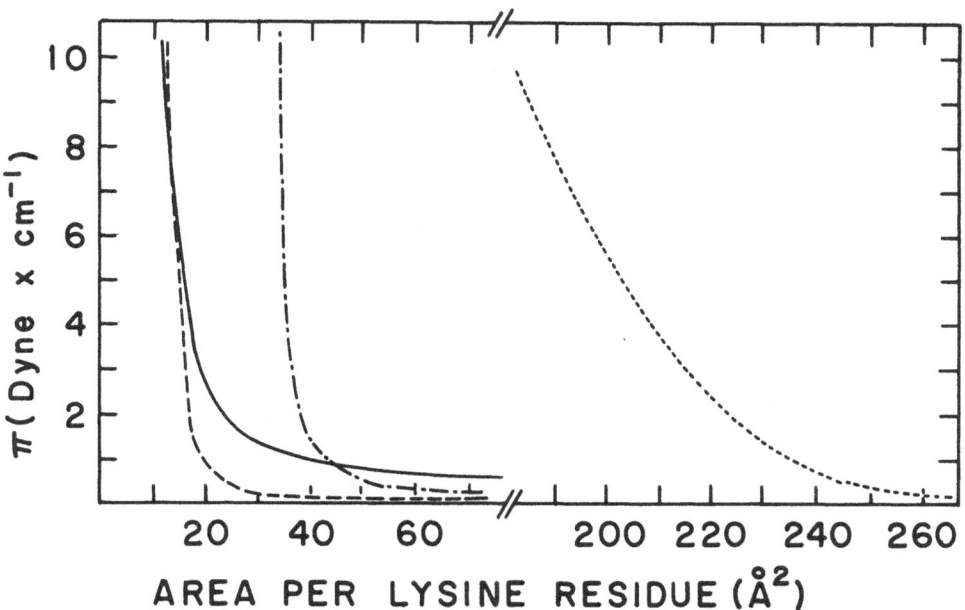

Fig. 5.   The dependence of surface pressure-πon the area-(A),
occupied by monolayers of copolymers of L-lysine, L-phenylalanine
on various substrate,

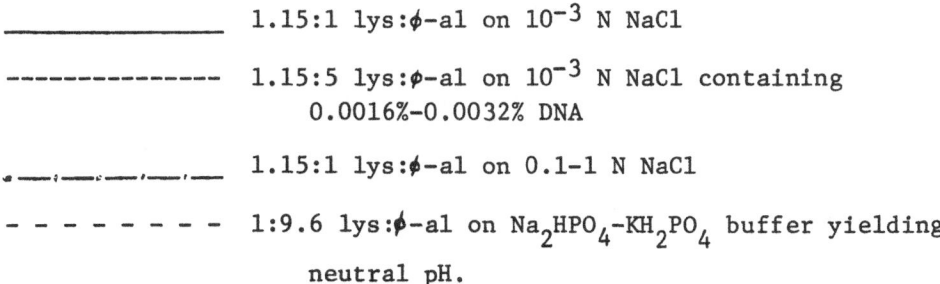

1.15:1 lys:ϕ-al on 10^{-3} N NaCl

1.15:5 lys:ϕ-al on 10^{-3} N NaCl containing
        0.0016%-0.0032% DNA

1.15:1 lys:ϕ-al on 0.1-1 N NaCl

1:9.6 lys:ϕ-al on Na_2HPO_4-KH_2PO_4 buffer yielding
        neutral pH.

double helix is a rigid structure having a surface charge density of 35 $Å^2$ per negative charge. The limiting area of CTACl is however about 90 $Å^2$ per positively charged ion. If, as assumed, the DNA is fully neutralized by CTA ions, they must penetrate into the solutions to various depths around the rigid DNA cylinder so that each of its negative charges is neutralized. Since the surface potential of the CTA/DNA monolayer is similar to that of CTACl on a high salt concentration substrate, the orientation of the dipoles and the packing of the long chain salt ions cannot be altered by this arrangement. Interaction of CTA without dipole orientation can be visualized only with those charges on the rigid DNA rod which do not point towards the solution, as shown in Figure 3. In this arrangement there is no rigid binding, and the CTA dipoles can "slide" around the cylindrical structure without change in orientation. It is very likely that the structure of charged monolayers, where the ionic components penetrate into the solution to various depths, is not unique for the case of polymeric "counter-ions", but may be a general feature of charged monolayers.

The lack of stoichiometry in the interaction of DNA with a polylysine monolayer, as well as steric effects in the CTA/DNA monolayers, hint very strongly that the DNA attached to the surface retains its double helical structure. This rigid structure eliminates the possibility of complete neutralization of the monolayer by the interacting DNA.

ACKNOWLEDGMENTS

The authors acknowledge the help of Mr. Mei-Marom who performed part of the surface pressure measurements of copolypeptides.

This work was sponsored by the National Institute of Health under research grant No. GM-08-519 and under research agreement No.615134.

This paper is abstracted from the Ph.D. thesis submitted by M.A. Frommer to the Feinberg Graduate School of the Weizmann Institute of Science.

REFERENCES

1. Frommer, M.A. and Miller, I.R., J. Phys. Chem., **72**, 2862 (1968).

2. Frommer, M.A. and Miller, I.R., Biopolymers, **6**, 1461 (1968).

3. Miller, I.F., J. Mol. Biol., **3**, 229 (1961).

4. Miller, I.R., J. Mol. Biol., **3**, 357 (1961).

5. Winsten, W.A., Biopolymers, 2, 337 (1964).

6. Miller, I.R., Biochim. et Biophys. Acta., 103, 219 (1965).

7. James, I.W. and Mazia, D., Biochim. et Biophys. Acta., 10, 367 (1953).

8. Kleinschmidt, A., Rutter, H., Hellmann, W., Zahn, R.R., Doctor, A., Zimmermann, E., Rubner, H. and Al-Ajwady, A.M., Z. Naturforschg., 146, 770 (1959).

9. Cheesman, D.F., Biochim. et Biophys. Acta., 11, 439 (1953).

10. Frommer, M.A. and Miller, I.R., Rev.Sci.Instr., 36, 707 (1965).

11. Frommer, M.A. and Miller, I.R., J. Colloid and Interface Sci., 21, 245 (1966).

12. Aubel-Sadron, G., Beck, G., and Ebel, J.P., Biochim. Biophys. Acta., 53, 11 (1961).

13. Ebel, J.P., Aubel-Sadron, G., Weil, J.H., Beck, G., and Hirth, L., Biochim. Biophys. Acta., 108, 30 (1965).

14. Davies, J.T., Proc. Roy.Soc., A 208, 224 (1951).

15. Zubay, G. and Wilkins, M.H.F., and Blout, E.R., J. Mol. Biol., 4, 69 (1962).

16. Bonner, J. and Ts'o, P.O.P. (editors), "The Nucleohistones", Holden-Day Inc., San Francisco, (1964).

17. Miller, I.R. and Katchalsky, A., "Proc. 4th Intern. Cong. Surface Activity", J.Th.G. Overbeek Ed., Vol. 2, p.275, Gordon and Breach Sci. Pub., New York (1967).

18. Trurnit, H.J., J. Colloid Sci., 15, 1 (1960).

19. Cumper, C.W.N. and Alexander, A.E., Trans. Farad.Soc., 46, 235 (1950).

20. Isemura, T. and Yamashita, T., Bull. Chem.Soc. Japan, 32, 1 (1959); 35, 929 (1962).

21. Mishuck, E. and Eirich, F.R., J. Poly.Sci., 16, 397 (1955).

22. Davies, J.T., Biochem. J., 56, 509 (1954).

23. Davies, J.T., Trans. Farad. Soc., 49, 949 (1953).

THE EFFECT OF MODIFIERS ON THE INTRINSIC PROPERTIES OF BILAYER

LIPID MEMBRANES (BLM)

H. Ti Tien

Department of Biophysics, Michigan State University

East Lansing, Michigan

INTRODUCTION

Black lipid membranes (BLM), as model systems for studies of biological membrane function, offer an opportunity for carrying out measurements which may pertain to cellular mechanisms of many important biological activities. For example, the mechanisms of nerve excitation, of water permeation, of active transport, and of energy transduction are just a few of the problems being actively investigated with the use of BLM. A number of comprehensive reviews covering the earlier studies on BLM have been published recently (1-4).

Although biological membranes are known to be complex and highly variable both in structure and function, it seems probable that there is a common construct basic to all of them (5). The strongest experimental evidence in support of such a supposition is provided by the electron microscopy of cellular membranes and organelles, which has led to the "unit membrane" hypothesis (6). Under the electron microscope, all the natural membranes thus far examined are in the order of 100 A in thickness and are generally interpreted to be as consisting of a lipid bilayer of the Gorter-Grendel type with adsorbed protein or nonlipid layers (7,8). The structure of BLM is frequently depicted to be similar to that of a Gorter-Grendel bimolecular lipid leaflet at a water-oil-water biface (9). This paper will attempt to present a unifying view that modified BLM are useful experimental models for at least five basic types of biological membranes. These types are the plasma membrane of erythrocyte, the nerve membrane of axon, the cristae membrane of mitochondrion, the thylakoid membrane of chloroplast, and the rod outer segment sac membrane of retina. Schematic and

135

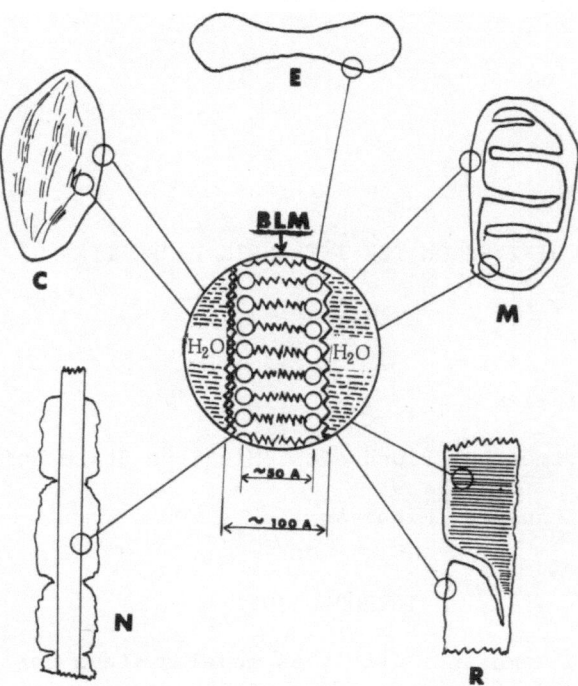

<u>Figure 1</u>. Schematic illustration of the five basic types of
biological membranes as they are generally visualized under the
electron microscope and their molecular interpretation according
to the bimolecular leaflet model (BLM). <u>E</u> erythrocyte (plasma
membrane). <u>C</u> chloroplast (thylakoid membrane). <u>M</u> mitochondrion
(cristae membrane). <u>N</u> nerve axon (myelin). <u>R</u> rod outer segment
membrane.

highly idealized pictures illustrating these basic units of life
processes are shown in Fig. 1. In describing these model systems,
the effects of modifiers on the intrinsic properties of unmodified
BLM will be stressed.

BLM VS. ULTRATHIN LAYER OF LIQUID HYDROCARBON

The intrinsic properties of an unmodified BLM generated from
either phospholipids or oxidized cholesterol in an alkane solvent
are strikingly similar to those expected of a thin layer of liquid
hydrocarbon of equivalent thickness. A comparison of the properties
of the unmodified BLM with the extrapolated properties of a 100 A
layer of liquid hydrocarbon is summarized in Table 1. It is
evident from the data given that an unmodified BLM appears to be a

Table 1. COMPARISON OF PROPERTIES OF UNMODIFIED BLM AND A
          LAYER OF LIQUID HYDROCARBON OF EQUIVALENT THICKNESS

PROPERTY	UNMODIFIED BLM (EXPERIMENTAL)	LIQUID HYDROCARBON (EXTRAPOLATED)
Thickness (Angstroms)	40-130	100
Resistance (Ω-cm^2)	10^7-10^9	10^8
Capacitance (μF-cm^{-2})	0.3-1.3	1.0
Breakdown Voltage (V/cm)	10^5-10^6	10^6
Dielectric Constant	2-5	2-5
Refractive Index	1.4-1.6	1.4-1.6
Water Permeability (μ/sec)	8-24	35
Interfacial Tension (dynes/cm)	0.2-6	50
Potential Difference per 10 fold Concentration of KCl (mV)	0	0
Electrical Excitability	None	None
Photoelectric Effects	None	None

poor model for the biological membrane. For instance, it is well
known that biological membranes are ion-selective. In the case of
nerve membrane, electrical "excitability" is one of the most unique
features.

   In an attempt to modify the intrinsic properties of the BLM,
literally several hundreds of compounds were evaluated in the
beginning (10). Among these the following groups of materials have
been tried: common proteins, enzymes, surfactants, fermentation
products, vitamins, tissue extracts (eg. retina), and a variety of
organic and inorganic compounds. This broad and preliminary testing of materials has led to the discovery of a modifier of uncertain
composition (still not known to date) termed "excitability inducing
material" (or EIM) which not only dramatically reduced the BLM
resistance but induced electrical "excitability" as well (10,11).
Beginning with EIM, a number of modifying agents (or modifiers)
has been discovered which, when present in the BLM, impart new
properties that are of biological interest. Since a detailed
review of these BLM modifiers will be given elsewhere (12), only
a general classification is given in Table 2. The present paper
will be concerned with the effects of modifiers on four different
BLM systems, which have been studied recently in this laboratory.

Table 2.   CLASSIFICATION OF BLM MODIFIERS

CATEGORIES	EXAMPLE
A. Those Altering the Basic Electrical Properties	EIM[a], KI
B. Those Conferring Ion Specificity	Valinomycin, I_2
C. Those Inducing Electrical Excitability	EIM[a], Alamethicin(13)
D. Those Changing the Mechanical Properties	HDTAB[b], Various Proteins
E. Those Generating Photoelectric Effects	Chlorophylls, Retinenes, Phthalocyanines, Various Organic Dyes, Inorganic Ions

[a]EIM - Excitability Inducing Material (10)
[b]HDTAB - Hexadecyltrimethylammonium Bromide

Table 3.   COMPOSITION OF BLM-FORMING SOLUTIONS

MAJOR LIPID COMPONENT OR SOURCE	MODEL FOR	REFERENCE
A. Brain lipids	Plasma membrane	(10)
Phospholipid and Cholesterol	" "	(14)
Erythrocyte extract	" "	(15)
E. Coli extract	" "	This work[a]
B. Brain lipids	Nerve membrane	(10)
Oxidized Cholesterol + DAP[b]	" "	(16)
C. Phospholipid + Cholesterol	Mitochondrial membrane	(17)
DAP[b] + Oxidized Cholesterol	" "	(16)
D. Chlorophylls + Phospholipid	Thylakoid membrane	(18)
Chloroplast extract	" "	(18)
E. Carotenoid pigments & Phospholipid + oxidized Cholesterol	Visual receptor membrane	(19)

[a]Details given in this paper
[b]DAP - Dodecyl acid phosphite (16)

In Table 3 the major component or source of lipids which has been used for the generating of BLM as models for various biological membranes is given. Further details concerning these BLM-forming solutions may be found in the published literature.

E. COLI BLM (Reconstituted Plasma Membrane)

Unlike most mammalian cells, E. Coli contain little lecithin or cholesterol. In spite of this difference in lipid composition, E. Coli also develop a high K^+ and low Na^+ interior when the nutrient solution is low in K^+ and high in Na^+. It was decided to study the effect of modifiers on BLM formed from E. Coli extract. The simple procedure for obtaining BLM-forming solution is shown in the flow chart (Fig. 2).

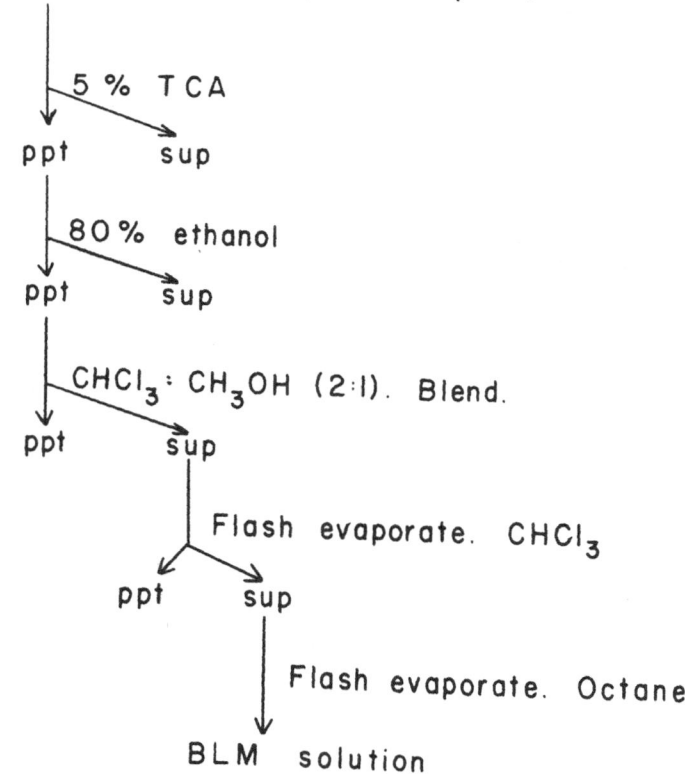

Figure 2.   Procedure for preparation of E. Coli BLM-forming solution.

E. Coli, strain K-12, were grown at 37°C in DIFCO culture medium with 3g/l sucrose added. The cells were harvested in the logarithmic phase of growth and treated first with 5% cold TCA, the 80% ethanol. The treated cells were dissolved in 20 ml/gm chloroform:methanol (2:1) and the mixture was blended in a Waring blender for 5 minutes. After centrifugation the supernatant was flash evaporated and the residue was redissolved in chloroform and filtered. This supernatant was flash evaporated and the final residue was dissolved in a hydrocarbon solvent. Dodecane was used in preliminary experiments; however it was found that black membranes formed faster and more smoothly in octane, and this solvent was subsequently used in all extractions.

A description of the apparatus has previously been given in detail (20). Briefly, membranes were formed on a 1.5 mm hole in a thin teflon sleeve which separated two chambers, each filled with aqueous solution. Temperature was maintained with a constant temperature bath connected to polyethylene or glass tubing which coiled around twice through the outer chamber. By keeping the inner chamber stirred, equilibration of temperature could be achieved in a short period of time. Calomel electrodes with agar-saturated KCl bridges were inserted in each chamber to measure electrical properties. For measurements of resistance and capacitance a small current at known potential (30-40mV) was passed across the membrane and through a standard resistor placed in series with the membrane. The potential drop across the BLM was measured with a Keithley-610 electrometer and the time course of this potential was recorded on a Bausch and Lomb VOM-5 recorder for capacitance determinations. Resistance was calculated from an application of Ohm's law and capacitance was calculated from the well-known DC discharge technique. For measurements of transmembrane potential all electronic components were switched out of the circuit except the electrometer and recorder. Lipid solution was applied to the teflon sleeve using a trimmed sable brush (10). At 37°C thinning was complete within two minutes.

The DC electrical resistance of E. Coli BLM was 10^7-$10^8\Omega$-cm^2. After correcting for area, usually 1 mm^2, the corrected resistance value was 10^5-$10^6\Omega$cm^2. The resistance was slightly increased in more dilute salt solutions, roughly a two-fold increase for every ten-fold dilution in the range of 10^{-3}-10^{-1}M salt solution. The change in resistance observed with changing temperature is shown in Fig. 3. From a plot of log R versus 1/T using the Arrhenius equation, the activation energy was calculated to be 18.6 Kcal/mole.

The capacitance of the E. Coli BLM was .75μF/cm^2 in 0.1N NaCl at 37°C. Variations in this parameter have not been studied in detail.

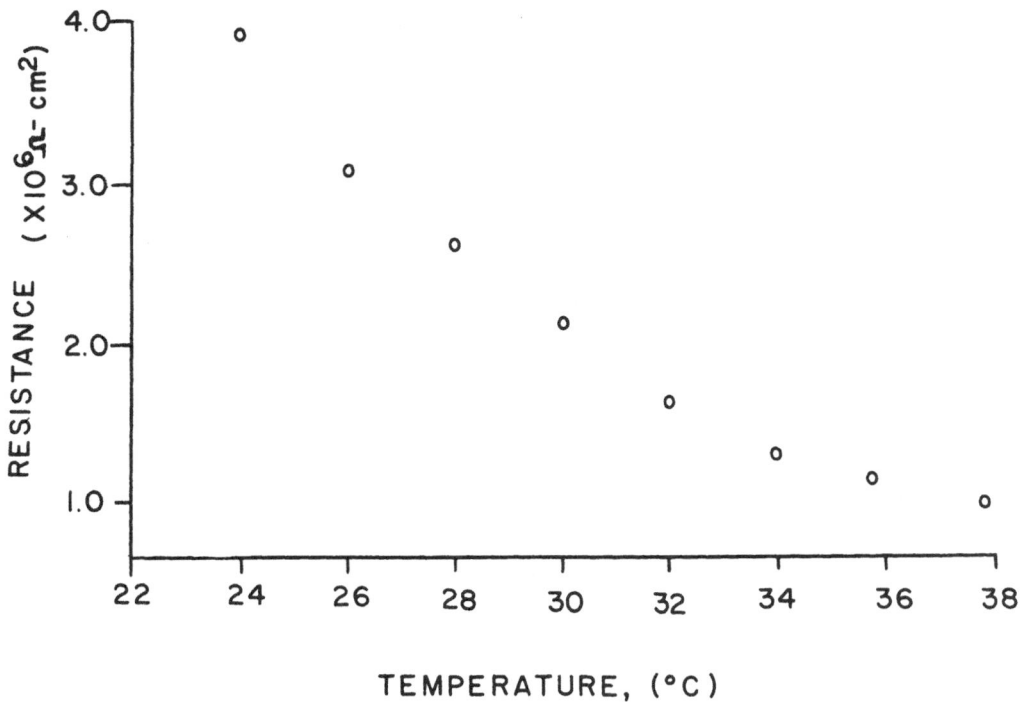

<u>Figure 3.</u> Temperature dependence of <u>E. Coli</u> BLM resistance.
Activation energy, 18.6 Kcal/mole.

The Effect of EIM

Two compounds which have a dramatic effect on most black
membrane systems were tested on <u>E. Coli</u> BLM. Both iodine and EIM
caused drops in resistance down to the level of $5x10^3 \Omega-cm^2$. The
effect with iodine was immediate, while the addition of EIM took
1-2 minutes to show its effect. The lowering of BLM resistance
by EIM (added to one side) as a function of time is shown in
Fig. 4. It is interesting to note that the membrane became
asymmetric.

The Effect of Surfactants

In an effort to determine qualitatively the net charge on
the BLM surfaces, various anionic and catonic surface active
materials were added to one side of the membrane and the transient
transmembrane potential was measured. The cationic materials
tested were hexadecyltrimethylammonium bromide (HDTAB) and
protamine sulfate; the anionic agents were sodium dodecyl sulfate

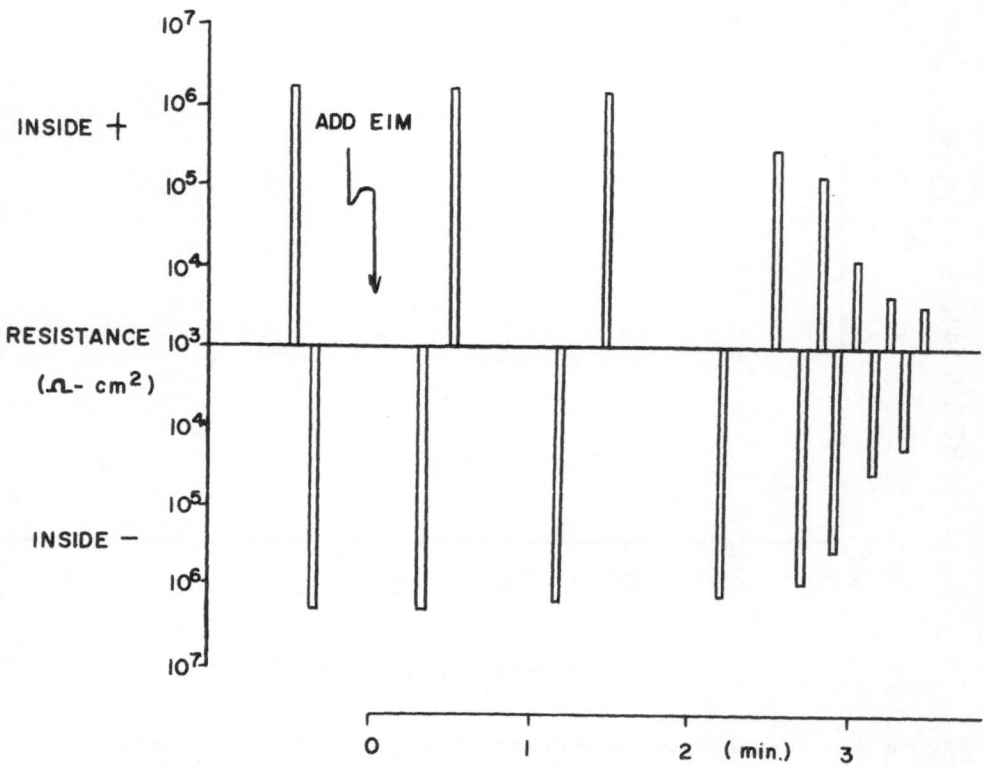

<u>Figure 4.</u> The effect of EIM on membrane resistance as a function
of time measured with both polarities. Note the membrane
resistance is asymmetrical.

and casein. Of these compounds, both of the positively charged
surface active materials produced large transmembrane potentials
when added to the aqueous solution on one side of the membrane.

The results of a set of experiments in which HDTAB was used
is shown in Fig. 5. Addition of HDTAB to one side of the BLM
produced a transmembrane potential which was positive on that side.
The potential developed a few seconds after the material was added,
the short delay being due to the amount of time required for the
stirring motion to bring the material to the membrane interface.
The potential developed across the membrane was transient and began
decaying in an exponential manner after ten seconds or so. The
total time course of the potential was in the range of 1-3 minutes.
The transmembrane potential increased with increasing concentrations

of HDTAB until the detergent caused the membrane to rupture.
Similar results were produced with protamine sulfate except that
the potential developed was much smaller.

The effect of these cationic surface-active materials was
much more pronounced when the salt concentration in the aqueous
solution was low. The negatively charged materials had little
effect, although casein was not tested at high concentrations due
to its very limited solubility in neutral salt solution. It was
concluded from these experiments that the net interfacial charge
on the E. Coli BLM was negative.

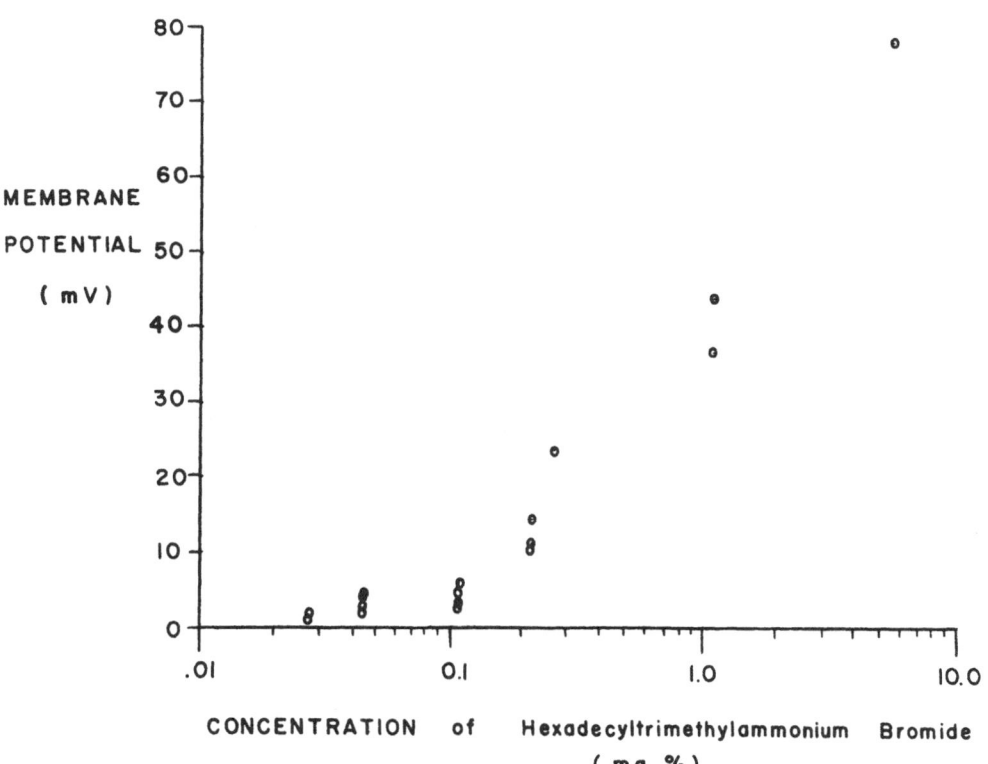

Figure 5. The effect of HDTAB on E. Coli BLM potential.

OXIDIZED CHOLESTEROL BLM

One of the difficulties in studying BLM is to find a lipid solution from which stable BLM can be formed. Oxidized cholesterol solution has been found to give very stable membranes (21), and it is by far the easiest to work with. The intrinsic electrical properties of oxidized cholesterol BLM are as follows: resistance --$5 \times 10^9 \Omega cm^2$, capacitance--$0.57 \mu F/cm^2$, and breakdown voltage-- 300 mV (16).

The effect of I_2 and KI on BLM was first observed by Lauger et al (22). They reported that the electrical resistance of lecithin BLM is lowered by 3 orders of magnitude when $KI-I_2$ was present in the bathing solution. Lauger et al. have also reported that a steady concentration potential of about theoretical value was observed, implying the BLM is highly specific for I^-.

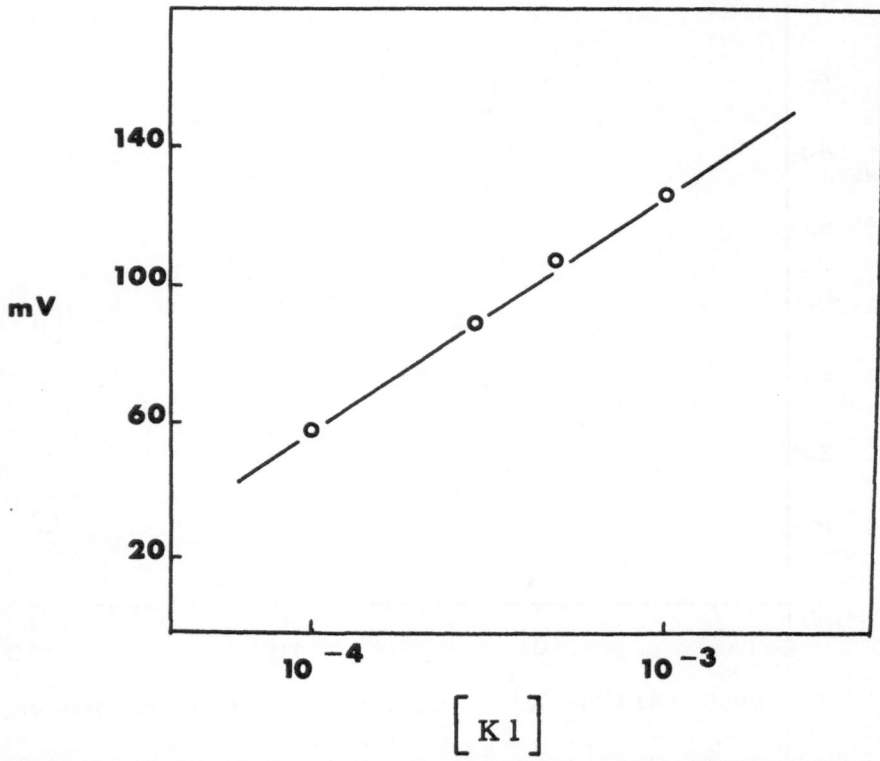

Figure 6. Typical response of an iodine-modified BLM to KI in 0.1 M KCl. The response to I^- depends only on the presence of I_2 and irrespective of BLM-forming solution.

In pursuing the effect of KI-I_2 on BLM, we have found that a
BLM can be made into an iodide-specific electrode by simply incor-
porating a minute amount of I_2 into the lipid solution prior to
BLM formation. For example, BLM formed from an oxidized cholesterol
solution containing 0.2 mg of I_2 (per 5 ml) responded ideally to
I^-. A typical curve is shown in Fig. 6. The presence of other
ions such as Cl^-, SO_4^{-2} or F^- did not interfere with the "electrode"
response to iodide ion.

In another series of experiments, we have measured the
current/voltage (I/V) curves of oxidized cholesterol BLM in 0.1 M
K I + 0.1 g I_2 with $Na_2S_2O_3$ added to one side of the BLM. The I/V
curve shown in Fig. 7 has been obtained with the aid of an
automatic recording polargraph (Polarecord Model E, Metrohm Ltd.).
The presence of $Na_2S_2O_3$ on one side of the BLM generated a potential
difference of about 45 mV. It is evident that the I/V curve is no

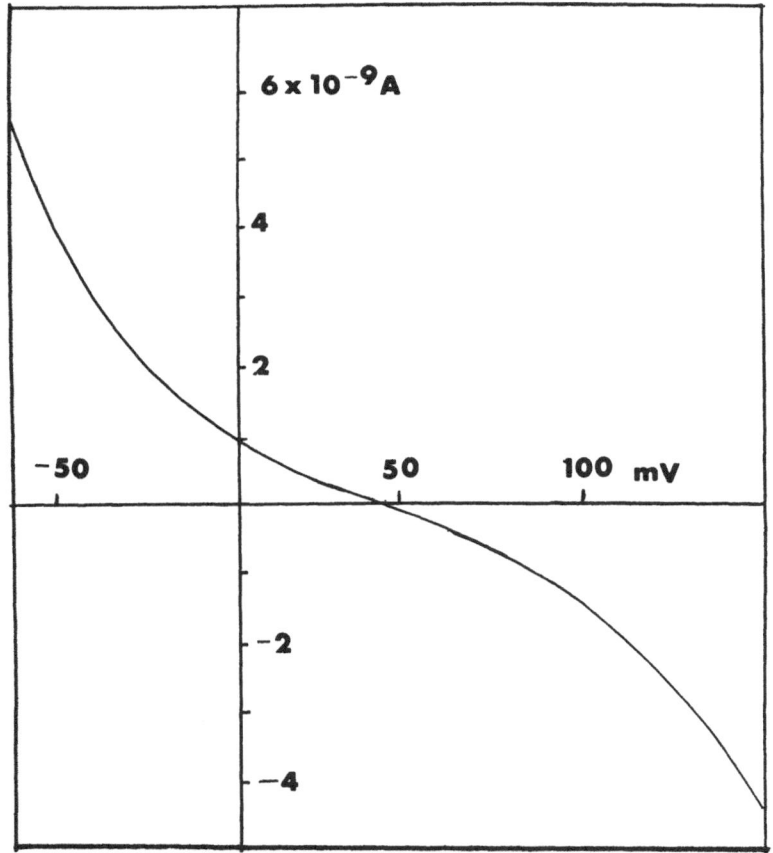

Figure 7. Current/voltage curve of an oxidized cholesterol BLM
in 0.1 N K I + $4x10^{-4}$ M iodine. In addition, the inner chamber
contained $5x10^{-3}$M $Na_2S_2O_3$.

longer linear but still entirely symmetrical. Our findings on
the effect of $KI-I_2$ on BLM are consistent with the observations
reported by other workers (22,23). The presence of I_2 in the BLM
facilitates the selective iodide transport, which in turn is
manifest in the development of a concentration potential in
accordance with the Nernst equation. A possible explanation for
the non-linear portion of the I/V curve will be given later.

CHLOROPLAST BLM (CHL-BLM)

The third type of BLM which has been studied is constituted
of spinach chloroplast extracts (18). The chloroplast black
lipid membrane (Chl-BLM) is of special interest in that upon
illumination with light, two basic photoelectric effects can be
induced. The photovoltaic effect has been found to be dependent
both on light intensity and wavelength. In addition, the electrical properties of Chl-BLM can be altered by a number of modifiers.
The results of some recent experiments are summarized in the
following paragraphs.

The Effect of Fe^{+3} and Ferredoxin

Unlike other BLM, Chl-BLM have been found highly specific to
H^+ below pH 8. For example, in the pH region 4-6, a potential
difference of theoretical value has been observed (i.e., 58 mV/pH).
This implies that the surface of Chl-BLM is negatively charged
(presumably due to sulfolipids and phospholipids). In the dark
the d.c. resistance of the membrane is ohmic when separating two
identical solutions. The unmodified Chl-BLM exhibits photo-
conductivity. When Fe^{+3} is added to one side of the Chl-BLM, a
"rectification" effect can be produced. The current/voltage
curve for an asymmetrical system containing $10^{-3}M$ Fe^{+3} on one
side is shown in Fig. 8A. Similar effect can be induced when
spinach ferredoxin is used in place of Fe^{+3} (see Fig. 8B).

The Effect of 2,4-dinitrophenol (DNP) and pH

The separation of charges by light in chloroplasts is
believed to be among the first steps involved in the primary
process of photosynthesis. Chemical compounds such as 2,4-
dinitrophenol (DNP) can uncouple electron transfer from energy
accumulation in the electron transport chain. The similar effect
of DNP on oxidative phosphorylation in the mitochondria is well
known (24). With the availability of BLM as model systems for
the biological membranes (see Fig. 1), various uncouplers including
DNP have been tested (16,25). In addition to DNP, we have
investigated the effect of carbonylcyanide-p-trifluoromethoxy-
phenolhydrazone (FCCP) on the electrical properties of Chl-BLM.

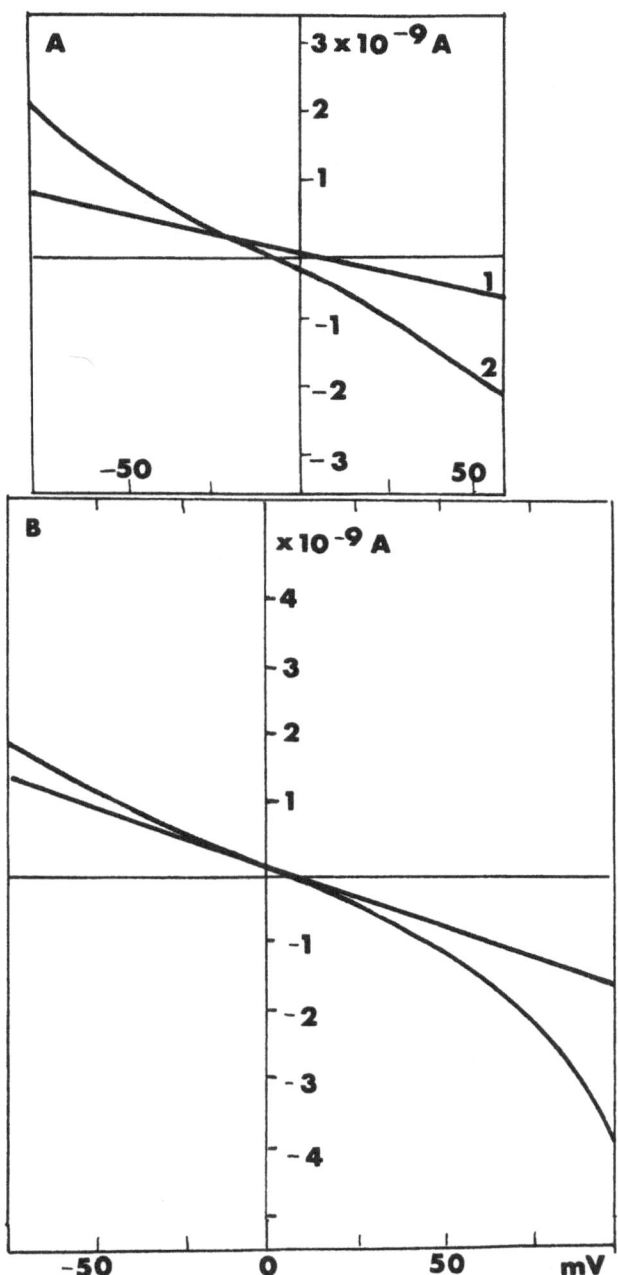

Figure 8. Current-voltage curves of chloroplast BLM in 0.1 M
citrus-phosphate buffer at pH 5.   A.   Inner chamber containing
10^{-3}M Fe^{+3}; curve 1 obtained in the dark, curve 2 with white
light.   B.   Inner chamber containing 0.1 mg ferredoxin preparation;
curve 1 in the dark, curve 2 with light.

In low concentrations (10^{-6}-10^{-5}M), both DNP and FCCP dramatically
lowered the Chl-BLM resistance, FCCP being twenty times more
effective.

The efficiency of FCCP and DNP on Chl-BLM has been found to
be pH dependent. Fig. 9A presents the results obtained with a
Chl-BLM as a function of pH (0.1M citrate-phosphate buffer). It
is interesting to note that in the absence of an uncoupler, the
electrical resistance of Chl-BLM itself exhibits a pH-dependency
(Fig. 9B).

The effect of these modifiers on non-linear I/V curves
(iodine, FCCP, and DNP) may be explained in terms of ion association
(26) in the lipid moiety and dissociation of the ionizable groups
of the Chl-BLM. At the BLM biface (i.e., the two co-existing
membrane/solution interfaces), dissociation of the ionizable
groups of the lipids must be governed by the pH of the bulk
solution and possibly also by the electric field. Thus, the
selectivity of the BLM would be controlled by the ionizable groups
at the biface whereas the specificity depends upon the solubility

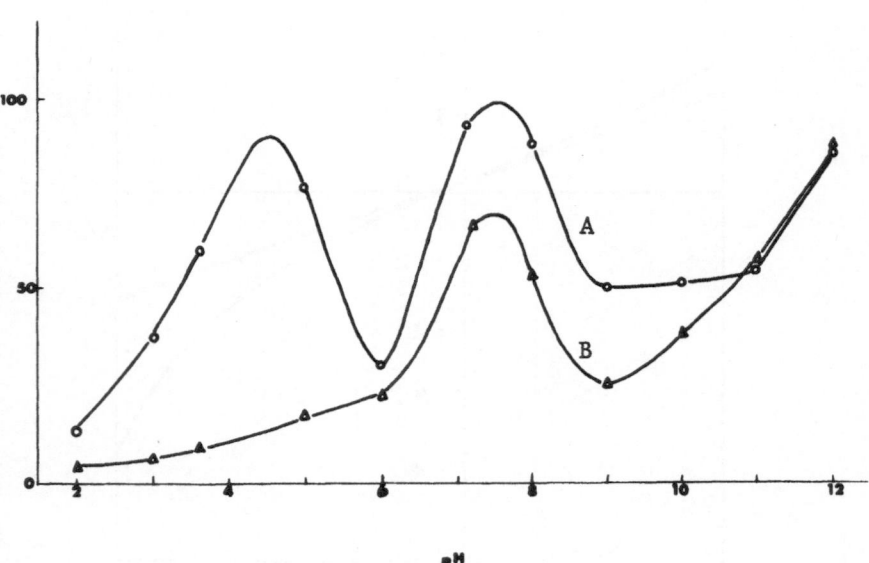

Figure 9. pH dependence of BLM current. BLM-forming solution
was prepared from spinach chloroplast extract. Aqueous solution=
0.1 M citrate-phosphate buffer; Applied voltage: 60 mV. Curve A
was with 2,4-dinitrophenol (DNP) as the modifier and curve B
without modifier. Applied voltage = 60 mV.

of lipid-soluble complex. According to Bjerrum's theory of ion association (26), the equation governing the following reaction

$$M^- + C^+ \rightleftharpoons M\text{---}C^+ \tag{1}$$

is given by

$$K^{-1} = \frac{4\pi N}{1000} \left(\frac{e^2}{DkT} \right)^3 Q(b) \tag{2}$$

where K is the dissociation constant for the reaction (1), and where N, D, R, T, and e have the usual significance. The quantity Q(b) is a definite integral, which is a function of (b) defined by the following equation

$$b = \frac{e^2}{DkTa} \tag{3}$$

where "a" is the distance between the ion pair. In principle one should be able to calculate the dissociation constant of the ion pair in the hydrophobic region of the BLM ($M^-\text{---} C^+$, where M^- stands for a negatively charged modifier, and C^+ is its counter ion). The Bjerrum theory of ion association has been quantitatively confirmed for aqueous solutions and non-aqueous systems by the work of Kraus and Fuoss (27). Whether the theory can be extended to BLM systems remains to be seen. The evidence at hand suggests that, when modifiers or ionizable species are present in the BLM system, the current/voltage curves are no longer linear (Figs. 7 and 8). It seems probable that the dissociation of ion-pairs under the condition of intense field ($\sim$1-5x10^5 V/cm) may be responsible for the non-linear I/V behavior. In fact, the production of additional charge carriers as a function of the electric field (the Wien effect) has been considered recently in conjunction with nerve excitation (28). The possibility that a similar mechanism is also operative in the BLM system under the influence of modifiers and electric field seems quite likely.

The Photoelectric Spectrum of Chl-BLM

As has been mentioned above, the observation of photoelectric effects provides ample evidence that a separation of charges (electrons and holes) by light is taking place in Chl-BLM. To study further the light-induced phenomena, we have now developed a quantitative technique to obtain the spectra of the Chl-BLM (29). The method is based upon the measurement of photoresponse of the BLM as a function of wavelength. The action spectrum of a Chl-BLM together with its absorption spectrum is shown in Fig. 10. Within the experimental error of detection instruments, the two spectra are practically identical. This finding is consistent with the observation that the spectral excitation curve of most organic photoconductors (semiconductors) is generally found to be very similar to that of the absorption spectrum.

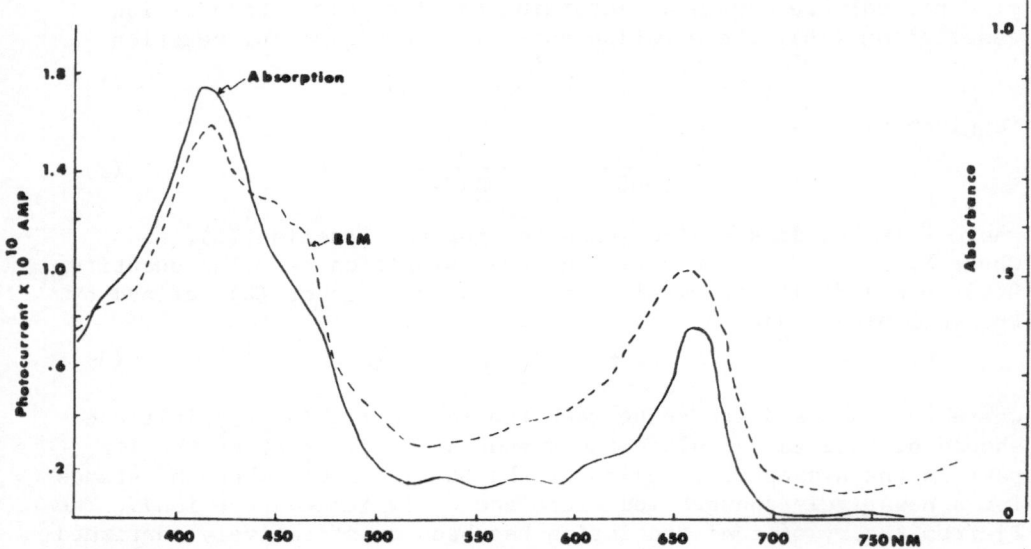

Figure 10. Absorption and photoelectric spectra of chloroplast extract (29).

CAROTENOID BLM

In the earlier studies BLM have been used as models for various types of biological membranes (Fig. 1). Attempts were made to incorporate visual pigments such as rhodopsin into the BLM but no significant results were obtained (30). More recently, β-carotene has been incorporated into a BLM system but no marked lowering of electrical resistance was observed (31). With availability of more stable BLM-forming solutions and improved experimental techniques, we have initiated a new series of experiments with the aim of constituting a different type of BLM as a model for the visual receptor membrane (19). The following paragraph summarizes our preliminary findings on black lipid membranes containing carotenoid pigments (or carotenoid BLM for short).

The experimental techniques used in the formation of the carotenoid BLM have been described previously (4). The lipid solutions used consisted of various carotenoids (all-<u>trans</u> retinal, 9-<u>cis</u> retinal, all-<u>trans</u> retinol, and β-carotene), phospholipids, and oxidized cholesterol dissolved in liquid alkanes. The dark d.c. resistance of the membrane is ohmic ranging from 10^6 to $10^7 \Omega$-cm^2, which is about 2-3 orders of magnitude lower than that of carotenoid-free BLM. Upon illumination with white light (DFG tungsten lamp, Sylvania) a maximum photo-emf of about 6 mV was observed in a BLM containing a mixture of all-<u>trans</u> retinal and β-carotene. All carotenoid BLM were found to be photoconductive. In addition, we have found that the photoresponses of these carotenoid BLM were more complex as the experimental conditions were altered. Depending upon the carotenoid pigments used and external factors (e.g. modifiers and applied voltage), the voltage/time curves can vary from a simple monophasic response to a typical biphasic wave-form not unlike those found in vertebrate retina (32). The various modes of responses which have been observed in our initial experiments are illustrated in Fig. 11.

The relevance of carotenoid BLM as an experimental model for the visual receptor membrane may be viewed from the standpoint of visual organelles (see Fig. 12). As has been revealed by the electron microscope, the outer segment membranes of rods and cones are highly organized (5). The structure of the outer segment membranes is believed to be similar to that of the unit membrane (6,34). Thus, the carotenoid BLM is one of the ideal systems with which not only initial energy transduction mechanism from photons to electrons and holes but also the triggering mechanism for ionic permeability across the membrane can be readily investigated. For example, Hagins et al (35) have reported that the initial membrane depolarization of the sacs of the rod outer segment is

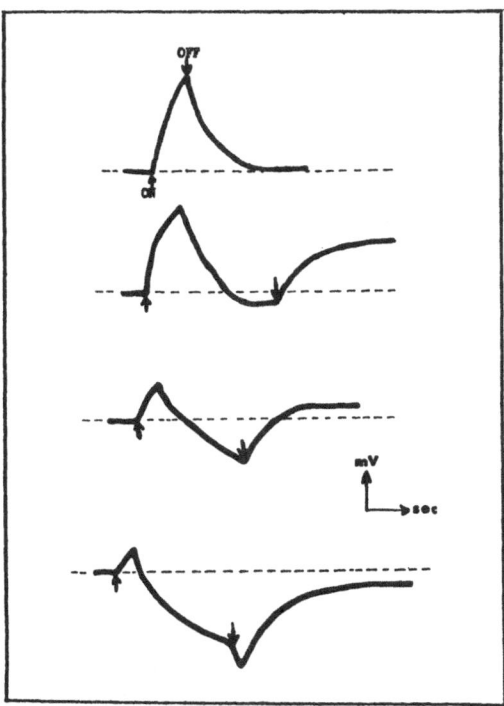

Figure 11. Modes of photoelectric response of a carotenoid BLM under different experimental conditions (19).

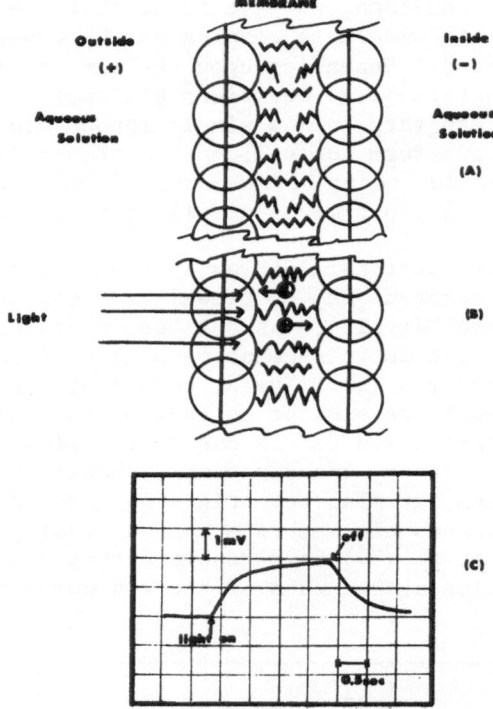

Figure 12. The effect of light on a carotenoid BLM containing β-carotene and retinal. (a) illustrating a hypothetic structure of the BLM in the dark, (B) generation of charge carriers by light, (C) an oscilloscope trace of a carotenoid BLM in response to illumination.

fairly localized while the outer plasma membrane depolarization occurs with a large latency and delocalization. We are at the present carrying out further experiments with the carotenoid BLM with the object of understanding the mechanisms of excitation and energy conversion. It is hoped that the results will be useful in analyzing the complicated electroretinogram (32,33).

SUMMARY

At the present time the molecular structure of biological membranes has not been definitely settled. Nevertheless, the general picture of membrane structure widely accepted today, is that based upon the bimolecular leaflet model. In view of this acceptance, a black or bilayer lipid membrane (BLM) upon suitable modification appears to be the closest approach to the biological membrane.

We have shown that an unmodified BLM in common salt solutions

(NaCl or KCl) exhibits characteristics expected of a liquid hydrocarbon layer of equivalent thickness. Upon the addition of modifiers, the intrinsic properties of a simple BLM can be drastically
altered. The BLM modifiers can be broadly divided into 5 groups:
(1) those altering the passive electrical properties, (2) those
changing the mechanical properties, (3) those conferring ion
selectivity, (4) those inducing excitability, and (5) those
generating photoelectric effects. The results of some recent
studies from this laboratory using a number of modifiers on a
variety of BLM systems as models for various biological membranes
have been described.

ACKNOWLEDGMENTS

The author acknowledges the invaluable collaboration of his
colleagues: Drs. A. L. Diana, S. P. Verma, N. T. Van, Mr. W. A.
Huemoeller and Mr. N. Kobamoto in the experimental work described
in this paper. The E. Coli cells were kindly supplied to us by
Professor Harold Sadoff. A sample of FCCP was provided by
Dr. W. W. Prichard of E. I. duPont Co. These investigations were
supported by Grant GM-14971 from National Institutes of Health.

REFERENCES

1. J. A. Castleden, J. Pharm. Sci., 58, 149 (1969).
2. A. D. Bangham, Progress in Biophysics, 18, 29 (1968).
3. L. Rothfield and A. Finkelstein, Ann. Rev. Biochem., 37, 480
    (1968).
4. H. T. Tien and A. L. Diana, Chem. Phys. Lipids, 2, 55 (1968).
5. F. S. Sjöstrand, Radiation Res. Suppl., 2, 349 (1960).
6. J. D. Robertson, Protoplasma, 63, 219 (1967).
7. E. Gorter and F. Grendel, J. Exptl. Med., 41, 439 (1925).
8. H. Davson and J. F. Danielli, "The Permeability of Natural
    Membranes", Cambridge, London, (1952).
9. H. T. Tien, J. Gen. Physiol., 52, 125s (1968).
10. P. Mueller, D. O. Rudin, H. T. Tien, and W. C. Wescott, in
    "Recent Progress in Surface Science", 1, 379 (1964).
11. P. Mueller, D. O. Rudin, H. T. Tien, and W. C. Wescott, Nature,
    194, 979 (1962).
12. H. T. Tien, in "The Chemistry of Bio-Surfaces", M. L. Hair,
    ed., Marcel Dekker, Inc., New York, to be published.
13. P. Mueller and D. O. Rudin, Nature, 217, 713 (1968).
14. T. Hanai, D. A. Haydon, and J. Taylor, J. Theoret. Biol., 9,
    422 (1965).
15. R. E. Wood, R. P. Wirth, and H. E. Morgan, Biochim. Bioph. 163, 171 (19
16. H. T. Tien and A. L. Diana, Nature, 215, 1199 (1967).
17. Bielawski, T., E. Thompson and A. L. Lehninger, Biochem.
    Biophys. Res. Commun., 24, 948 (1966).

18. H. T. Tien, W. A. Huemoeller, and H. P. Ting, Biochem.
     Biophys. Res. Commun., 33, 207 (1968).
19. H. T. Tien and N. Kobamoto, Nature, in press (1969).
20. H. T. Tien and A. L. Diana, J. Colloid Int. Sci., 24 287
     (1967).
21. H. T. Tien, S. Carbone, and E. A. Dawidowicz, Nature, 212,
     718 (1966).
22. P. Lauger, W. Lesslauer, E. Marti, and J. Richter, Biochim.
     Biophys. Acta, 135, 20 (1967).
23. A. Finkelstein and A. Cass, in "Biological Interfaces: Flows
     and Exchanges, Little, Brown & Co., New York, 1968. p. 145.
24. P. Mitchell, Biol. Rev., 41, 445 (1966).
25. E. A. Liberman, E. N. Mokhova, V. P. Skulachev, and
     V. P. Topaly, Biofizika, 13, 188 (1968).
26. N. Bjerrum, "Selected Paper", Emar Munksgaard, Copenhagen
     (1949). pp. 108-119.
27. R. M. Kraus and C. A. Fuoss, J. Am. Chem. Soc., 55, 1010
     (1933).
28. L. Bass and W. J. Moore, Nature, 214, 393 (1967).
29. N. T. Van and H. T. Tien, Biochim. Biophys. Acta., (1969)
     to be published.
30. D. O. Rudin and H. T. Tien, unpublished data (1961-1962).
31. R. B. Leslie and D. Chapman, Chem. Phys. Lipids, 1, 143
     (1967).
32. K. T. Brown, Vision Res., 8, 633 (1968).
33. W. L. Pak and R. A. Cone, Nature, 204, 836 (1964).
34. J. J. Wolken, J. Am. Oil Chemists' Soc., 45, 251 (1968).
35. W. A. Hagins, H. V. Zonada, and R. G. Adams, Nature, 194,
     844 (1962).

ASYMMETRIC PHOSPHOLIPID MEMBRANES: EFFECT OF pH AND Ca^{2+}

Shinpei Ohki and Demetrios Papahadjopoulos

Dept. of Biophysical Sciences, State University of New York at Buffalo and Dept. of Experimental Pathology Roswell Park Memorial Institute, Buffalo, New York

SUMMARY

Phosphatidylserine membranes (either as liquid-crystalline vesicles or as bilayers) in 0.1M NaCl and neutral pH, are very impermeable to cations or anions and exhibit high electrical resistance ($5 \times 10^7 \Omega cm^2$). However, the same membranes become unstable under conditions of asymmetric distribution of Ca^{2+} or H^+: Addition of Ca^{2+} to the aqueous salt solution only on one side of the membrane, produces a lowering of the d.c. resistance, and above a certain concentration results in breakage of the membrane. On the other hand, if Ca^{2+} is present on both sides, the membranes are stable and show very high electrical resistance ($5 \times 10^8 \Omega cm^2$) over a wide pH range. The above phenomena were not observed with membranes made of phosphatidylcholine. It is suggested that the instability of PS membranes observed in this study is due to the difference in surface energy between the two opposing sides of the bilayer. The biological implications of membrane instability following an asymmetric distribution of Ca^{2+} or H^+ is discussed.

INTRODUCTION

Phospholipid membranes have recently become the subject of intensive research as models for biological membrane function (4, 47). Since most biological membranes are complex mixtures composed predominantly of lipids and proteins, artificial membranes composed exclusively of lipids can serve only as very simplified models. However some of the properties of such artificial membranes bear striking similarity to those of the biological membranes (20, 19,

39, 6). Thus it appears that within certain limitations, artificially produced phospholipid membranes are very promising tools for studying "in isolation" certain molecular events relating to specific membrane functions.

Most of the published work on artificial phospholipid membranes has been performed either with neutral purified phospholipids or with ill-defined lipid-extracts from various natural sources (4, 47). However, it would appear that for studies relating to electrical excitation where binding of ions on fixed charges is an important factor, the phospholipids of choice would be some of the species that carry formal charges (17, 26, 31). Anionic phospholipids (PS, PA, PI, PG, diPG, triPI) are present in most membranes in widely varying proportions, although the percentage of each species within a given membrane is fairly characteristic (51, 3).

Recent work on the surface properties of anionic phospholipids indicates that they bind bivalent metals with high stability constants and even in the presence of high concentration of monovalent salts (2, 13, 26, 5). In contrast, it has been shown that unsaturated PC does not bind appreciable amounts of Ca^{2+} in the presence of 0.1M Na^+ or K^+ (15, 36, 40, 26, 11).

Studies concerned with the permeability properties of anionic (acidic) phospholipids indicate that these compounds are able to form bilayer structures with very high resistance to the diffusion of ions (17, 28, 34, 33). Membranes composed of PS either in the form of unilamellar vesicles or as planar bilayers are characterized by high electrical resistance, high capacitance and low diffusion of Na^+, K^+ and Cl^- (34, 31, 21, 22). However, bivalent metals appear to have a pronounced effect on these permeability properties. Thus, it has been shown that Ca^{2+} increases the diffusion rate of Na^+ and K^+ out of unilamellar PS vesicles (28, 34, 33). It has also been shown that PS bilayers formed in the presence of Ca^{2+} and Na^+ are more stable and have higher electrical resistance than those formed in the presence of only NaCl (21, 22).

This intriguing dual role for Ca^{2+} in its ability to produce both an increase and a decrease in the permeability of PS membranes has recently been investigated further by the present authors (31). The data obtained indicate that the stability of a phospholipid membrane is markedly effected by the asymmetric distribution of fixed charges and counterions on the two opposing sides of a phospholipid bilayer.

The present communication is a review of recent data concerning the effect of Ca^{2+} and H^+ on the stability of PS bilayers and an account of the biological implications of such a system.

RESULTS AND DISCUSSION

Effect of pH on Conductivity of Phospholipid Membranes:

Bilayer membranes composed of PS show a very high resistance (10^7 to 10^8 Ωcm^2) when made in 0.1 N NaCl and slightly acidic solutions, pH 3 to 7 (Fig. 1). This range of electrical resistance values is remarkably similar to those calculated (34, 33) from data of K^+ diffusion rates through PS vesicles (2 x 10^{-15} equiv/cm^2 sec = 1 x 10^8 Ωcm). The decrease in resistance with pH lower than 3.0 shown in Figure 1 could be explained by the protonation of the phosphate group which would render both phospholipids positively charged or alternatively by the permeability to protons. The decrease in resistance between pH 4 and 8 (also observed with PC bilayers (21)), is more difficult to explain in terms of ionization of specific groups on the molecules. Titrations of monomolecular films indicate no changes in surface potential

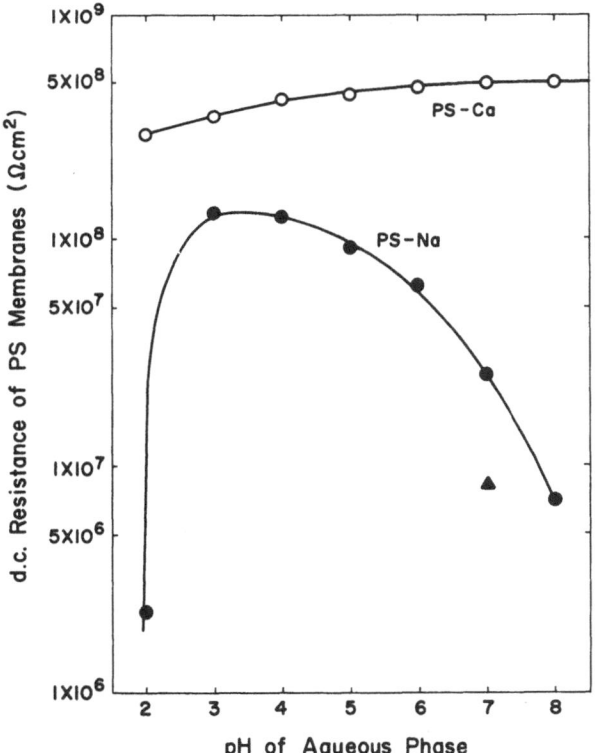

Fig. 1. Electrical resistance of PS membranes at different pH: The membranes were formed in the following solutions: in 0.1 M NaCl, ●; in 0.1 M NaCl and 1.0 mM $CaCl_2$, o; point ▲ indicates the resistance of a PS membrane formed in 0.1 M NaCl, with 1.0 mM $CaCl_2$ added to one side of the membrane after the formation.

between pH 3 and 11 for PC, and between pH 6 and 9 for PS (26).   A
possible explanation could be the presence of small amounts of fatty
acids which are either present from the beginning or are produced by
hydrolysis during the experiment. However, changes in resistance
due to ionization seem to be much more pronounced than changes in
ΔV. Thus, the ionization of the amino group in PS produces a change
of 40 mV in ΔV and the ionization of the secondary phosphate of PA
only 25 mV (26). Thus, a small degree of ionization (1%) would not
produce appreciable changes in ΔV (less than 1 mV) but it would be
expected to affect the conductivity. This would explain the instability of PS membranes above pH 8 although it is still difficult
to apply the same argument for PC with a quaternary ammonium group.

Figure 1 also shows the effect of small amounts of Ca (1mM)
in the presence of 100 mM NaCl on the resistance of PS bilayers.
In this case the resistance is very high (4-5 x 10^8 Ωcm^2) and shows
no change over a wide range of pH. This result is very similar to
the titration curves of PS monolayers in the presence of the same
concentrations of Ca^{2+} and Na^+ (26). Low concentrations of Ca^{2+}
have no effect on PC resistance, although an increase in resistance
is noted with high Ca^{2+} (0.1N) at high pH (21). It should be noted
here that PS bilayers in 0.1 NaCl show generally lower resistance in
the presence of EDTA. Obviously small amounts of contaminant bi-
and multi-valent metals stabilize the membrane and might explain
the higher resistance of PS compared to PC (21, 33). Small amounts
of bi- and multi-valent metals are usually present as contaminants
in the monovalent salts and are also extracted along with PS from
natural sources. Metal determinations performed by SPANG Micro-
analytical Laboratory indicated the following: For a sample of
bovine PS purchased from Applied Science Laboratories there was 204
parts per million Ca^{2+} and 4030 parts per million Mg^{2+}. For a
sample of bovine brain PS prepared in our laboratory, there was
100 parts per million Ca^{2+} and 40 parts per million Mg^{2+}.

Asymmetric Distribution of Ca^{2+}:

When $CaCl_2$ is added on one side of a PS membrane prepared in
0.1 N NaCl (pH 7.0) the electrical resistance of the bilayer drops
by a factor of more than two (Fig. 1). Above a certain concentration of Ca^{2+} the membrane becomes unstable and breaks. In a representative experiment where the resistance of PS was 6.8 x 10^7 Ωcm^2
As it has been noted in the previous section, when 1 mM $CaCl_2$ is
present on both sides, the resistance is much higher (3 x 10^8 Ωcm^2).
The concentration of $CaCl_2$ necessary to produce breaking is highly
dependent on the pH. As Figure 2 shows, while 1 mM is enough to
break the membrane at pH 7.8 it takes 5 mM $CaCl_2$ at pH 7.0 and even
higher at lower pH (curve a). These experiments were conducted in
the presence of 0.1 mM EDTA in order to remove any higher valency
metals either contributed as contaminants of the NaCl or extracted

with the PS fraction from natural sources. When EDTA is not included in the solution, the concentration of Ca^{2+} required for
breaking are usually higher (point b in Fig. 2). The presence of
Tris-Cl also has an inhibitory effect on the ability of Ca^{2+} to
break the membrane. Points 1, 2, 3, 4 in figure 2 were obtained
with 0.1, 1.0, 5.0 and 10 mM Tris-Cl respectively. EDTA was present
in these experiments at 0.1 mM concentration which rules out contamination by bivalent metals and points to the possibility of a
competitive interaction. Similar instability and breaking is observed with membranes of PS formed in 1 mM $CaCl_2$ solution (with 100
mM NaCl),equimolar amounts of EDTA are added on one side only. However, it should be noted here that membranes made of PC do not show
any instability at all with concentrations of Ca^{2+} up to 20 mM on
one side only.

The above results bear remarkable similarity to the effect of
Ca^{2+} on the permeability properties of PS vesicles (28, 34). The
addition of 1 to 2 mM CaCls to the outside solution of such vesicles
produces complete discharge of the Na^+ and K^+ trapped inside the

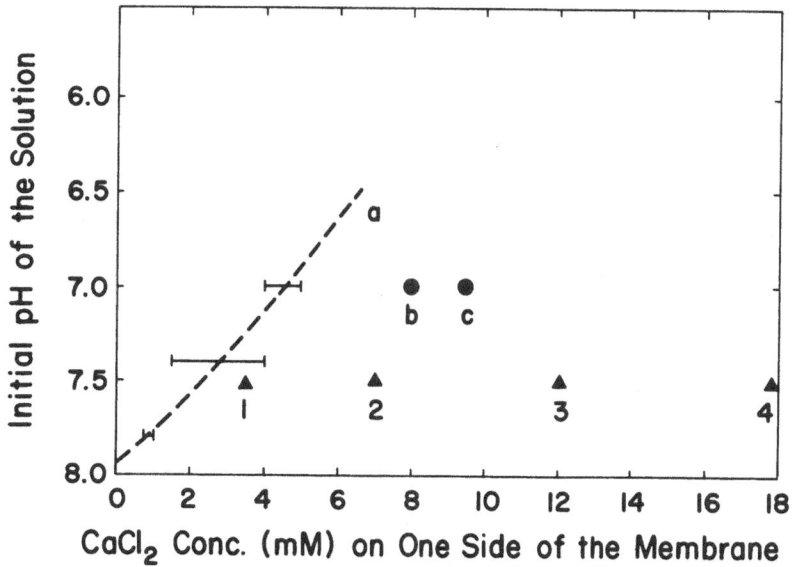

Fig. 2. Stability of PS membranes with asymmetric distribution of
Ca^{2+}. Membranes were formed in 0.1 M NaCl and 0.05 mM EDTA. Curve
a (----) indicates the amount of Ca^{2+} (in mM) added to one side of
the membrane before breakage. Points 1, 2, 3, 4 (▲) indicate the
presence of increasing amounts of Tris-Cl (0.2, 1.0, 2.0, 10.0 mM
respectively). Point b (●) indicates the amount of $CaCl_2$ needed
in the absence of EDTA (other conditions same as points of curve
a. Point c (●) indicates the effect of $CaCl_2$ on a PS-cholesterol
membrane.

vesicles within minutes. Smaller amounts of Ca^{2+} produce a slow
increase in K^+ or Na^+ diffusion (33). Using the same vesicles it
has been shown that at a concentration of 1mM of Ca^{2+} in 145mM KCl,
PS binds one equivalent of Ca^{2+} per molecule (5). Titration of PS
monolayers has shown (5, 26) that the same concentration of Ca^{2+}
produces a decrease in film pressure (increase in surface tension,
$\Delta\pi$ = 6 dynes/cm) and an increase in surface potential (40mV). In
spite of the binding of one equivalent of Ca^{2+} per PS molecule,
microelectrophoresis of PS vesicles in the presence of 1mM $CaCl_2$
(and 145 NaCl) indicates that the surface is still electronegative:
 potential = -35 mV. This was interpreted (26) as evidence for
the discharge of the amine proton during Ca^{2+} binding also shown
earlier by titrimetric techniques (1). The binding of Ca^{2+} to PS
(Fig. 3) is tentatively illustrated as a hexadentate complex involving four neighboring PS molecules in a linear polymetric

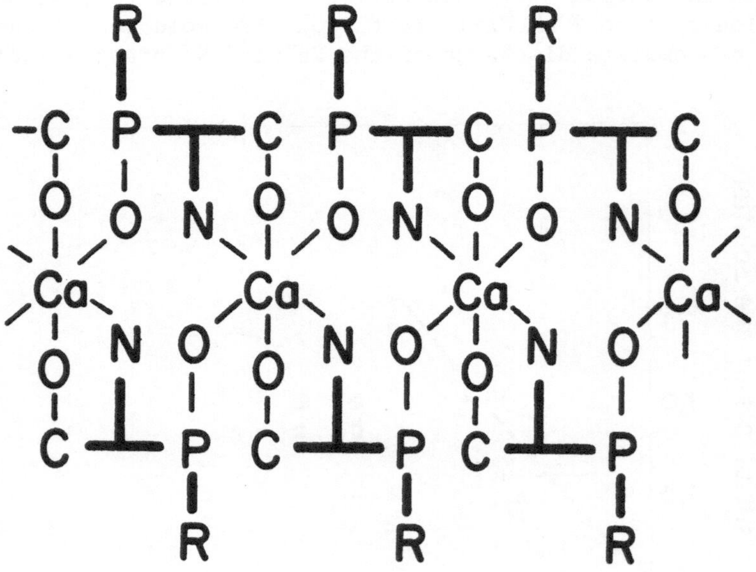

PS-Ca^{2+} complex, TOP VIEW

Fig. 3. Diagrammatic representation of PS-calcium complex. In
this illustration the phospholipid molecules are placed with the
fatty acid chains perpendicular to the plane of the page. The
letter R signifies the rest of the molecule (diglyceride) connected to the head group phosphorylserine. Each Ca^{2+} is bound to a
total of six groups (two phosphates, two carboxyls and two amines)
of four different molecules. The whole complex is a linear
polymeric arrangement.

arrangement. Not shown in figure 3 are the other counter-ions (Na^+ or K^+ which must be present as a diffuse layer adjacent to the PS-Ca^{2+} complex which is still electronegative. The instability of PS membranes following an asymmetric distribution of Ca^{2+} in respect to the two planes of the membrane was recently interpreted as the result of differences in the surface energy between these two planes (23).

Counterbalancing of Ca^{2+} with H^+

If PS membranes are formed at a certain pH and the pH of the solution facing only <u>one side</u> of the membrane is changed subsequently, a point of instability is reached (breaking) which is characteristic of the initial pH of the solution. Figure 4 illustrates some of the

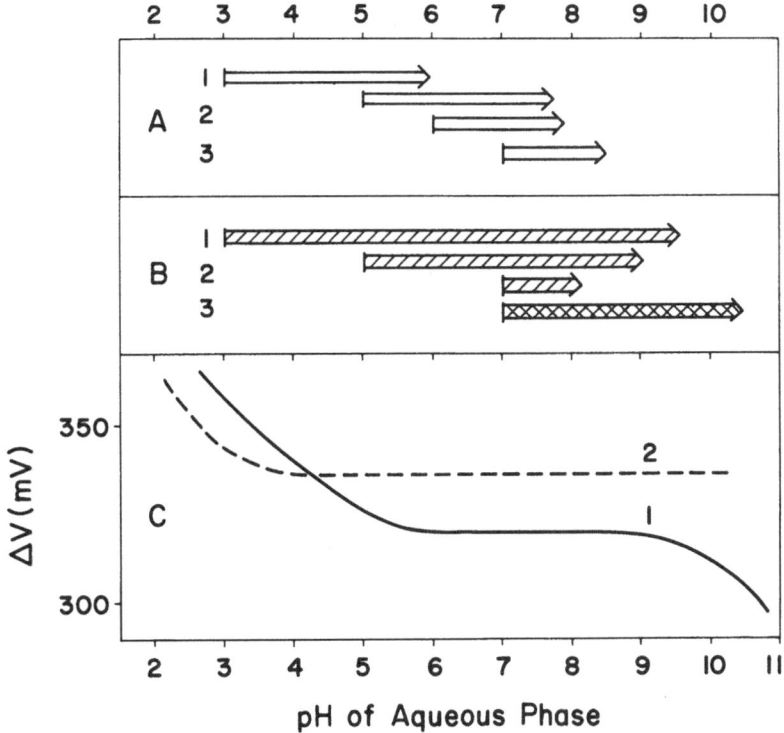

Fig. 4. Limits of stability of PS membranes under asymmetric distribution of H^+ and Ca^{2+}. Each horizontal bar represents the difference in pH between the two sides of the membrane, before the membrane breaks. The beginning of each bar is the initial pH and the end point of each bar the final pH of the solution on the outside aqueous compartment. A: experiments in 0.1 M NaCl only. B1 and B2: experiments in 0.1 M NaCl with 1 mM Ca^{2+} added to the outside before the change in pH. B3: experiments in 0.1 N NaCl and 1.0 mM $CaCl_2$. C: Titration of PS monolayers · 1: in 0.1 M NaCl, 2: in 0.14 M NaCl and 1.0 mM $CaCl_2$ (for details see ref.).

results indicating the stability of PS membranes under asymmetric
distribution of H$^+$. Figure 4A represents experiments without Ca^{2+}.
If a PS membrane is formed at pH 3.0, it reaches the point of instability (breaking) when the outside pH is increased to 6.0 (Fig. 4A,1).
However when the initial pH is higher (Fig. 4A, 2 and 3), the outside pH at which the membrane breaks is also higher. It is thus apparent that the stability of the membranes is defined by the difference in ionization of the three head-groups (phosphate, carboxyl and
amino) between the two sides of the membrane. At pH 3.0 PS should
be near the isoelectric point (Fig. 4C, curve 1) while at pH 6.0
each molecule carries one extra negative charge per molecule. This
difference of one charge per molecule (zero vs. one) produces instability as shown in A1. If both sides are charged with one charge
per molecule (at pH 6 or 7) the membrane is stable. However when
the amine starts losing its protons above pH 6 or 7, a fraction of
the extra charge per molecule is enough to produce breaking (Fig.
4A, 2 and 3).

The remarkable ability of Ca^{2+} in stabilizing asymmetric membranes is seen in Fig. 4B. Here the membranes were made at pH 3.0.
Then 1 mM CaCl$_2$ was added outside only, followed by NaOH to increase
the pH. At this point the PS membranes were neutral on one side
(pH 3.0) and charged on the other (pH 3.0-9.5) but with Ca^{2+} binding
on the charged groups. Such membranes were stable up to pH 9.5.
This stability up to high pH decreases as initial pH becomes higher
and H$^+$ do not counterbalance Ca^{2+} (Fig. 4B,2). When Ca^{2+} is present
on both sides, PS membranes formed at pH 7.0 can be titrated on one
side with NaOH up to pH 10.5 (or with HCl down to pH 2.5) before
they become unstable (Fig. 4B, 3). This stability over a wide pH
range is also found upon titration of PS monolayers in the presence
of Ca^{2+} (Fig. 4C, curve 2).

Effect of Cholesterol, Alcohols, Local Anaesthetics:

The presence of cholesterol in the membrane forming solution
(1/1 W/W or 1/2 mole/mole PS/cholesterol) has generally a stabilizing effect on the membranes. It increases electrical resistance and
decreases capacitance (32) of bilayer membranes. When mixed with PS
in 1/1 molar proportion before the formation of sonicated vesicles,
it reduces the diffusion rate to both Na$^+$ and Cl$^-$ and increases the
activation energy for the diffusion of these ions (32, 27). Cholesterol also has an inhibitory effect on Ca^{2+} induced permeability
changes. As shown in Figure 2 (point c) it takes twice as much Ca^{2+}
to break the membrane compared to the amount needed to break the
membranes in the absence of cholesterol. The same effect was noticed
with PS vesicles (32). This is probably due to the dilution of surface charge by the neutral molecules of cholesterol. A similar
phenomenon has also been noticed with mixtures of PS and PC (33).

General and local anaesthetics have been used in conjunction with studies on the permeability of PC multilamellar vesicles. It has been reported (6) that alcohols tend to increase, while local anaesthetics tend to decrease the diffusion rate of K^+ through such model membranes. More recently, we have studied the effect of alcohols and local anaesthetics on the ability of Ca^{2+} to induce permeability changes in PS membranes. Both groups of anaesthetics seem to have a pronounced inhibitory effect on Ca^{2+} action. When PS membranes (bilayers) are made in 0.1 N NaCl solution containing 0.1 mM Butanol, addition of Ca^{2+} on one side has no effect up to 20 mM, the highest concentration studied. Presumably, this is again the effect of "dilution" of the PS molecules at the surface of the membrane by butanol molecules. Capacitance and resistance of the membrane ($C \sim 0.4 \mu F/cm^2$, $R \sim 1 \times 10^7$ cm^2) indicate that the bilayers are still intact. The effect of nupercaine on PS vesicles is shown in Figure 5. Nupercaine alone (0.5 mM) has no effect on permeability (Fig. 5, A and B) $CaCl_2$ alone (1mM) increases the diffusion rate of

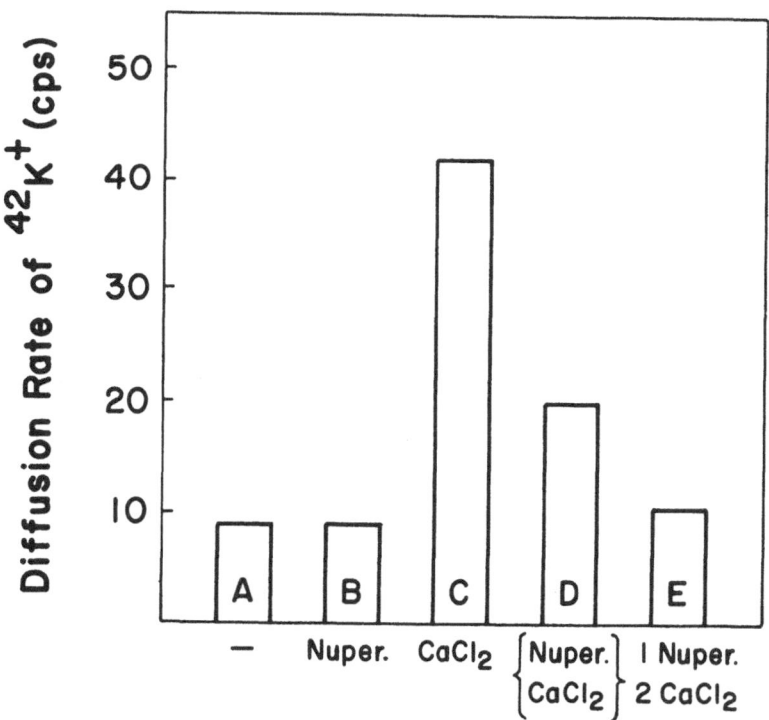

Fig. 5. Effect of $CaCl_2$ on PS vesicles in the presence of Nupercaine. Diffusion rate of ^{42}K from PS vesicles in 0.14 M KCl and 10 mM Tris-Cl at pH 7.4. A: no further additions. B: 0.5 mM Nupercaine added. C: 1.0 mM $CaCl_2$ added. D: Same concentrations of Nupercaine and $CaCl_2$ added simultaneously outside the vesicles. E: Same amounts as above but with Nupercaine added 1 hour before $CaCl_2$.

K^{42} out of vesicles (Fig. 5, C). When nupercaine is added simultaneously with $CaCl_2$, the increase in $^{42}K^+$ diffusion rate is limited
(Fig. 5, D). When nupercaine is added first, followed by $CaCl_2$ the
effect of Ca^{2+} is completely inhibited (Fig. 5, E). Local anaesthetics have been shown to decrease the negative charge on the surface of phospholipid vesicles (6) producing a small decrease in
$^{42}K^+$ diffusion. However the effect shown here probably represents
inhibition of Ca^{2+} binding, with subsequent failure of Ca^{2+} to produce permeability changes. Inhibition of Ca^{2+} binding to phospholipids has been shown earlier in two-phase systems (8) and monolayers (42). The results presented here, represent further evidence
implicating changes in permeability properties as a result of Ca^{2+} -
local anaesthetic antagonism. This subject is now being actively
pursued in this laboratory.

Theoretical Consideration of Asymmetrical Membranes:

To understand the instability of the asymmetric membranes mentioned, we have considered the following membrane model. The membrane is composed of a monomolecular layer, each molecule of which,
is a long chain with an equal length and has a dissociable polar
head group at both ends of the molecule. The molecules lie with
their molecular axes normal to the plane of the membrane such as in
a unimolecular crystal, while the head groups coordinate in two
outer surfaces facing the aqueous phases. The head group is able
to chelate with a metal ion according to both the concentration of
metal ions and the degree of dissociation of the polar group. It
is assumed that if a polar head group chelates with a metal ion,
the polar group loses its net charge, or the polar group does not
receive any electrostatic interaction from the other charged polar
groups.

Let us denote the position of the head groups on the surface
of the membrane as "sites". We may denote the number of total sites
on one surface of the membrane by N. When N_1 sites out of N are
chelated with metal ions, the number of the sites which are not
occupied by the metal ions is

$$N - N_1 = N_2 \tag{1}$$

We may define the average probability with which a polar group
is chelated with the metal ion as follows:

$$X = \frac{N_1}{N} \tag{2}$$

From Eqs. (1) and (2), we have

$$N_1 = NX \left. \vphantom{\begin{matrix}N_1 \\ N_2\end{matrix}} \right\} \qquad (3)$$

$$N_2 = N(1-X)$$

As extreme cases, if all sites are chelated with the metal ions, $X = 1$, and if all sites are not chelated, $X = 0$.

Let us suppose in the following case: On one surface of the membrane, each site is chelated with a metal ion with a probability X_o. In other words, the number of chelated sites is NX_o and the number of unchelated sites is $N(1-X_o)$ on the surface, and on the other surface of the membrane, there is no chelated site; $X = 0$. Then, total energy for this system is given by

$$F(X_o) = F^{bulk} + F^{surface}(X_o) \qquad (4)$$

where F^{bulk} is the free energy of the inside phase of the membrane, $F^{surface}$ is the free energy of both surfaces of the membranes which interact with water phase. $F^{surface}$ may be expressed by

$$F^{surface}(X_o) = \left[E(X_o)^{sites} - (ST)^{sites} + F_o^{surface} \right] \qquad (5)$$

where E^{sites} is the electrostatic interaction energy due to the dissociated polar groups; $(ST)^{sites}$ is the entropy due to the mixing of the sites; and $F_o^{surface}$ is the free energy of the surface excluding the contribution from the polar groups. If we assume that $F_o^{surface}$ does not change whether or not the polar groups chelate with the metal ions, $F^{surface}$ varies only with the variation of the probability X_o. In the above case, let us suppose that NX molecules out of NX_o chelated molecules on one side of the membrane turn over to the other side of the membrane, or NX sites out of NX_o on one side exchange with NX unchelated sites on the other side. Then the probability of unchelation for a site on the one side of the membrane will be $(1-X_o+X)$ and the probability of unchelation of a site on the other side of the membrane will be $1-X$. The energy of electrostatic interaction due to charged polar groups is

$$E^{sites}(X_o) = E_1^{sites}(X_o) + E_2^{sites}(X_o) \qquad (6)$$

where E_1 and E_2 are the energies of electrostatic interaction at side 1 and the other side 2, due to all polar groups on the surface of side 1 and side 2 of the membrane. E_1 and E_2 may be expressed by

$$E_1 = \frac{1}{2} \ Neq \ \psi_1 (1-X)$$

$$(7)$$

$$E_2 = \frac{1}{2} \ Neq \ \psi_2 (1-X_o +X)$$

where ψ_1 and ψ_2 are the electrostatic potentials at a site on side 1 with surface charge densite σ and on side 2 with surface charge densite σ', respectively. ψ_1 and ψ_2 are given as follows (24)

$$\psi_1 = \frac{4\pi}{\varepsilon_o \kappa} \ \frac{\sigma +\sigma' + \sigma' \ \frac{\varepsilon_o \kappa h}{\varepsilon_1}}{2 + \frac{\varepsilon_o \kappa h}{\varepsilon_1}}$$

$$(8)$$

$$\psi_2 = \frac{4\pi}{\varepsilon_o \kappa} \ \frac{\sigma +\sigma' + \sigma \ \frac{\varepsilon_o \kappa h}{\varepsilon_1}}{2 + \frac{\varepsilon_o \kappa h}{\varepsilon_1}}$$

where ε_o and ε_1 are the dielectric constants of aqueous and lipid phases, respectively, and σ is the average charge density due to dissociated polar groups. If the area per molecule is A and the net charge per molecule is eq, we have

$$\sigma = \frac{eq}{A} \ (1-X)$$

$$(9)$$

$$\sigma = \frac{eq'}{A} \ (1-X_o +X)$$

$$(10)$$

Then, the total energy due to the electrostatic interaction of the polar groups is

$$E(X_o) = E_1(X_o) + E_2(X_o)$$

$$= \frac{2\pi e^2 q^2}{\varepsilon_o \kappa A} \ \frac{1}{2+ \frac{\varepsilon_o \kappa h}{\varepsilon_1}} \left\{ (1-X_o +X)^2 +(1-X)(1-X_o +X) \right.$$

$$+ \frac{\varepsilon_o \kappa h}{\varepsilon_1} (1-X_o +X)^2 + (1-X)^2 +(1-X)(1-X_o +X)$$

$$\left. + \frac{\varepsilon_o \kappa h}{\varepsilon_1} (1-X)^2 \right\}$$

$$(11)$$

The partition function of the total system is

$$Z = e^{-F/kT} = e^{\frac{ST^{sites}}{kT}} \; e^{-E\frac{sites}{kT}} \; e^{-F_o\frac{surface}{}} \; e^{-\frac{F^{bulk}}{kT}}$$

$$= {}_NC_{NX_o} \; {}_{NX_o}C_{NX} \; e^{-\frac{E^{sites}}{kT}} \; e^{-\frac{F_o^{surface}}{kT}} \; e^{-\frac{F^{bulk}}{kT}}$$

$$= {}_NC_{NX_o} \; {}_{NX_o}C_{NX} \; e^{-\frac{E^{ele}}{kT}} \; e^{-\frac{F^o}{kT}} \tag{12}$$

where ${}_NC_{NX_o} \; {}_{NX_o}C_{NX}$ is the number of various distributions of the sites, and $F^o = F^{bulk} + F_o^{surface}$.

The free energy of the membrane is

$$F = -kT \log Z = E^{site}(X_o) - kT \log \left\{ {}_{NX_o}C_{NX} \; {}_NC_{NX_o} \right\} + F_o \tag{13}$$

Using the approximation $N! \cong N \log N$ for $N!$, we have

$$F = E^{site}(X_o) = kT \, N \left[\log \frac{1}{1-X_o} - X_o \log \frac{X_o - X}{1-X_o} - X \log \frac{X}{X_o - X} \right] + F_o \tag{14}$$

Since we assume that F_o is not affected with the change of polar groups, in order to express the free energy change of the membrane it is sufficient to know the relative free energy $(F^{rel}/N \equiv \frac{F-F_o}{N})$ per molecule due to the change of the polar groups. The relative specific free energy is defined as follows:

$$\frac{F-F_o}{N} \cong F^{rel} = \epsilon(X_o) - kT \left[\log \frac{1}{1-X_o} - X_o \log \frac{X_o - X}{1-X_o} - X \log \frac{X}{X_o - X} \right] \tag{15}$$

where $\epsilon(X_o) = \frac{E(X_o)^{sites}}{N}$

For given X_o, the stationary state of the system can be determined by

$$\frac{\partial F^{rel}}{\partial X} = 0 \tag{16}$$

With the value X_1 which satisfies Eq. (16), the relative free energy per molecule will be a minimum. We call this state $F(X_1)$ a

stable state of the membrane. The difference ΔF between the
specific free energies at $X = 0$ and $X = X_1$, is expressed in terms of
X_o and q with Eqs. (11), (15) and (16).

$$\Delta F^{rel} = F^{rel}(X_o) - F^{rel}(X_1) \tag{17}$$

The numerical values of F with respect to the probability of initial
chelation X_o of a site and the degree of dissociation q of the polar
groups can be calculated. For the case of $X_o = 1$, the numerical
value of ΔF is shown with various values of q, in Fig. 6.

According to Salem (37, 38), the London van der Waals dispersion energy per molecule in lipid monolayer films is given by

$$W_{dis} = \frac{n}{2} \; \frac{A}{4\ell^2 D^4} \quad \rho \left(3 \; \tan^{-1} \rho + \frac{\rho}{1+\rho^2} \right) \; , \quad \rho = L/D \tag{18}$$

where A is the coefficient of the dispersion interaction between
two basic molecular units (CH_2 hydrocarbon units), n is the number
of the nearest neighbor hydrocarbon chains, ℓ is the length of a
CH_2 unit, D is the mutual distance of two hydrocarbon chains, L is
the length of a hydrocarbon chain and N_c is the number of hydrocarbon molecules.

Since the phospholipid is composed of two aliphatic hydrocarbon
chains, if we use 60 $Å^2$ for the area per lipid molecule, the mutual
distance of two nearest hydrocarbon chains D is estimated to be 5.48
$Å$ for a square lattice packing. If we apply the formula of Eq.
(18) to the bilayer membrane with the following numerical values:

> A: 1340 $Å^6$ Kcal/mole for CH_2-CH_2 (Ref. 38)
>
> ℓ: 1.26 $Å$
>
> N_c :   18
>
> D : 5.48 $Å$ for a square lattice

Thus, the cohesive energy of a hydrocarbon chain in the lipid molecular layer due to the dispersion forces of neighboring hydrocarbon chains is

$$W_{dis} = 4.04 \times 10^{-13} \text{ ergs.}$$

In the square lattice molecular arrangement, if energy is
added in amounts greater than $\frac{W_{dis}}{2}$ (which corresponds to the
energy to break two nearest neighboring chains) to a lipid molecule,
the membrane can break. That is, if ΔF is greater than $W_{dis}/2$,
the membrane may break. As shown in Fig. 6, for a given concentration of chelation there is a critical degree of dissociation q_c of

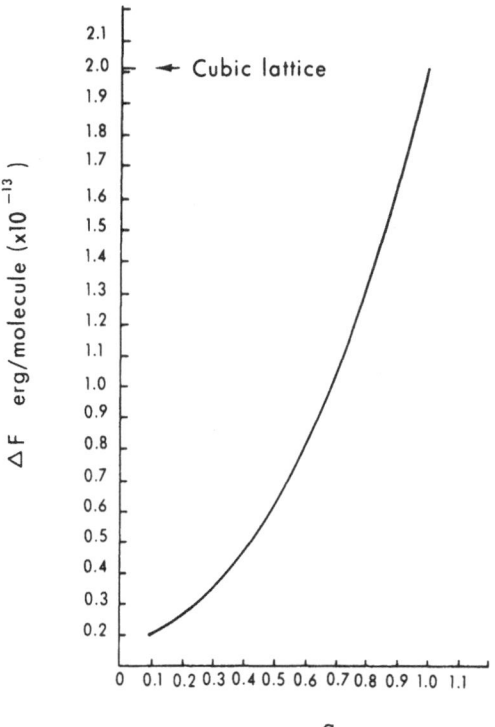

Fig. 6. Relative free energy per lipid molecule with respect to
the net charge per molecule. An arrow mark shows the energy necessary to break the membrane with a square lattice packing.

the polar groups which is enough to break the membrane.

$$q_c = 0.95 \quad , \quad X_o = 1 \tag{19}$$

It is appreciated that the analysis given here is oversimplified
and contains a number of assumptions (such as the nonspherical character of CH_2 groups, orientation effect, and the neglect of
microscopic dielectric constant) which will be valid only in extreme
cases. However, the treatment is qualitatively adequate to indicate
the significance of the instability of the asymmetrical membrane.
The instability of the asymmetrical membrane with divalent metals
(e.g. Ca^{++}) may be greatly related to the excitation phenomena of
biological membranes, as elaborated in the following section.

Another factor which could also be involved in producing instability is entropy changes following the chelation of Ca^{2+} with
polar groups of PS. The calculation similar to the one mentioned
above, taking into account the fixing of PS "sites" by chelation
indicates the enhancement of the instability of the asymmetric

membrane.   The detail analysis is in progress.

Analogies to Biological Membranes:

The importance of bivalent metals on the structure and function
of biological membranes is well recognized. For example, it has
been shown that Ca^{2+} has a pronounced effect on the permeability of
red cells (14a), epithelial cell junctions (18) and squid axon (9)
as well as of many other membranes. In addition, Ca^{2+} is a partici-
pant in mitochondrial function (16), cell-contact phenomena and
cell deformability (52), and activation of prothrombin by thrombo-
plastin(29). All these reactions involve membranous structures
containing both phospholipids and proteins. The observations presented here concerning the effect of Ca^{2+} on the permeability of
PS membranes seem highly relevant to the effect of Ca^{2+} on whole
biological membranes.

Several investigators have considered the possibility of phospholipids being implicated in nerve excitation process. Goldman
(10) has speculated on the re-orientation of the dipoles of the
phospholipid head-groups as a mechanism accounting for the nerve
axon properties. Tobias (48) has proposed a theory of excitation
involving Ca^{2+} binding on PS and its replacement by K^+ during action
potential. This later postulation was based mainly on the observation that Ca^{2+} increased the electrical resistance of PS-impregnated
millipore filter membranes (50). However Tobias' theory did not ·
involve any re-orientation of the whole phospholipid molecules or
the effect of asymmetry. Tasaki and Singer (44) have formulated
a "macromolecular" theory of the mechanism of action potential involving the binding of divalent metals on membrane negative sites.
The actual molecular site for these events was not discussed by
these authors, but in view of the observations described in this
paper, the PS-Ca^{2+} complex appears as a reasonable possibility.
Recent evidence on birefringence charges of whole nerve membranes
(7) and spectroscopic changes of dye molecules incorporated into
the membrane (43) during electrical stimulation suggest the occurrence of molecular re-orientation within the axon membrane during
excitation. Cohen, Keynes and Hille (7) have speculated that the
observed changes in light scattering and birefringence might reflect
conformational changes of the protein molecules. However changes
in birefringence can also be accounted for by a re-orientation of
phospholipid molecules (inversion) as a response to the instability
created by asymmetric distribution of fixed negative charges or
Ca^{2+} (31).

Based on the observed instability of asymmetric PS membranes
and various other experimental evidence from other laboratories, we
propose the following mechanism for nerve excitation: Let us

suppose that the plasma membrane of a nerve is "variegated" in composition. The neutral phospholipids and proteins constitute the majority of the membrane, interdispersed with small areas (patches) rich in PS or other acidic phospholipids with a high binding constant for Ca^{2+} (Fig. 7A). At the resting state, the membrane areas ("sites") containing high proportions of PS are "balanced" by binding to Ca^{2+} on the outside surface of the membranes and to the positively charged groups of a protein on the inside of the membrane (Figure 7B). Removal of the stabilizing factor (protein) from the inside surface by electrical impulse, produces an asymmetry (31) characterized by instability. As a result of the difference in surface energy between the two planes of the membrane, PS molecules or clusters of molecules "invert" in order to equilibrate the difference in surface energy. During this reorientation, the membrane undergoes profound changes in conductivity and quickly reverts back

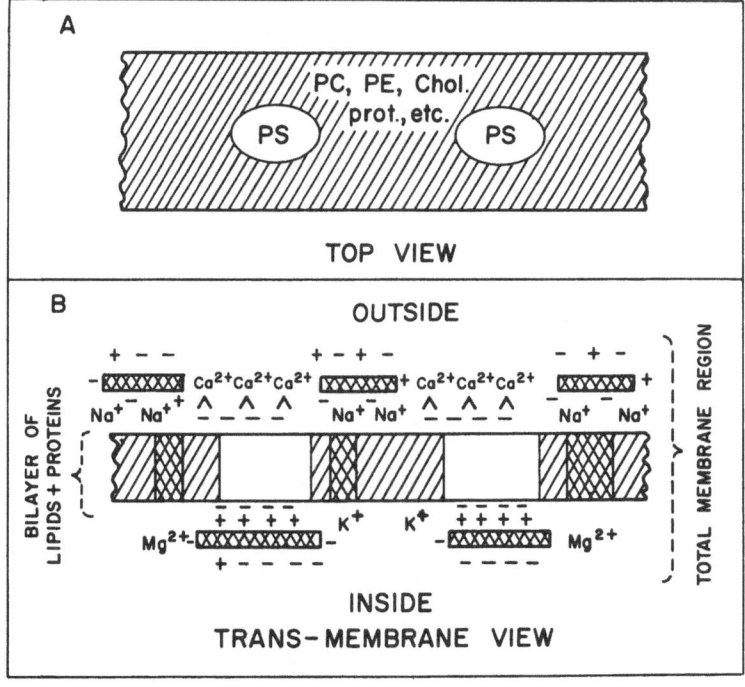

Fig. 7. Diagrammatic representation of a nerve axon membrane. A: Top view showing areas rich in PS surrounded by areas composed of the rest of (neutral) phospholipids, cholesterol and proteins. B: Trans membrane view showing phospholipids present in a central area of bimolecular thickness with proteins present both in the bilayer region and at the water interface. Calcium ions are shown bound to the PS areas.

to a stable state. During the "inversion", Ca^{2+} is transported into the interior, along with Na^+, and K^+ is transported out. Ca^{2+} diffuses into the cytoplasm and the internal PS molecules bind again to the basic groups of the "stabilizing" protein. During the Ca^{2+} induced high permeability stage, Na^+ influx would increase first, predominantly due to the potential difference across the membrane, followed by K^+ efflux due to diffusion. This may correspond to the early Na^+ and late K^+ conductance postulated by Hodgkin and Huxley (14). The areas of the membrane composed of the neutral phospholipids and proteins with low binding for Ca^{2+} would not be affected under these conditions.

There is experimental evidence to support the suggestion that a protein involved in the mechanism of the action potential is situated "inside" and not "outside" the axon membrane (49, 35, 45). Tasaki and Singer have calculated the extra ionic flux accompanying the electrical excitation of squid axon as $3-6 \times 10^{-8}$ moles, $cm^{-2} \cdot sec^{-1}$ (44). A very relevant question would be whether the amount of PS present in the nerve axon membrane is enough to transport the calculated number of ions. From various data presented by Ansel and Hawthorn (3) and also more recently by Sheltawy and Dawson (41) it appears that non-myelinated axons have a relatively high percentage of cephalin, of which approximately 40% is PS. Based on total dry weight, a value of 5% can be calculated for the PS content of non-myelinated nerve. Assuming a value of 55 $\AA^2$ area per molecule of PS (26) it can be calculated that a bilayer membrane consisting 100% of PS would contain 550×10^{-8} moles $\cdot$ cm^2. Taking into account the value of 5% PS content within the axon membrane, the amount of PS molecules present within a cm^2 would be 22×10^{-8} moles $\cdot$ cm^2, more than enough to accommodate the observed ionic flux. This calculation assumes an equal distribution of all the membrane components w/w within the membrane surface, which is probably not the case. It is difficult to state with any certainty the magnitude of the error involved in this assumption because of lack of direct evidence on the fine structure of biological membranes. However it would probably be within the order of magnitude. In any case, if one assumes a modified Danielli-Davson structure with protein both in the region of the bilayer, and also at the water interface, the percentage (per unit area) of each liquid component within the bilayer would be higher than the percentage based on total dry weight.

The membrane model presented in figure 7 should be considered only as a very schematic representation of some of the membrane features that seemed relevant for the discussion. It illustrates proteins as present both in the bilayer region and also adsorbed at the water interface.

Acknowledgement:

This work was supported partly by grants from the U. S.
National Institute of Neurological Diseases and Stroke, Grant
No. 1 RO1 NB08739-01 (S. O.); and from the American Heart Associa-
tion, Grant No. A.H.A. 68698 (D. P.).

References:

1. Abramson, M. B., Katzman, R., and Gregor, H. P., J. Biol. Chem.
    239, 70 (1964).
2. Abramson, M. B., Katzman, R., Gregor, H. P. and Curci, R.,
    Biochemistry, 5, 2207 (1966).
3. Ansell, G. B. and Hawthorne, J. N., Phospholipids, Elsevier
    Publ. Co., Amsterdam, N. Y., 1964.
4. Bangham, A. D. in "Progress in Biophysics and Molecular Biology"
    (edit. by J. A. V. Butler and D. Noble), Pergamon Press,
    Oxford, 1968, p. 29.
5. Bangham, A. D. and Papahadjopoulos, D., Biochim. Biophys. Acta,
    126, 181 (1966).
6. Bangham, A. D., Standish, M. M., and Miller, N., Nature, 208,
    1295 (1965).
7. Cohen, L. B., Keynes, R. D. and Hille, B., Nature, 218, 438,
    (1968).
8. Feinstein, M. B., J. Gen. Physiol., 48, 357 (1964).
9. Frankenhaeuser, B. and Hodgkin, A. L., J. Physiol., London,
    137, 218 (1957).
10. Goldman, D. E., Biophys. J., 4, 167 (1964).
11. Hauser, H., and Dawson, R. M. C., European J. Biochem., 1,
    61 (1967).
12. Hauser, H. and Dawson, R. M. C., Biochem. J., 109, 909 (1968).
13. Hendrickson, H. S., and Fullington, J. G., Biochem. 4, 1599(1965).
14. Hodgkin, A.L. and Huxley, A.F., J. Physiol., 117, 500 (1952).
14a. Hoffman, J. F., Circulation, 26, 1201 (1962).
15. Kimizuka, H., Nakahara, T., Vejo, H., and Yamauchi, A.,
    Biochim. Biophys. Acta, 137, 549 (1967).
16. Lehninger, A. L., Ann. N. Y. Acad. Sci., 137, 700 (1966).
17. Lesslauer, Richter, J., Lauger, P., Nature, 213, 1224 (1967).
18. Loewenstein, W. R., J. Colloid Sci., 25, 34 (1967).
19. Mueller, P. and Rudin, D. O., Nature, 217, 713 (1968).
20. Mueller, P. and Rudin, O. D., J. Theoret. Biology, 18, 222 (1968).
21. Ohki, S., J. Coll. Interf. Sci., 30, 413 (1969).
22. Ohki, S., Biophysical J., 9, 1195 (1969).
23. Ohki, in F. M. Snell, et al. Eds., "Physical Principles of
    Biological Membranes", Gordon and Breach, Science Publ.
    Inc., in press.
24. Ohki, S. and Aono, O., J. Coll. Interf. Sci., in press.
25. Ohki, S. and Goldup, A., Nature, 217, 458 (1968).

26. Papahadjopoulos, D., Biochim. Biophys. Acta, 163, 240 (1968).
27. Papahadjopoulos, D., in preparation.
28. Papahadjopoulos, D. and Bangham, A. D., Biochim. Biophys. Acta, 126, 185 (1966).
29. Papahadjopoulos, D., and Hanahan, D. J., Biochim. Biophys. Acta, 90, 436 (1964).
30. Papahadjopoulos, D. and Miller, N., Biochim. Biophys. Acta, 135, 624 (1967).
31. Papahadjopoulos, D. and Ohki, S., Science, 164, 1075 (1969).
32. Papahadjopoulos, D. and Ohki, S., J. Amer. Oil Chem. Soc., in press (Symposium on Model Membranes, San Francisco, April 1969).
33. Papahadjopoulos, D. and Ohki, S., Symposium on "Ordered Fluids and Liquid Crystals", Advances in Chemistry, Number 43, American Chemical Society, in press.
34. Papahadjopoulos, D., Watkins, J. C., Biochim. Biophys. Acta, 135, 639 (1967).
35. Rojas, E. and Luxoro, M., Nature, 199, 78 (1963).
36. Rojas, E. and Tobias, J. M., Biochim. Biophys. Acta, 94, 394 (1965).
37. Salem, L., Molec. Phys., 3, 441 (1960).
38. Salem, L., J. Chem. Phys., 37, 2100 (1962).
39. Rothfield, L. and Finkelstein, A., Annual Reviews of Biochem. p. 463 (1968).
40. Shah, D. O. and Schulman, J. H., J. Lipid Res., 8, 227 (1967).
41. Sheltawy, A. and Dawson, R. M. C., Biochem. J., 100, 12 (1966).
42. Skou, J. C., Acta. Pharmacol. Toxicol., 10, 325 (1954).
43. Tasaki, I., Carnay, L., Sandlin, R. and Watanabe, A., Science, 163, 683 (1969).
44. Tasaki, I. and Singer, I., Ann. N. Y. Acad. Sci., 137, 792 (1966).
45. Tasaki, I., and Takameaka, T., Proc. Nat. Acad. Sc., 52, 804 (1964).
46. Tasaki, I., Watanabe, A. and Lerman, L., Amer. J. Physiol., 213, 1465 (1967).
47. Tien, H. Ti, and Diana, A., Chem. Phys. Lipids, 2, 55 (1968).
48. Tobias, J. M., Nature, 203, 13 (1964).
49. Tobias, J. M., J. Gen. Physiol., 43, Suppl. 57 (1960).
50. Tobias, J. M., Agin, D. P. and Pawlowski, R., J. Gen. Physiol., 45, 989 (1962).
51. Van Deemen, L. L. M., in "Progress in the Chemistry of Fats and Other Lipids", Vol. 8, part 1, Pergamon Press (1965).
52. Weiss, L., J. Cell. Biol., 35, 347 (1967).

DISSOCIATION OF FUNCTIONAL MARKERS
IN BACTERIAL MEMBRANES

Martin S. Nachbar and Milton R. J. Salton

Department of Medicine and Department of Microbiology

New York University School of Medicine, New York

INTRODUCTION

While much is known of the gross structural, chemical and biochemical properties of biological membranes, (1-4) little is yet known of their existing structure - function relationships, (5) intermolecular architecture, (6) inherent informational content or possible dynamic changes occurring in response to extra - and intracellular stimuli. There exists no universally accepted model of membrane architecture and, indeed, several patterns may exist, not only in different membranes, but in portions of the same membrane.

The elucidation of these problems would seem to require the combined knowledge of gross changes in membrane architecture as determined by alterations of (1) protein and lipid conformations and interactions as described from IR, NMR, CD and ORD spectroscopy and (2) the study of the properties of and the interrelationships between the individual building blocks.

The membrane system we have chosen to discuss is that of the microorganism Micrococcus lysodeikticus (ML) (Fig. 1).

An outer cell wall composed of a peptidoglycan, an outer plasma membrane and an inner mesosomal membrane structure (comprised predominantly of lipid and protein) and the cell sap are evident. In common with other bacteria, ML lacks the

175

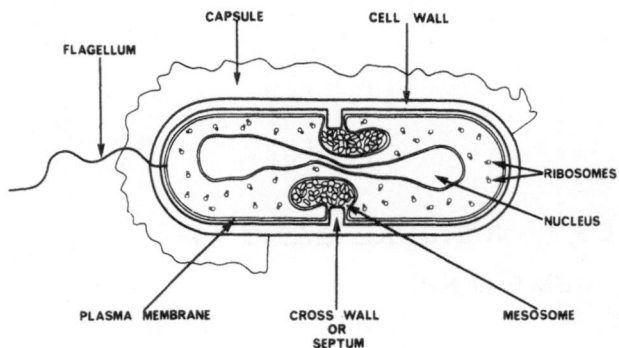

<u>Fig. 1</u>   A diagrammatic representation of
the anatomy of a gram-positive bacterium
indicating surface appendages, surface layers
and internal structures.   The discontinuity of
the capsule indicates that not all species
possess this surface component (Ref. 16).

intracellular membranous organelles i. e. - mitochondria, golgi
apparatus, endoplasmic reticulum, nuclear membrane, etc.
which are found in higher biological forms.

The membrane systems of ML appear relatively simple
but they must be so organized as to be able to perform various
complex functions such as ionic and organic transport or translocation, electron transport and oxidative phosphorylation, wall
and membrane synthesis, nuclear anchoring and separation
during cell division.   The challenge is to know which components
of the membrane perform these specific functions and whether
these components are scattered randomly over the membrane or
whether a more highly organized and probably more efficient
distribution of activities is to be found.

The membrane systems of this organism are easily
obtained.   The outer wall can be removed by digestion with the
enzyme lysozyme and the membranes obtained by differential
centrifugation.   As shown by Salton (5) successive washing of the
particulate or membranous fraction by 0. 1M Tris buffer pH 7. 5
results in progressively less release of protein into the supernatant washes.   A minimum seems to be reached after about 3-4
washes.   This suggests that any components released after this

stage are probably located in the membrane.

Examination of washed, sonicated membranes reacted against protoplasmic antisera in double diffusion tests shows a progressive reduction of reactive material with successive washes. Little or no reactivity with cytoplasmic antisera after the 3rd wash is demonstrated.

These results would tend to militate against much entrapment of cytoplasmic material in membrane bound vesicles. However, upon sonication of the membrane after 5 washes and a subsequent high speed centrifugation, the supernatant revealed the presence of an enzyme thought to be cytoplasmic (TPN specific Isocitric Dehydrogenase). Without sonication, this enzyme could not be found in the supernatant fraction after the third wash. (9) (Fig. 2). Therefore some cytoplasmic trapping within membrane bound vesicles cannot be ruled out.

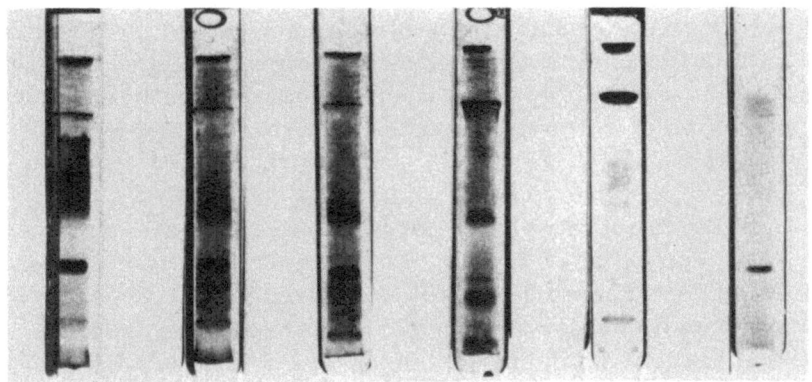

Fig. 2   Disc electrophoresis of washes and supernatant fraction of sonicated membrane stained for various enzyme activities (8). From left to right gels represent washes 1-5, 0.03M Tris pH 7.5, gel on far right is supernatant fraction of sonicated membrane after 5 washes, stained only for TPN-specific Isocitric Dehydrogenase (9). Comparable zone of staining is seen in washes 1-3, faintly in wash 4 and not in wash 5. Standard 7% separating gel run at pH 9.5, stacking gel pH 8.9 (17, 18).

The gross chemical composition of the membrane
(Table 1) reveals a preponderance of protein with a protein/lipid
ratio of about 2:1, very little carbohydrate and a small but
constant amount of RNA.

TABLE 1

CHEMICAL COMPOSITION OF Micrococcus lysodeikticus MEMBRANES

| | PROTEIN | | LIPID | CARBOHYDRATE | RNA |
	LOWRY	BIURET		ANTHRONE	ORCINOL
% (by weight)	49	64.5	23.8	4.2	1-2

What are the functional markers that characterize this
membrane fraction? We will define a marker as any component
(i. e. protein, lipid, lipoprotein, glycoprotein, polysaccharide,
etc.) which is found in significant amounts in the membrane
fraction after washing and which can be recognized by its enzymatic, antigenic, chemical or physio-chemical properties.
Phospholipids, carotenoids and menaquinones have been shown to
reside exclusively in the membrane (10) fraction after separation
from cytoplasm.

Individual proteins may not behave in this all or none
manner. As can be seen in Fig. 3, the distribution of enzyme
activities between membrane and cytoplasm varies over a wide
range from those exclusively found in the cytoplasm such as
DPN-TPNH transhydrogenase to those found entirely in the
membrane fraction such as phophatidic acid-cytidyl transferase
(Nachbar - unpublished). A large bipartite region can also be
seen. Most of those enzymes associated predominantly with the
cytoplasm can be released by repeated washing, i. e. in 0.03-
0.1M Tris buffer pH 7.5,and these have arbitrarily and perhaps
wrongly excluded from consideration as "true" membrane
markers. It should be realized that these proteins may be
associated with the membrane by weak ionic or hydrophobic
bonds which are disrupted by even the mildest of procedures.

As can be seen the "true" markers (i. e. those firmly
attached to the membranes) belong to the electron transport and

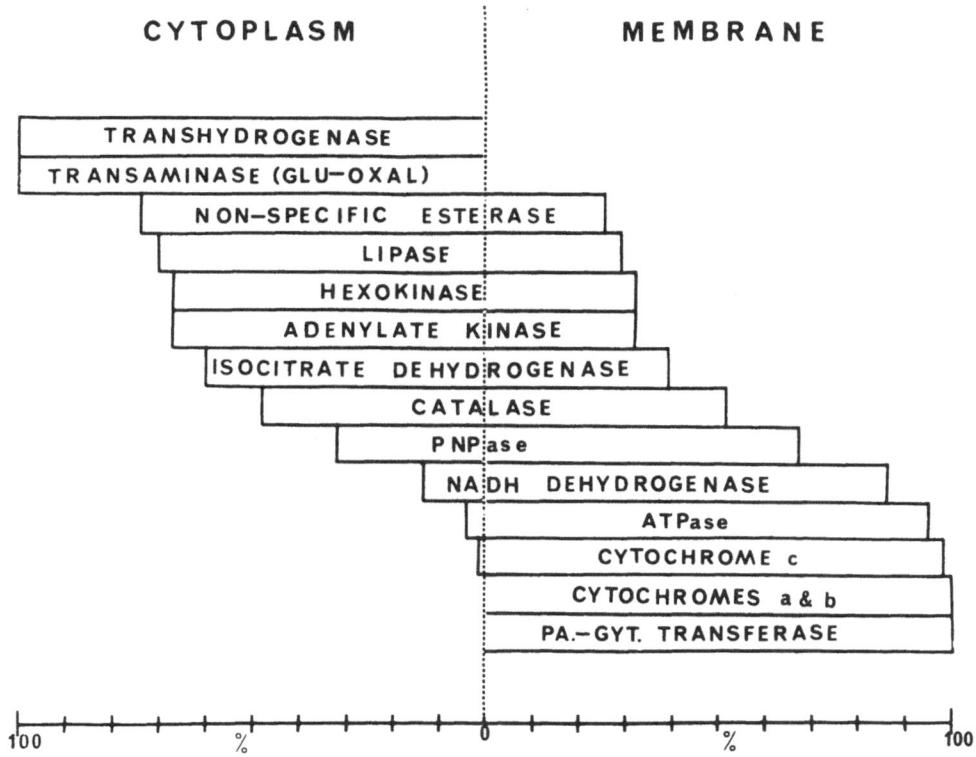

Fig. 3   Histogram of distributions of enzymes between
cytoplasm and membrane after separation of cytoplasm and
membrane fractions by 30' centrifugation at 0°C at 30,000 xg
(Nachbar - unpublished).

oxidative phosphorylation chains, a finding almost universal for
all biological membranes, and to enzymes concerned with
phospholipid synthesis as recently found by Dr. August De Siervo
(unpublished) in our laboratory. Two other systems suspected of
being present but not yet identified are the cell wall synthesizing
enzymes and components necessary for specialized transport.

The problem which now arises is that of the dissociation
of individual components of the membrane while retaining some
recognizable characteristics. Table 2 demonstrates the various
methods we have thus far employed, the bonds believed to be
involved and enzymes partially purified by these techniques.
They differ in degree of severity but with each, certain markers

TABLE 2

METHODS OF MEMBRANE DISSOCIATION AND COMPONENTS ISOLATED

PROCEDURE	BOND	MAIN COMPONENTS RELEASED AND ISOLATED
1. Low Ionic Strength Environment	Weak Hydrophobic Bonds Ionic Bonds Via Cations	ATPase
2. Chelation	Ionic Bonds	NADH Dehydrogenase, Cardiolipin, ? Glycolipid
3. Detergents	Hydrophobic Bonds	Cytochromes, Succinic Dehydrogenase
4. Organic Solvents	Hydrophobic Bonds	ATPase, Cytochromes
5. Mechanical Disruption	Weak intermolecular Bonds Hydrogen Bonds	ATPase, Cytochromes

can easily be recognized.

What follows is a more detailed description of the procedures employed and results obtained.

1) Low Ionic Strength Environment

Dr. Emilio Munoz, while working in our laboratory
demonstrated that a Ca^{++}-dependent ATPase could be released
from ML membranes by lowering the ionic strength of the wash
solution. Membranes were washed four times in 0.03M Tris
buffer pH 7.5 and then with 0.003M Tris pH 7.5. The results
may be seen in Fig. 4.

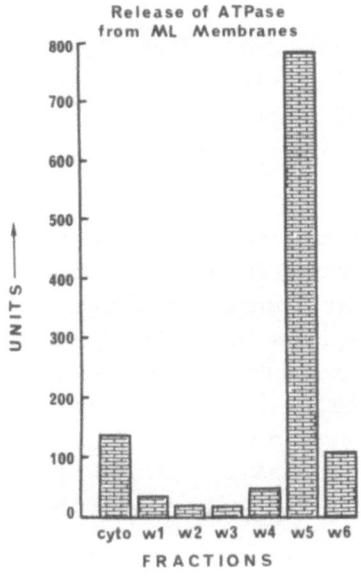

Fig. 4   Histogram showing
release of ATPase from ML
membranes.
cyto=cytoplasm, w1-w4=0.03M
Tris pH 7.5 washes, w5=first
0.003M Tris wash, w6=second
0.003M Tris wash. Assayed
according to Munoz (11).

This treatment results in a marked release of the enzyme
as measured by total activity and, in addition, the specific activity of the 5th and 6th washes are 4 to 5 times that of any
preceding wash. The specificity of the procedure and identification of the enzymes are demonstrated in Fig. 5. The complexity
of the regular washes can be seen and the selectivity of the low
ionic strength or "shock" wash can be appreciated. Enzymatic
staining of the gel shows the major component of the "shock"
wash to have the ATPase activity. Dr. Munoz has further purified this enzyme. The details are beyond the scope of this paper.

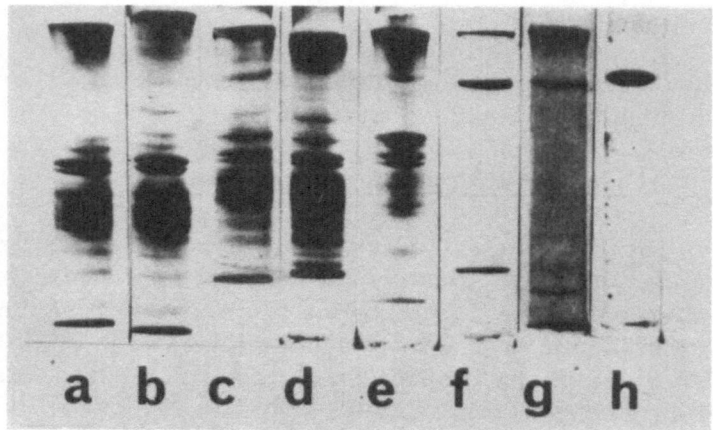

<u>Fig. 5</u>   Disc electrophoresis showing aniline
black staining of proteins in a) cytoplasm;
(b-g) six consecutive washes of the membranes,
and enzymatic staining for ATPase (h).   Method
of Weinbaum and Markman (19).

The morphology of the membranes before and after
treatment by low ionic strength buffer demonstrates some interesting changes.   Fig. 6a shows the particle studded membrane
prior to low ionic strength treatment.   Fig. 6b demonstrates the
apparent disappearance of the particles after treatment.   The
electron micrograph of purified ATPase (Fig. 6c) demonstrates
a particulate structure similar to those seen in previous slides.
The calculated size of the smaller membrane particles and of
purified ATPase are about 100 angstroms.   We believe that the
two are the same.

In summary, ATPase dissociation from membrane is
dependent upon prior removal of poorly defined materials (cation,
protein, lipids) with subsequent release upon exposure to a low
ionic strength environment.   Hydrophobic and/or ionic linkages
seem to be the most important bonds in this enzyme's association
with the membrane.

2) <u>Cation Depletion by Chelating Agents (EDTA)</u>

Selective release of NADH dehydrogenase activity
(Nachbar - unpublished) may be achieved in the following manner.
After three to four washes with 0.03M Tris pH 7.5, the mem-
branes are washed again in same buffer but this time made

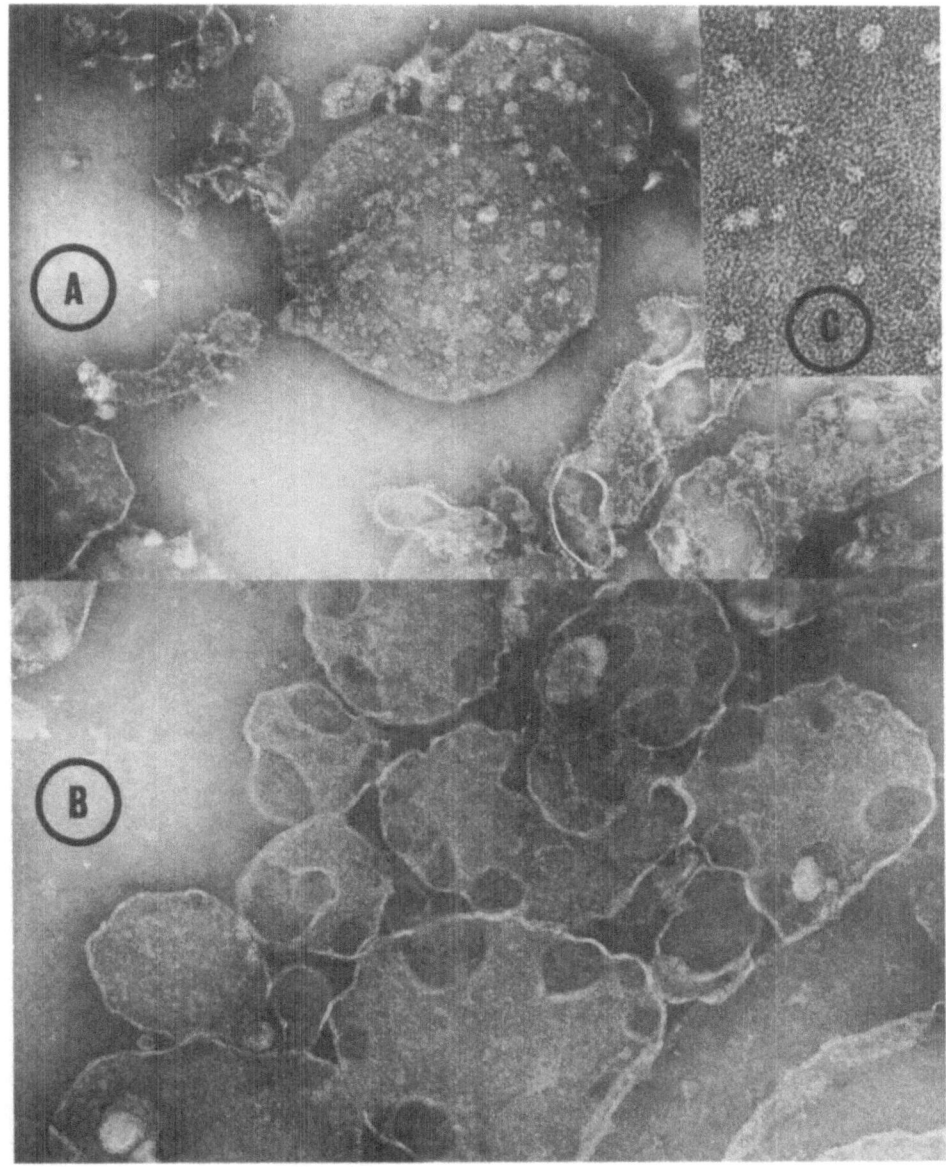

Fig. 6a-c   Negatively stained membrane fragments and ATPase
preparations. (a) membrane fragments prior to "shock wash"
showing associated particles x 95,000, (b) membrane fragments
after shock wash showing loss of particles x 95,000, (c) active
ATPase preparation after purification, (Ref. 12) showing spherical
particles of approximately 100 Å diameter x 260,000.

0.005M with respect to EDTA. NADH dehydrogenase activity is assayed using 2,6-Dichlorophenolindophenol as the electron acceptor. The units are arbitrary and an O. D. change of 0.1 in 30 seconds at room temperature using a Bausch and Lomb Spectrometer at 600 nm is defined as 1 unit of activity. As can be seen in Fig. 7, treatment with EDTA results in a dramatic release of the activity. Less than 15% of the original activity remains in the membrane. In addition to the large amount of activity released, the specific activity of the EDTA wash is 3 to 4 times that of the original membrane and the preceding washes.

Fig. 7   Histogram showing release of NADH dehydrogenase from ML membranes. cyto=cytoplasm; mem nw= membranes before washing; mem 5w=membranes after 5 washes, w1-w3=consecutive washes with 0.03M Tris pH 7.5; w4=0.03M Tris pH 7.5 and 0.005M EDTA wash; w5=0.003M Tris pH 7.5 wash.

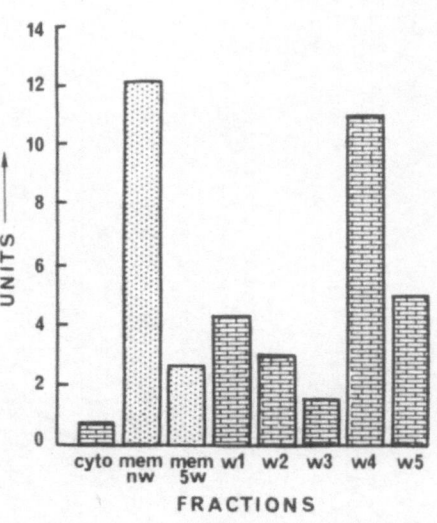

Release of NADH Dehydrogenase from ML Membranes

          Assay modified from Saurge (13). Assay mixture contain 6.0 μmoles NADH, 100 μmoles sodium phosphate buffer pH 7.5, 0.1 μmoles 2,6-Dichlorophenolindophenol in a final volume of 2.95 ml assay at room temperature read at 600 nm. Zero time reading after addition of indicator. The sample is then added (0.05 ml) with rapid mixing and the mixture read at 30 seconds. A change of 0.1 O.D. units in 30 seconds is equal to 1 unit of activity. The best range is a decrease of between 0.1 and 0.5 O.D. units. Corrections made for control (no enzyme).

          Morphological studies demonstrate the vesicular nature of the "soluble" EDTA wash (Fig. 8).

          In a sucrose density gradient of 15-30% run at 25,000 rpm for 24 hrs. at 4°C in a Spinco SW 27 head, particles of activity of 11-12s, 21-22s and a pellet were observed.

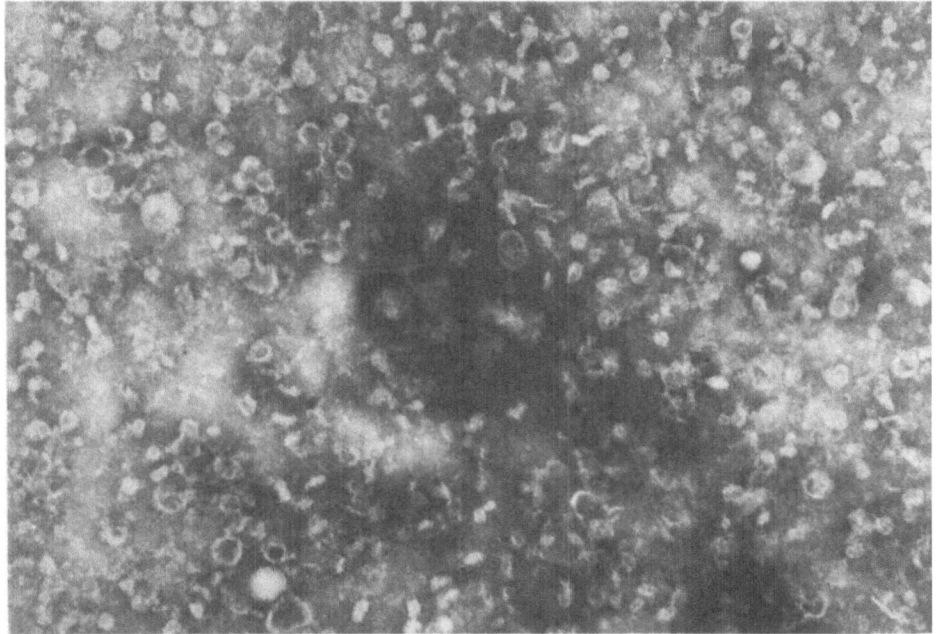

Fig. 8   Negatively stained EDTA wash showing vesic-
ular nature of the wash with smallest vesicles of the
order of 300 Å in diameter. (x 100,000). Note the
relative uniformity of preparation.

Enzymatic staining of the EDTA wash for NADH dehydrogenase in a 7% polyacrylamide gel after electrophoresis
(17,18) using Triphenyltetrazolium as the indicator, demonstrates
2-3 bands of activity, two in the middle of the gel and one faster
moving band. We are not yet sure this represents a group of
enzymes or one enzyme in several polymeric forms. (Fig. 9)

An unexpected finding was the apparent selective release
of lipid components by EDTA. The protein-lipid ratio of the
membrane is of the order of 2:1. The EDTA wash showed a much
higher value for the lipid and in addition, selective release of one
phospholipid, cardiolipin. (Table 3)

We are tentatively interpreting these results to mean that
there may be very specific binding affinities between the various
phospholipids and certain membrane proteins, which are highly
dependent upon the presence of divalent cations.

Fig. 9   Disc gel electrophoresis of EDTA wash
stained for NADH dehydrogenase activity.   Assay
mixture contains 6.0 μmoles NADH, 3.0 μmoles
triphenyltetrazolium chloride, 100 μmoles 0.03M
Tris pH 7.5, 100 μmoles colbalt chloride in a final
volume of 3.4 ml.   The mixture was allowed to
stand 10-30 minutes at 37°C. Note three bands of
activity. Positive results appear as red bands.

       The selectivity of EDTA can be explained in one of three
ways: (1) specific lipids and NADH dehydrogenase exist at different sites on the membrane but are each dependent upon divalent
cations for association to the membrane (2) lipids and proteins
are located in the same region of the membrane but are not
intimately associated and EDTA releases the entire lipid-protein
"patch" (3) lipid and protein are intimately associated in a large
lipoprotein complex which attaches to the membrane via divalent
cations.   Combinations of these may also exist.   In any event, the
findings indicate a heterogeneous makeup of the membrane both
for protein and lipid and are perhaps evidence against a simple
repeating unit structure for the membrane organization.

TABLE 3

LIPID AND LIPID-PROTEIN RELATIONSHIPS

OF THE MEMBRANE AND EDTA WASHES

PREPARATION	LIPID/ PROTEIN RATIO	CARDIOLIPIN/ PHOSPHATIDYL GLYCEROL
Membrane	0.5	2.2
EDTA Wash	1.3	8.9

3) Detergents

Early work by Salton (8) demonstrated the isolation of cytochromes b, c, a by a combination of ionic and/or non-ionic detergents and salt fractionation. However, this method, while it achieved the separation of the cytochromes from most of the lipids and approximately 2/3 of the proteins left the residue without any obvious structural organization; more recent work with membranes washed with 1% deoxycholate achieved a more remarkable result.

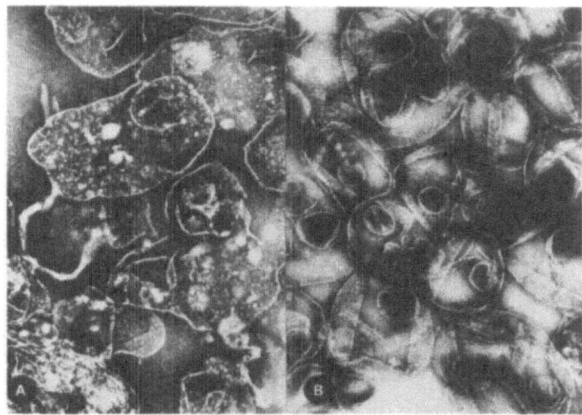

Fig. 10   Electron micrographs of M. lysodeikticus membrane fractions negatively stained with 2% ammonium molybdate A, control membrane after 6 washes with 0.05M Tris B, DOC-insoluble residues after 6 successive washes with 1% DOC in 0.05M Tris. A and B x 86,000.

The membranes are washed until the supernatant is devoid of yellow color (most of the phospholipids and carotenoids having been removed). These residues contain 10-15% of the initial membrane protein and 3-5% of the original lipid. The absorption spectra of these sheets show that characteristic of the cytochromes. In addition, succinic dehydrogenase activity is found to reside in this residue. The DOC-soluble phase spectra shows little cytochrome absorption but strong carotenoid absorption. Both NADH dehydrogenase activity and ATPase activity are recovered in the soluble phase and not in the residue.

The DOC-treated lipid depleted membranous sheets can be studied (Salton - unpublished) further for their response to other dissociating agents.   A neutral detergent Triton-X-100 and 8M urea are not very effective in solubilizing this material.   Complete solubilization is achieved with SDS at concentrations of 0. 1-0. 5%, with Nonidet P-40, 1% with 1% Triton-X-100 solutions containing 0. 04M $AlCl_3$ and with 10. 0M acetic acid.   The protein concentration for all experiments is 5 mg/ml.   Dialysis to remove dissociating reagents usually results in rapid reaggregation into amorphous sheets but membranous sheets are generated in the SDS dissociated fractions dialysed against 0. 05M Tris pH 7. 5 containing Mg^{++}.   Precise conditions for optimal reaggregation into membranous sheets requires further study.

Release of cytochrome c from the other cytochromes of this system can be achieved by dissociating this fraction with 0. 5% SDS followed by $(NH_4)_2 SO_4$ precipitation (20%).   Cytochrome c is released in a soluble form, the a and b cytochrome fraction being precipitated.   The major constituents of the fraction seem therefore to be held together by very strong hydrophobic interactions between proteins and to a lesser extent by electrostatic or ionic bonds.

It is possible that with this technique we have skinned the membrane almost down to the bone.   The residue may represent its skeleton upon which the other lipids and proteins are deposited.

4) Solvents

A variety of organic solvents have been employed in attempts to solubilize protein components.   Of these n-butanol (8) and 3-Pentanol (Nachbar-unpublished) have been the most useful.   Extraction of sonicated membranes in a water phase by a saturating solution of n-butanol at 0°C results in removal of (a) almost all of the lipid in a random fashion - greater than 95% (b) solubilization of many enzyme proteins in an active form e. g. ATPase and the recovery of the cytochromes at the interface. However, NADH dehydrogenase activity is almost completely destroyed.

3-Pentanol extraction preserves 30-40% of the NADH dehydrogenase activity while removing > 90% of the lipids.   After three extractions, no further lipid can be removed.   This solvent appears to selectively remove lipids from the membrane.   Analysis

of the remaining phospholipids in the soluble phase by the
techniques of P^{32} labelling, methanol-chloroform extraction of
lipids (20) followed by silica gel paper chromatography using
system 2 of Wurthier (21) and isotope counting demonstrated
a marked relative enrichment in this fraction of cardiolipin.

SUMMARY AND CONCLUSION

In summary, we have attempted to disect the membrane
of one organism Micrococcus lysodeikticus into its functional
components, using first very mild procedures and progressing
to more drastic measures. We have succeeded in partially
purifying a few enzymatic markers. We also believe we have
demonstrated a great range of strength of binding affinities of
lipids to lipids, lipids to proteins, and proteins to proteins. Our
data suggest but do not yet prove that there are specific differences
between lipids in their participation in various lipo-protein complexes and/or that there may exist a mosaic of lipid or lipoprotein
aggregates in different regions of the membranes. Evidence
points to a regional distribution of specific proteins, as might be
expected for reasons of functional efficiency and not a randomly
associated collection of protein molecules bound together by
haphazard associations.

We would hasten to point out that our findings may well be
peculiar to this organism and generalizations to other membrane
systems should be appropriately tempered.

Acknowledgements:

This work was supported in part by a National Science
Foundation Grant (GB 7250) and by the P. H. S. General Research
Support Grant (FR 05399).

Senior author's (M. S. N.) work done while Public Health
Service Training Fellow, P. H. S. Grant GM00466.

We wish to thank Mr. William Winkler, Mrs. Marion T.
Schor, Mrs. Marjorie Schmitt and Mr. Charles Harman for their
technical assistance and Dr. John H. Freer for the electron
micrographs used in this study.

References

1.   Korn, E. D. , Fed. Proc., 28, 6, 1969.
2.   Korn, E. D. , Ann. Rev. of Biochem. , 38, 263, 1969.
3.   Chapman, D. and contribution by D. F. H. Wallach, in
      Recent Physical Studies of Phospholipids and Natural
      Membranes in Biological Membranes: Physical Fact and
      Function, D. Chapman, Ed. , p. 125, Academic Press,
      New York, 1968.
4.   Salton, M. R. J. , in Protides of the Biological Fluids,
      15, 279, 1967.
5.   Racker, E. and A. Bruni, in Membrane Models and the
      Formation of Biological Membranes, L. Bolis and
      B. A. Pethica, Eds. , p. 138, North-Holland Publ. Co. ,
      Amsterdam, 1968.
6.   Engelman, D. M., in Membrane Models and the Formation
      of Biological Membranes, L. Bolis and B. A. Pethica,
      Eds. p. 203, North-Holland Publ. Co. , Amsterdam, 1968.
7.   Rothfield, L. M. , Weiser M. and A. Endo, J. Gen.
      Physiol. , 54, 27s, 1969.
8.   Salton, M. R. J. , Trans, N. Y. Acad. Sci. , 29, 764, 1967.
9.   Bancroft, J. , An Introduction to Histochemical Technique,
      p. 224, Appleton-Century-Crofts, New York, 1967.
10.   Salton, M. R. J. , in Microbial Protoplasts and L-Forms,
      L. B. Guze, Ed. , p. 144, Williams and Wilkins Co. ,
      Baltimore, 1968.
11.   Munoz, E. , Nachbar, M. S. , Schor, M. T. and Salton,
      M. R. J. , Biochem. Biophys. Res. Commun. , 32, 539, 1968.
12.   Munoz, E. , Salton, M. R. J. , Ng, M. H. , Schor, M. T. ,
      European J. Biochem. , 7, 490, 1969.
13.   Saurge, N. , Biochem. J. , 67, 146, 1957.
14.   Salton, M. R. J. , Freer, J. H. , Ellar, D. J. , Biochem.
      Biophys. Res. Commun. , 33, 909, 1968.
15.   Nachbar, M. S. - in preparation.
16.   Salton, M. R. J. , in The Future of the Brain Sciences,
      S. Bogoch, Ed. , p. 1, Plenum Press, New York, 1969.
17.   Ornstein, L. , Ann. N. Y. Acad. Sci. , 121, 321, 1964.
18.   Davis, B. J. , Ann. N. Y. Acad. Sci. , 121, 404, 1964.
19.   Weinbaum, G. and Markman, R. , Biochim. Biophys. Acta,
      124, 207, 1966
20.   Bligh, E. G. and Dyer, W. J. , Canad. J. Biochem.
      37, 911, 1959.
21.   Wuthier, R. E. , J. Lipid Research, 7, 544, 1966.

RNA IN THE CELL PERIPHERY

E. Mayhew and L. Weiss

Department of Experimental Pathology

Roswell Park Memorial Institute, Buffalo, New York

All mammalian cells so far studied carry a net negative surface charge, at neutral pH, as measured by cell electrophoresis.
However the nature of the fixed ionogenic groups responsible for
the electrokinetic behavior of cells has not as yet been fully
elucidated.

This paper is divided into three sections; (1) a presentation
of evidence that ionogenic groups specifically susceptible to
ribonuclease are present at the periphery of some cell types; (2)
a discussion of the dynamic state of the cellular electrokinetic
surface, and (3) speculation on the possible importance of this
dynamism on cellular properties.

(1) Ribonuclease-susceptible anionic groups

Evidence that there are ionogenic groups within the cell
periphery susceptible to ribonuclease comes mainly from measurements of cell electrophoretic mobility.

The original observation (Weiss and Mayhew, 1966) was made
that ribonuclease reduced the electrophoretic mobility of two types
of cultured human cells. The results showed that this reduction
was independent of, and additive to, the known effects of neuraminidase on these cells (Table 1). This was taken as evidence that
ribonuclease interacted with sites separate from those acted upon by
neuraminidase. Analysis of cell supernatants showed that ribonuclease does not release any neuraminic acids from cells.

On the one hand, loss of cell surface net negativity could
result from fission of phosphodiester bonds, indicating the

191

Table 1

Electrophoretic Mobility of RPMI No. 41 cells
after treatment with HBSS,
Ribonuclease and/or Neuraminidase

1st treatment	2nd treatment	Electrophoretic* mobility	%Decrease in mobility
HBSS	---	$-1.09 \pm .021$	---
HBSS	HBSS	$-1.11 \pm .022$	---
Neuraminidase	Neuraminidase	$-0.65 \pm .023$	40.4
Neuraminidase		$-0.63 \pm .022$	43.2
Ribonuclease	---	$-0.78 \pm .021$	28.4
Ribonuclease	Ribonuclease	$-0.77 \pm .019$	30.6
Ribonuclease	Neuraminidase	$-0.43 \pm .020$	61.3
Neuraminidase	Ribonuclease	$-0.45 \pm .017$	59.5

*$\mu.\text{sec}^{-1}. \text{v}^{-1} \text{cm}. \pm$ standard error of mean.

Table 2

Effect of Ribonuclease on Electrophoretic mobility of Cells

Cells	Electrophoretic mobility[1]	%Change in mobility[2]	p value[3]	RNA at cell Periphery[4]
Chicken Erythrocytes	-0.83	-3 to +3	p>0.5	-
Mouse Erythrocytes	-1.19	-3 to +3	p>0.5	-
Human Erythrocytes	-1.09	-3 to +3	p>0.5	-
Human Peripheral Blood				
Lymphocytes	-0.87	-10 to -15	p<0.01	+
Polymorphs	-0.80	-6 to -10	0.02>p>0.01	?
Monocytes	-0.57	-3 to -10	0.1> p>0.05	?
Platelets	-0.96	-3 to +3	p>0.5	-
Acute Lyphocytic Leuk.[5]	-1.42_6	-8 to +5	p>0.1	?
Chronic Lymphocytic "[5]	-1.18_6	+2 to +10	p>0.5	-
Human Cultured				
RPMI No. 41	-1.10_6	-10 to -35	p<0.01	+
Burkitts Lymphoma	-1.00_6	-3 to +3	p>0.2	-
Normal Lymphoid	-1.25_6	-11 to -20	p<0.01	+
Mouse				
Peritoneal exudate[5]	-0.77	-5 to -10	0.1>p>0.05	?
Thymocytes	-0.89	-6 to -10	0.02>p>.01	?
Liver	-0.84	-3 to -15	0.02>p>01	?
Lymph node	-1.10	-3 to +3	p<0.1	-
Mouse Tumour				
Ehrlich ascites	-1.06	-10 to -30	p<0.01	+
S37	-1.00	-10 to -25	p<0.01	?
Cultured Mouse				
Ehrlich Ascites	-0.95	-10 to -25	p<0.01	+
1210	-0.83	-6 to +9	p<0.01	+
929 Fibroblasts	-0.98	+3 to -15	.05>p>0.02	?

1. μ sec^{-1}. volts^{-1} cm. at 25° pH 7.2-7.4 in PBS or HBSS.

2. The extreme ranges of change after ribonuclease treatment in different experiments are shown.

3. Probability value at mean variation in mobility.

4. − no peripheral RNAse-susceptible sites detectable.
   ? RNAse susceptible sites possibly present.
   + RNAse susceptible sites present under optimal conditions.

5. Nucleated cells.

6. Measured at 37°.

Table 3

Effect of Ribonuclease A and T_1 on electrophoretic
mobility of Ehrlich ascites tumour cells

	Electrophoretic Mobility $\mu.sec^{-1}. v^{-1} cm.$	%Decrease in mobility
Control	-1.10 ± .014 (374)*	---
Ribonuclease A	-0.87 ± .021 (185)	19.8
Ribonuclease T_1	-0.96 ± 0.22 (245)	13.0

*mean ± standard error (number of determinations)

Table 4

Effects of active and inactive ribonuclease on
the electrophoretic mobility of Ehrlich ascites cells[1]

Final Concentration Enzyme mg/ml	Electrophoretic Mobility[2]	
	Ribonuclease A	Inactivated Ribonuclease A
0	1-11 ± .026	1.11 ± .026
10^{-3}	0-96 ± .021	1.05 ± .024
10^{-2}	0.89 ± .023	1.08 ± .030
10^{-1}	0.88 ± .020	1.10 ± .027

1.cells treated 30 mins at 37°

2.μ sec.$^{-1} v^{-1}$ cm.

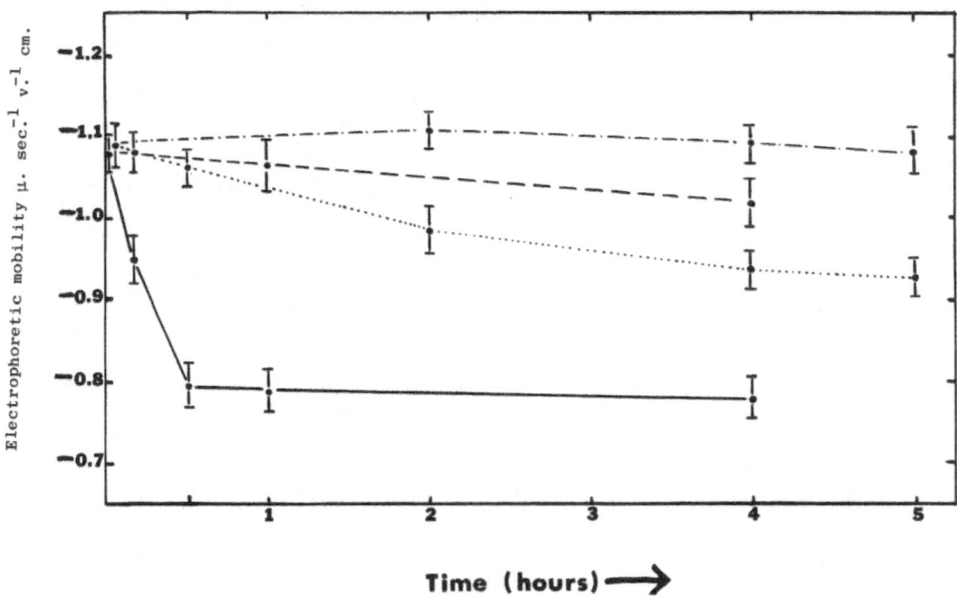

Figure 1. Electrophoretic mobility of Ehrlich ascites cells
treated with ribonuclease at 10° or 37°, and measured
at those temperatures.
———— 37° Control, ————— 37° ribonuclease-treated
—·—·—·— 10° Control, ... 10° ribonuclease-treated

presence of RNA in the cell periphery. On the other hand, as
ribonuclease is positively charged at neutral pH, at which most
electrophoretic measurements are made, non-specific binding of
this enzyme to the cell surface could also reduce net surface
negativity.

Several pieces of evidence suggest that non-specific adsorption of ribonuclease is not the cause of the changes in electrophoretic mobility. Table 2 shows the effect of ribonuclease on the
electrophoretic mobility of some cell types. Although all these
cells carry net negative surface charge, only about a third of the
various types have their mobilities appreciably reduced by ribonuclease; and of the RNAse-susceptible cells the reductions range
from amounts barely detectable by cell electrophoresis (approximately 5%), to more than 30%. These results suggest that any
adsorption of RNAse to cell surfaces is not nonspecific.

The mobility of Ehrlich ascites cells is reduced both by
ribonuclease T_1 and by ribonuclease A (Table 3). (Ribonuclease
A from bovine pancreas specifically cleaves phosphodiester bonds
between $3'$ and $5'$ positions of ribose groups in pyrimidine ribonu-
cleosides (Brown and Todd, 1955) whereas ribonuclease T_1 from
Asperigillus oryzae specifically cleaves phosphodiester bonds
between $3'$ guanylic acid groups and $5'$ hydroxyl groups of adjacent
nucleotides in RNA (Egami et al, 1964).) As at neutral pH, ribonuclease T_1 is negatively charged, it is therefore unlikely that
the reduction in cell surface charge produced by both ribonucleases
T_1 and A can be due to adsorption. The more likely explanation is
that both enzymes act by cleaving their specific substrates within
the cell periphery. Comparative studies have also been made of the
effects of ribonuclease A, and ribonuclease A which was specifically
inactivated by Barnard and Stein's method (Barnard and Stein, 1959)
(Table 4). The inactivated enzyme differs from native RNase in
that histidine residue 119 is carboxymethylated and inactivation
produces only minor conformational changes visible by small angle
X-ray diffraction at 3Å resolution (Kartha, personal comm.) and
very minor changes in net charge (Crestfield et al, 1963). The
inactivated enzyme produced no significant change in cell electrophoretic mobilities. Again these collected data are against the
proposition that adsorption of RNase causes the loss of net cell
surface negativity. The possible slight effect at high concentrations was probably due to slight residual traces of active ribonuclease in the inactivated enzyme preparation.

In another series of experiments, the effects of temperature
of incubation were observed. The mobilities of cells were determined at different times after treatment with ribonuclease at $10°$
or $37°C$ (Figure 1). Whereas the mobility of cells treated at $37°$
decreased rapidly with maximum reduction occurring after about 20
minutes incubation, the mobility of cells treated at $10°$ decreased

at only approximately 1/5th of this rate. From consideration of Gibbs' adsorption equation it is expected that adsorption at 37° would occur slightly less rapidly than at 10°. As the rate of reaction of ribonuclease against isolated RNA is approximately 8 times faster at 37° than at 10°, the rate of decrease of electrophoretic mobility at the two temperatures is thus more consistent with an enzyme reaction than with adsorption.

More phosphates and nucleotides are liberated from cells treated with ribonuclease than from cells treated with inactive ribonuclease or no enzyme at all (Weiss, Mayhew, 1966). Ribonuclease may well penetrate into cells, act on intracellular substrate and release enzymatic products, as well as acting at the cell periphery. Another possibility is that activity of ribonuclease at the cell periphery causes permeability changes, resulting in leakage of various materials. Thus, analyses of liberated enzymatic degradation products do not necessarily give information on cell surface chemistry.

RNA which leaks from cells could adsorb to their surfaces, and experiments have been made to determine whether the surface RNA is in fact a leakage product. Cells were washed 20 times before treatment with enzyme and their mobilities measured (Table 5). Ribonuclease treatment resulted in a marked reduction in mobility both after 1 or 20 washings. However the control mobilities also decreased slightly from 1 wash to 20 washes, suggesting that some of the ribonuclease-susceptible material can be removed by washing, but most cannot. In another series of experiments, a cell suspension was ultrasonically homogenized and then mixed with either controls, or cells previously treated with ribonuclease. The mobilities were again measured (Table 6), and no changes in mobility were observed. These experiments indicated that cellular debris, containing nucleic acids, does not affect cellular electrophoretic mobility. In cultures where a high percentage of dead cells was present, the amount of detectable peripheral RNA is usually lower than in "healthy" cultures. If adsorbed leaking RNA was the source of cell surface RNA, a higher amount of peripheral RNA would have been expected in "unhealthy" cultures. Other possible sources of peripheral RNA are mycoplasmal or viral contaminants. However, RNase susceptible Ehrlich ascites cells have been shown to be free of mycoplasma (Weiss and Mayhew, 1969) and cell surface viruses (Horoszewicz, personal communications).

The total experimental evidence strongly suggests that in some cell types there are sites in the cell periphery specifically susceptible to ribonuclease, which in view of the specificity of the enzyme, may be taken to indicate the presence of RNA in this region.

Table 5

The Electrophoretic mobility of Ehrlich ascites
cells washed different number of times before
treatment with ribonuclease

Number of washes before treatment	Electrophoretic mobility*	
	Controls	Treated with Ribonuclease A
1	1.12 ± .030	0.83 ± .021
5	1.05 ± .031	0.85 ± .022
10	0.98 ± .027	0.81 ± .024
15	1.01 ± .028	0.80 ± .021
20	0.96 ± .027	0.78 ± .021

$*\mu \ sec^{-1}v^{-1}cm.$

Table 6

The Effect of Ehrlich ascites cell homogenates on the
Electrophoretic Mobility of HBSS and Ribonuclease treated Ehrlich
Ascites cells

1st treatment	Electrophoretic mobility	2nd treatment	Electrophoretic mobility
HBSS	-1.06	Medium only 30 min.	-1.08
HBSS	-1.06	Cell homogenates 30 min.	-1.10
Ribonuclease	-0.82	Medium only 30 min.	-0.88
Ribonuclease	-0.82	Cell homogenates 30 min.	-0.89

$*\mu \ sec^{-1}v^{-1}cm.$

Following ribonuclease treatment, the loss of anionic groups from the cell periphery can come about in two main ways (Fig. 2). Some ionized phosphates of RNA could be in the cellular electrokinetic surface, and be lost into the environment on enzyme treatment. Cellular RNA is usually bound to proteins, mucopolysaccharides and other materials and a second possibility therefore, is that the RNase susceptible anionic groups at the cellular electrokinetic surface are not part of the actual RNA molecule.

(2) The dynamic state of the cellular electrokinetic surface

(a) Changes with growth rate. The electrophoretic mobilities of untreated cells or cells treated with neuraminidase and ribonuclease were measured in cultures where the growth rate was varied.

Figure 3 shows that when the growth rate of cultures was slowed by not adding fresh medium, the mean electrophoretic mobility of the control cells decreased; their mobility returned to normal one day after adding fresh medium. The mean mobilities of neuraminidase treated cells ran approximately parallel to the controls, and corresponded to a reduction in net negativity of 20-25% throughout the experiment. Fig. 4 shows the effects of ribonuclease. Although there was initial drop of about 25%, after five days without replenishing the medium, ribonuclease did not appreciably decrease the mobility of cells at all. However, within one day of adding fresh medium, decreases in mobility following RNase-treatment were observed. These results indicate that the density of ribonuclease-susceptible ionized groups at the cell surface can be modified by changes in cultural conditions related to changes in growth-rate. In other experiments, the growth-rate was varied by culturing cells in media containing different amounts of calf serum. It can be seen (Table 7), that in the presence of the highest concentration of calf serum, where multiplication was most rapid, RNase-susceptibility was the greatest, and vice versa. In contrast to the variable effects of RNase on cell electrophoretic mobility, neuraminidase-susceptibility did not vary significantly as a function of growth rate.

An attempt was made to follow effects of growth rate changes in vivo (Mayhew, 1968), by determining the mobilities of Ehrlich ascites cells at different times after inoculation into mice. The mobilities of cells rose rapidly in the first day after inoculation, and then declined slowly (Table 8). The percentage reduction in the neuraminidase-treated cells remained constant, irrespective of age of tumors. However, after ribonuclease treatment the mobilities fell to a constant value, indicating that the density of ribonuclease-susceptible ionogenic groups at the surfaces of these cells changed with time. The most ribonuclease-susceptible sites were found in the younger tumours, where growth rate is highest. Thus, these results, although not as clear-cut as in the in vitro studies,

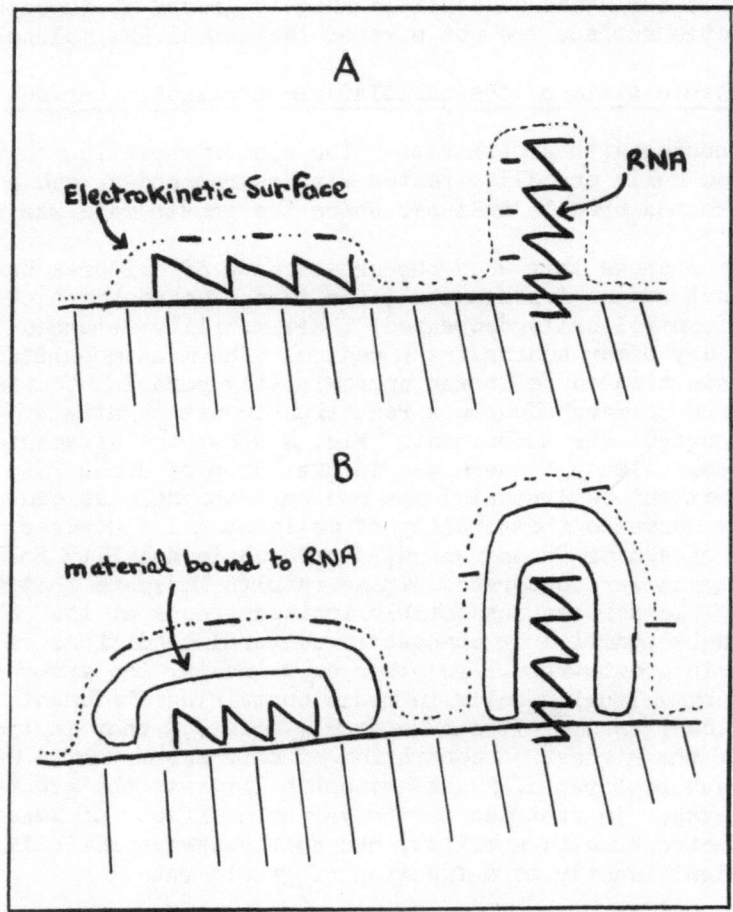

Figure 2. RNA in cell periphery: possible localization. A)
Ionogenic groups of RNA are in electrokinetic surface.
B) Ionogenic groups of unknown material bound to RNA
are in electrokinetic surface.

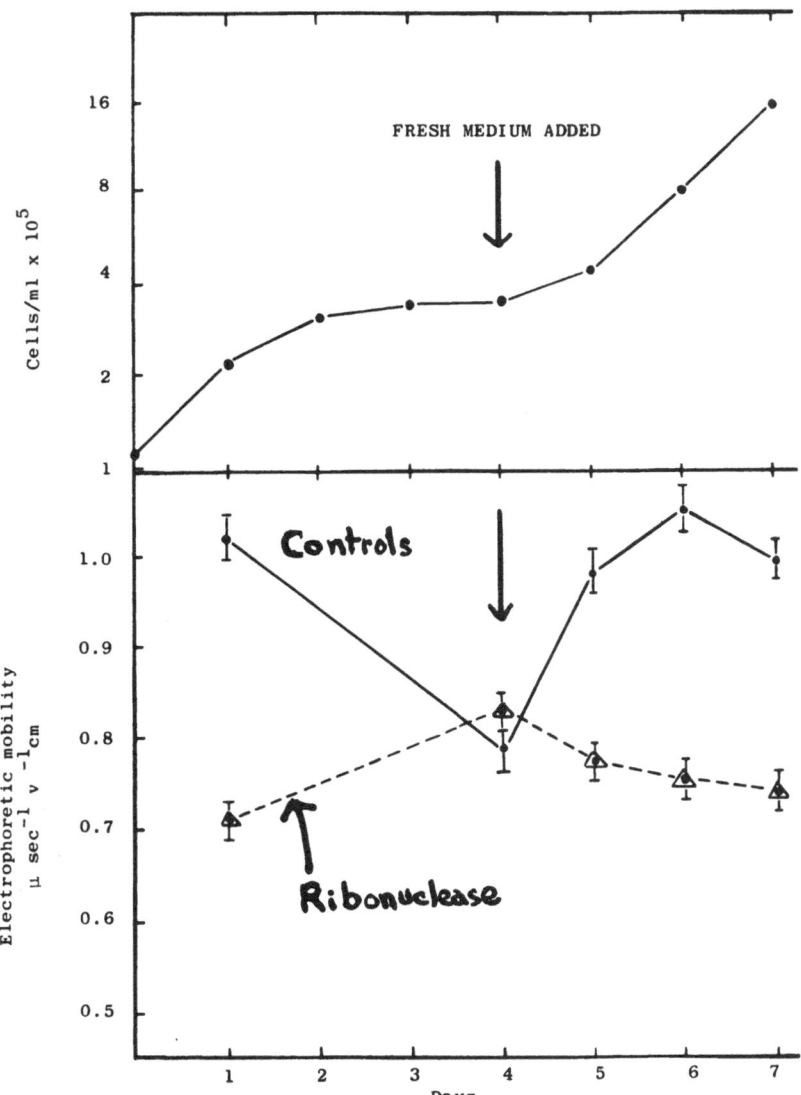

Figure 3. The effect of growth rate changes on electrophoretic
           mobility of RPMI No. 41 cells.
                ──────── Controls, Δ────Δ ribonuclease-treated,
           ⊕──────⊕ neuraminidase-treated.
           Top graph shows the growth of cells. Cells were not
           fed from day 0 to day 3 but were refed on day 4.

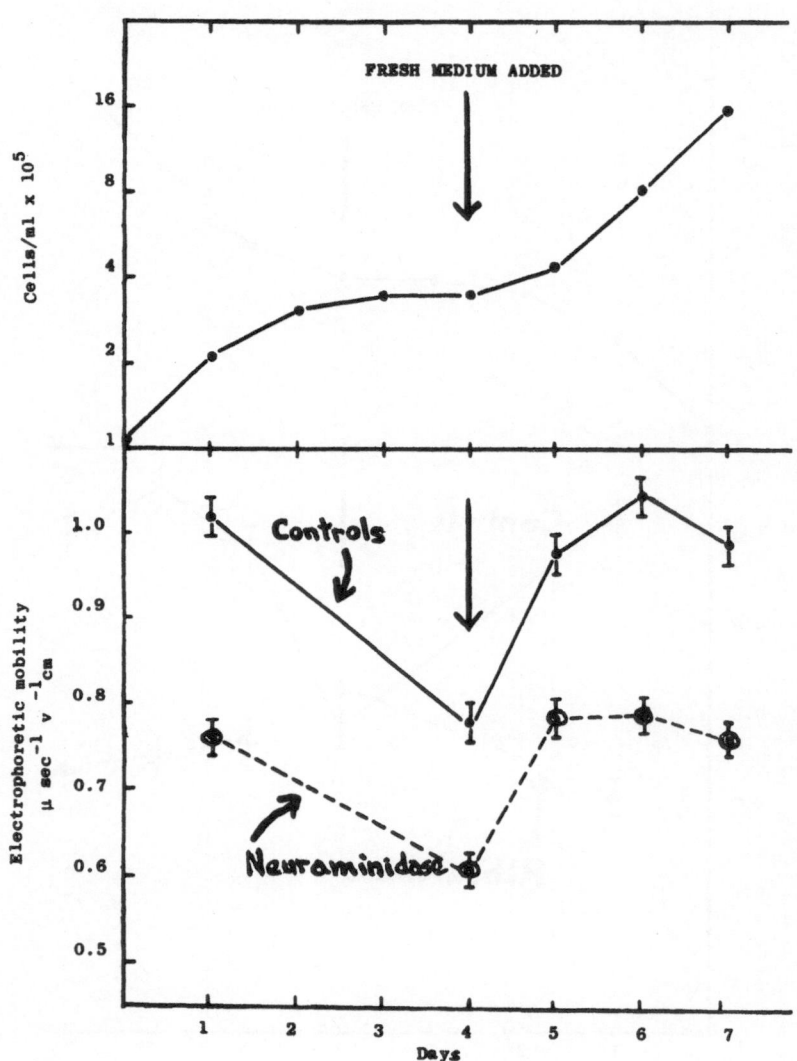

Figure 4. The effect of growth rate changes on electrophoretic
           mobility of RPMI No. 41 cells.
                ——————— Controls, △----△ ribonuclease-treated,
           ⊖------⊖ neuraminidase-treated.
           Top graph shows growth of cells. Cells were not fed
           from day 0 to day 3 but were refed on day 4.

Table 7

Electrophoretic Mobility of RPMI No. 41 cells grown in different
calf serum concentrations, with or without treatment
with Ribonuclease and/or Neuraminidase

	Calf Serum 5% Calf Serum	%Reduction	0.5% Calf Serum	%Reduction
HBSS	1.04	--	0.91	--
Neuraminidase	0.79	24	0.70	22
Ribonuclease	0.79	24	0.79	13
Neuraminidase/Ribonuclease	0.53	48	0.64	30
Mean Doubling Time	25 Hrs.		51 Hrs.	

$$* \mu \ sec^{-1}v^{-1}cm.$$

Table 8

Effect of neuraminidase and ribonuclease on
electrophoretic mobility
of Ehrlich ascites cells grown in mice

Days after tumour inoculation	Controls	Electrophoretic Mobility*	
		Neuraminidase treatment	Ribonuclease treatment
0	1.02 + .030	0.66 + .020	0.64 + .018
1	1.23 + .034	0.76 + .019	0.63 + .017
2	1.21 + .033	0.84 + .024	0.59 + .019
3	1.18 + .025	0.85 + .022	0.66 + .020
4	1.12 + .024	0.75 + .017	0.64 + .019
5	1.10 + .032	0.76 + .019	0.67 + .019
7	1.00 + .026	0.65 + .020	0.65 + .022
11	0.96 + .024	0.66 + .021	0.67 + .017
14	0.84 + .023	0.62 + .020	0.66 + .019
16	0.78 + .026	0.56 + .019	0.59 + .017

$$*\mu \ sec^{-1}v^{-1}cm.$$

support the concept that growth-rate is directly related to the
density of ribonuclease-susceptible groups in Ehrlich ascites cells.

This linkage of electrophoretic mobility to cellular metabolic
processes is a subtle one, as short-term experiments lasting
approximately 30 minutes, have failed to reveal changes in mobility
associated with severe depression of cellular oxygen-consumption,
anaerobic glycolysis or uncoupling of oxidative phosphorylation
(Weiss and Ratcliffe, 1968).

(b) Changes with cellular life cycle. Cells at specified
stages of the cell-life cycle may be obtained in parasynchronous
cultures and it has been shown that changes occur in the mobility
of cells during their mitotic cycle (Mayhew, 1966).

Fig. 5 shows an experiment where synchronized growth was
induced in a cultured human cells by double thyimidine treatment
(Peterson and Anderson, 1964). The top graph shows the growth of
the cells after release from the second thymidine block. No cell
division occurred for about 7 hours, then wave of division lasting
to 10-11 hours took place. No cells were detected in mitosis for
about 6 hours, a wave of mitosis then occurred, which reached a
peak about 8 hours after release, and then declined. The electrophoretic mobilities of these cells, with or without treatment with
ribonuclease and/or neuraminidase, are shown in the bottom graph.
The mobilities of the control cells reached a peak at 8 hours and
declined afterwards. After neuraminidase-treatment the mobilities
fell to a constant level irrespective of time, whereas after ribonuclease treatment there was a slight maximum at 8 hours. These
results indicate that the increases in mobility during the mitotic
cycles were due mainly to an increased surface density of sialic
acid moieties; if there was a variation in the surface density of
ribonuclease-susceptible sites it was not detectable in these
experiments.

(3) Functional aspects of RNA at the cell periphery

The experimental data show that in the cells examined by us,
the density of ribonuclease-susceptible anionic groups at their
electrokinetic surfaces varies with growth-rate. The question
arises of the functional significance of both the peripheral RNA,
and of changes in its surface density.

In addition to the processes discussed earlier, it should also
be remembered that changes in the physicochemical nature of the
cell periphery can be brought about by enzymes released from cells.
Examples have been given of non-lethal autolytic changes produced
by released lysosomal enzymes (Weiss, 1967), and among these
enzymes in RNase. Thus, any of the multitude of factors causing
lysosomal activation, can possibly affect the status of the RNA in

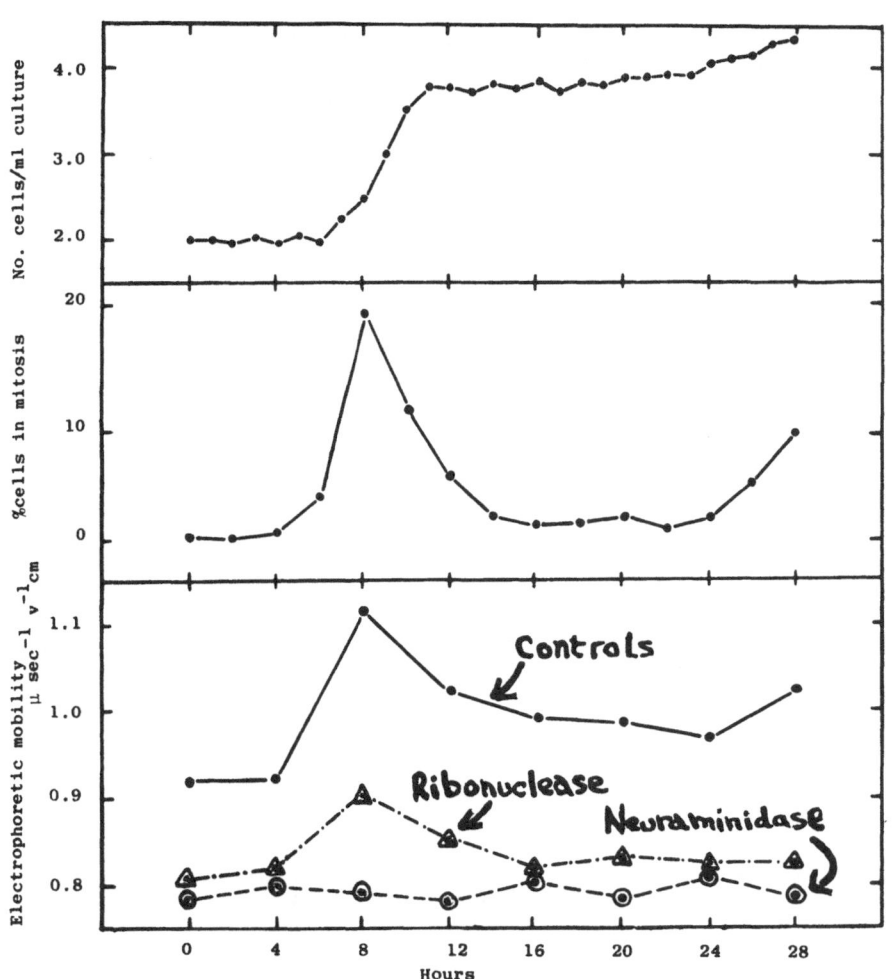

Figure 5. Electrophoretic mobility of RPMI No. 41 cells in
           parasynchronous growth, with or without treatment
           with ribonuclease or neuraminidase.

the peripheral region of cells.

Previous work has shown that the rigidity of the peripheries of some cells was dependent on sialic acid, possibly due to mutual electrostatic repulsion between its ionized carboxyl groups (W eiss, 1965). When macrophages were treated with neuraminidase, the reduction in net surface negativity and increased deformability were associated with enhancement of phagocytic activity (Weiss, Mayhew, and Ulrich, 1966). However, ribonuclease produces no change in the deformability of a number of cells (Weiss, 1968) and in addition RNase-susceptible groups have not been detected at the macrophage surface (Table 2).

It has often been suggested that ion-binding to carriers is a necessary preliminary step to transmembrane ion movements (Rosenberg and Wilbrandt, 1957). However, studies on Ehrlich ascites tumour cells, have not revealed an effect of ribonuclease-treatment of either net flux of Na+ or K+, or on unidirectional K+ fluxes (Weiss and Levinson, 1969).

The ionized moieties in the cell periphery may well influence drug adsorption. The charged groups themselves may be the actual receptor or part of the receptor complex; they may orient the drug to invoke drug/receptor specificity as in hapten/antibody interactions; they may affect drug/receptor interactions by altering the conformation of the cell surface and finally, the ionogenic groups may offer steric hindrance to drugs which can react with underlying receptors. Preliminary, unpublished work, suggests that RNase-treatment changes the susceptibility of Ehrlich ascites cells to antitumour drugs.

The observation that whereas normal, human circulating lymphocytes have their electrophoretic mobilities significantly reduced by RNase, whereas circulating lymphoid cells from patients with acute or chronic lymphocytic leukaemia do not, raises the question of whether this difference is capable of diagnostic or chemotherapeutic exploitation(Weiss & Sinks, 1969).

Bennett et al (1969) have studied RNA in the peripheries of rapidly proliferating mouse lymphoid cells, by transplanting lymphopoietic cells into irradiated recipients. Where as normal lymph-node cells had no detectable peripheral RNA, allogeneic and syngeneic (transplanted) lymphnode cells did, on the evidence that following RNase-treatment, their mean electrophoretic mobilities were reduced by 14.2 and 10.3%, respectively. These observations may be possibly related to those of Mowbray and his colleagues (1969) that pancreatic RNase has immunosuppressive activity, and may be capable of exploitation in transplantation.

The nature of the RNA in the peripheries of some cells is

not known. Speculations have been made on the possible implications if it has genetic informational content (Weiss, 1967), and
how this informational material can be transferred from the
periphery of one cell to another, where it can be endocytosed and
"used".

Contact between the charged surfaces of cells has been
treated extensively in terms of lyophobic colloid theory (Curtis
1967, Weiss, 1967). The work presented here suggests that in any
consideration of electrostatic interactions between two opposing
cell surfaces, the metabolic state of the cells must be considered.
If cell contact interactions are in fact dependent on peripheral
RNA, then these interactions must also be related to other aspects
of cellular activity at that time.

Acknowledgement   This work was supported in part by American
Cancer Society Grant P403B.

Literature Cited

Barnard E.A. and W.D. Stein, 1959, The histidine residue in the
active centre of ribonuclease 1. A specific reaction with bromo-
acetic acid, J. Mol. Biol. 1, 339-349.

Bennett M., Mayhew E., and L. Weiss 1969, RNA in the periphery of
rapidly proliferating mouse lymphoid cells J. Cell. Physiol.
(in press).

Brown D.M. and A. R. Todd, 1955, Evidence on the nature of the
chemical bonds in nucleic acids. In "Nucleic Acids" Vol. 1,
p.409-446 Eds. E. Chargaff and J. N. Davidson, Academic Press,
New York.

Curtis A.S.G., 1967, The Cell Surface, Its Molecular Role in
Morphogenesis, Logos, Academic Press, London.

Crestfield, A. M., Stein W.D. and S. Moore, 1963, Alkylation and
Identification of the Histidine Residues at the active site of
ribonuclease J. Biol. Chem. 238, 2413-2419.

Egami F., Takahashi K. and T. Uchida, 1964, Ribonuclease in Taka-
Diastase; properties, chemical nature and applications. Prog. in
Nucleic acid Res. and Mol. Biol. Vol. 3, Academic Press, New York.

E. Mayhew 1966, Cellular Electrophoretic mobility and the mitotic
cycle J. Gen. Physiol., 49, 717-725.

E. Mayhew, 1967, Effect of ribonuclease and neuraminidase on the
electrophoretic mobility of tissue culture cells in parasynchronous
growth. J. Cell. Physiol., 69 305-310.

E. Mayhew, 1968, Electrophoretic mobility of Ehrlich ascites car-
cinoma cells grown in vitro or in vivo Cancer Res. 28, 159-1595.

Mowbray J.F., Boylston A. W., Milton J. D. and M. Weksler M. 1969,
Studies on the mode of action of immunosuppressive ribonuclease
Antibiotica et Chemotherapia, 15, 384-392.

Petersen D. F. and E. C. Anderson, 1964, Quantity Production of
synchronized mammalian cells in suspension culture, Nature, 203,
642-643.

Rosenberg T., and W. Wilbrant, 1957, Uphill transport induced by
counterflow J. Gen. Physiol., 41, 289-296.

Weiss L. and C. Levinson, 1969, Cell Electrophoretic Mobility and
Cationic Flux, J. Cell. Physiol., 73, 31-36.

Weiss L. and E. Mayhew, 1966, The presence of ribonucleic acid
within the peripheral zones of two types of mammalian cell
J. Cell. Physiol. 68, 345.

Weiss L. and Mayhew E. 1969, Ribonuclease-susceptible charged
groups at the surface of Ehrlich ascites tumour cells Int.
J. Canc. (in press).

Weiss L, and Ratcliff T. 1968, Effect of Chemotherapeutic and
other agents on cellular electrophoretic mobility. J. Nat. Cancer
Inst. 41, 957-966.

IMMUNOLOGICAL REACTIONS CARRIED OUT AT A LIQUID-SOLID INTERFACE

WITH THE HELP OF A WEAK ELECTRIC CURRENT

Alexandre Rothen and Christian Mathot

Rockefeller University

New York, New York

We showed more than twenty years ago that monolayers of proteins spread on the water surface of a Langmuir trough and transferred to metalized glass slides were capable of adsorbing specifically corresponding antibodies (1,2). These layers were completely surface denatured, about 8 Å thick per monolayer and in
spite of their loss of secondary and tertiary structures they had
kept their immunological properties to interact specifically with
antisera. The test consisted in smearing on a metalized glass
slide coated with the transferred protein layers a drop of dilute
antiserum. After 10' or so, the slide was washed and dried and
the thickness of the layer adsorbed from the antiserum was measured optically with an ellipsometer, an apparatus developed in
our laboratory, permitting the determination of a thickness within
a fraction of one Angstrom (3,4). The thickness adsorbed was about
60 Å as compared to approximately 10 Å when the serum was a normal
or a heterologous one. Spread films of serum bovine albumin exhibited a unique property not shared by any other protein films
investigated, namely the thickness of the layer adsorbed from a
homologous antiserum was proportional to the number of monolayers
of albumin transferred to the slide. The adsorbed antibody layer
was roughly 80 Å thick per double layer of albumin and 240 Å
could easily be adsorbed on four underlying double layers of albumin. Protein molecules could also be adsorbed directly on a
slide from a dilute solution. In this case the protein molecules
were not surface denatured and a layer of antibodies as thick as
80 Å could be specifically immobilized on the subjacent adsorbed
layer of antigen. The antigenic layers could also be made of
substances other than proteins. Similar experiments carried out
with polysaccharides from Types I, III and VIII pneumococcus gave
analogous results. In this case the thickness of the adsorbed

209

layer of antibodies was very large (up to 600 Å) whereas the underlying layer of polysaccharide was only 4 to 5 A thick (2).

        Such results showed that a method for quantitative determina-
tions of immunological reactions might be developed based on the
measurements of the thickness of specifically adsorbed layers of
antibodies. Theoretically at least, the concentrations of either
antigens or antibodies could be determined. In actual practice the
method proved to have serious limitations because the antigen solutions had to be relatively pure and concentrated and the antisera
could not be greatly diluted. However, when an electric current
with the proper polarity was applied during the antigen and antibody adsorption periods, the slide upon which the deposition took
place being one of the electrodes, clear cut differentiation between specific and non-specific adsorption could be obtained with
very crude antigenic preparations or with highly diluted antisera.
These facts were the basis for a new technique to detect immunological reactions which we called "The Immunoelectroadsorption
Method". It was successfully applied to the detection of eight
arthropod-borne viruses. The antigen solutions were crude extracts
of infected new-born mouse brain. In one case immunological reactions could still be detected with an antiserum diluted 1 to 10^5
(5). This method was also used to follow the formation of antibodies in mice against Friend's virus. It was possible to detect
the appearance of antibodies already two days after infection (6)
and to investigate the influence of the strain of mice on antibodies
formation (7). Using pure substances as antigens, we were able to
determine the smallest concentration of antigen still detectable
with a good rabbit antiserum. For instance the smallest detectable
concentration of bovine serum albumin dissolved in water was 10^{-8} g/ml
(8). It was important to know the influence of foreign material,
protein or otherwise, present in the solutions of antigen on the
sensitivity of the method. This information was especially needed
if the method was to have wide applications when only very impure
antigen preparations were available. The smallest measurable concentration of bovine albumin was ≈10^{-7} g/ml, when the albumin was
dissolved in a 2% guinea pig serum solution named "carrier". In
this case the ratio by weight of foreign protein to bovine albumin
was roughly 10.000. It is remarkable that the specific adsorption
of antibodies was so little influenced by the presence of an over-
whelming number of foreign protein molecules present in the solution
of the antigen used for the first adsorption.

        The determination of circulating human and bovine growth hor-
mones in the physiological range was successfully achieved with the
immunoelectroadsorption method (9). The limit of sensitivity was
2×10^{-10} g. of growth hormone per ml of the carrier solution. For
an assay of growth hormone the serum to be tested was diluted with
a carrier until the thickness adsorbed subsequently from a rabbit
immune serum was the same as that observed when the slide was

treated for the first adsorption with the carrier only. The concentration of growth hormone in two sera, one from an acromegalic
and the other from a hypopituitary subject were found to be
12×10^{-9} g/ml and 0.8×10^{-9} g/ml respectively.

EXPERIMENTAL

The immunoelectroadsorption method was applied to the problem of
the interaction between the polysaccharides mainly from type III
and VIII pneumococci and their corresponding rabbit antisera. The
results obtained are presented in this communication.

Chromium plated slides were used throughout and the intensity
of the current was always kept at 300μA. See references (5) and
(8) for technical details. The polysaccharides were dissolved
either in pure water or in a carrier consisting of a 2% guinea pig
serum solution in order to test the influence of foreign material.
The results obtained with solutions of polysaccharide from type
III pneumococcus in a 2% guinea pig serum have been summarized in
Table I. The antiserum solution was a rabbit serum diluted 1 to
10 in veronal buffer.

The tests were conducted as follows: small test tubes were
filled with 0.5 ml of the antigen solutions. The chromium plated
slides connected to either the negative or positive pole of a D.C.
source giving a constant current were inserted into the tubes along
with a platinum wire 0.07 cm in diameter connected to the other
pole of the D.C. source. After 1' the slides were washed, dried
and the thickness adsorbed measured. The slides were then always
connected to the positive pole and immersed in 0.5 ml of the immune
serum for 1'. After washing and drying, the increase in thickness
adsorbed was measured. The figures in column 3 of fig. I are the
thicknesses adsorbed from the antiserum in A units. After each
test carried out with a sample of the antigen solution, a control
test was conducted in a similar way using the same slide but with
no polysaccharide in the 0.5 ml of the carrier. Tests and their
corresponding controls have been bracketed in column 3. The thickness adsorbed from the antiserum in the control experiment was
smaller than that observed when some polysaccharide was present in
the carrier solution. The difference in the thickness adsorbed in
the test and the corresponding control was a measure of the specific
adsorption due to the polysaccharide (see column 4). It was important to use the same slide for both test and control, since the
optical thickness adsorbed from an immune serum varied from slide
to slide. It is apparent from the table that when the slides were
positively charged during the deposition of the polysaccharide, no
specific adsorption of antibodies could be observed for concentrations smaller than 10^{-4} g/ml of polysaccharide. However when the
slides were negatively charged specific adsorption of antibodies

could still be observed when the titer of the polysaccharide solution was as small as $\times 10^{-7}$ g/ml. At first sight, it would thus appear that the polysaccharide was positively charged. However, when the polysaccharide was dissolved in pure water no specific adsorption of antibodies could be observed if the slide had been negatively charged for the adsorption of the polysaccharide at very low concentrations (10^{-8} g/ml). In contrast, if the slide had been positively charged a specific adsorption could still be observed with aqueous solutions of polysaccharide as dilute as 10^{-13} g/ml. These results showed that the polysaccharide was negatively charged. Therefore, we are inclined to believe that when the slides were negatively charged in the presence of a carrier the adsorption of the protein of the carrier was so reduced as to permit the detection of an immunological reaction at greater dilutions of the antigen solutions.

Table I

Sign of charge on slide for antigen deposition	Concentration polysaccharide in carrier solution	Thickness adsorbed from antiserum in $\overset{\circ}{A}$	Specific thickness of adsorbed antibodies in $\overset{\circ}{A}$
+	2.1×10^{-4}	166	
			55
+	0	111	
+	2×10^{-5}	103	
			0
+	0	105	
–	2×10^{-5}	164	
			41
–	0	123	
–	1×10^{-5}	140	
			24
–	0	116	
–	1×10^{-6}	126	
			10
–	0	116	
–	1×10^{-7}	132	
			12
–	0	120	

Specific adsorption of antibodies as a function of concentration and polarity of the slide for the first electrodeposition.

When working with extremely dilute solutions of polysaccharide

the nature of the metalized surface played a very important role.
This had also been previously observed in the system growth hormone
antigrowth hormone rabbit serum. Certain batches of slides metalized apparently under the same conditions did not permit to observe
an immunological reaction at the greatest dilutions. However, an
adequate surface could be obtained if a metalized slide yielding
no results was coated with five monolayers of Ba stearate prior
to the electroadsorption of the antigen (8). It was important
to use the slides within 4-6 hours after the deposition of the
stearate layers. No positive results could be obtained with slides
coated 20 hr before they were used for an IEA test, that is the
thickness observed after the electroadsorption of the immune serum,
was the same on the part of the slide coated with antigen and
carrier and the part coated with carrier only. However, if the
slides were kept in a test tube in a water vapor saturated atmos-
phere they could be used at least 24 hr after their preparation.
Slides freshly coated with five layer of stearate lost immediately
their property for identification of immunological reactions if
heated at 50 for 5'. The same thing happened if the slides were
placed under vacuum for 10 min.

Most interesting results were obtained when we investigated
the reactions between the polysaccharide from types III and VIII
pneumococci and their respective rabbit antisera. For the experiments summarized in Table II the adsorption of the antigen as
well as of the antibodies were carried out without electric current.
It is known that a certain amount of cross reaction takes place
between these two systems. This is clearly evident from the first
row of the Table. The adsorption period of the polysaccharides
was 1' and that of the antisera 10' - the antisera were diluted
one to ten in veronal buffer pH 7.5.

The figures in columns 2,3,5 and 6 of the table are the thickness in A units adsorbed from the antisera against polysaccharide
III or VIII. It is evident that homologous reactions could still
be differentiated from the heterologous in the concentration
range of 10^{-6}g/ml of the polysaccharide. It is known that a certain amount of cross reaction takes place between these two systems. This is clearly evident from the first row of the table
which shows that when the polysaccharide solutions were concentrated (10^{-3}g/ml) very thick layers could be adsorbed from the
immune sera. These thicknesses would vary within large limits
depending on whether the slide had been dried or not after the
adsorption of the antigen. Layers nearly twice as thick were
adsorbed if the slides were kept wet after the adsorption of the
polysaccharide until they were immersed into the immune serum
solutions. This observation, which has great significance for
the mechanism of the adsorption will be more fully discussed in
a forthcoming paper.

Table II

Antiserum against polysaccharide from Type III pneumococcus			Antiserum against polysaccharide from Type VIII pneumococcus		
conc. poly g/ml	poly III Å	poly VIII Å	conc. poly g/ml	poly III Å	poly VIII Å
3×10^{-3}	278	180	3×10^{-3}	89	158
4×10^{-6}	55	43	2×10^{-4}	35	92
4×10^{-7}	30	27	4×10^{-6}	35	-
4×10^{-8}	-	23	8×10^{-7}	24	36
			1.6×10^{-10}	-	23

Specific adsorption of antibodies as a function of concentration of polysaccharides solutions without the use of an electric current. The figures under the headings poly III and poly VIII which refer to the type of polysaccharide used as antigen, are the thicknesses in Å adsorbed from antisera against either type III or VIII pneumococcus.

Table III

Rabbit antiserum against pneumococcus III			Rabbit antiserum against pneumococcus VIII		
conc. poly g/ml	poly III Å	poly VIII Å	conc. poly g/ml	poly III Å	poly VIII Å
3×10^{-12}	80	50	3×10^{-10}	98	122
3×10^{-13}	74	55	3×10^{-11}	94	115
3×10^{-13}	78	50	3×10^{-13}	108	120
3×10^{-14}	60	56	6×10^{-16}	77	86
0	60		6×10^{-18}	75	77

Specific adsorption of antibodies with the use of a current as a function of the concentration of the polysaccharides solutions. The figures under the headings poly III and poly VIII which refer to the type of polysaccharide used in the first electroadsorption, are the thicknesses in Å adsorbed from antisera against Type III or VIII pneumococcus.

The data of Table III show that when a current was used for the deposition of the antigen and antibodies an immunological reaction was still detectable with concentrations in antigen as small as 3×10^{-13} g/ml, seven orders of magnitude smaller than the limit attained without current, The slides were positively charged for both adsorption periods which lasted 1'.

Dilutions of the antiserum. All experiments reported so far were performed with slides sparsely coated with antigen and treated with relatively concentrated antiserum solutions (1 to 10 dilution). We investigated the effect of the dilution of the antiserum on the immunological reaction carried out on slides fully coated with antigen. The antigen solution contained 10^{-3} g/ml of polysaccharide, and no current was used for the adsorption period which lasted 1'. The adsorbed layer was roughly 5A thick. The antiserum adsorption lasted 2', with the slides positively charged as usual. Specific adsorption of 36 A, 17 A and 10 A were recorded on slides treated with rabbit antiserum (against the polysaccharide from type III pneumococcus) diluted 1/4000, 1/40,000 and 1/400,000 respectively. Such specific adsorption could not be obtained without current.

DISCUSSION

It is most remarkable that such minute quantities of poly-saccharide could be detected by this method. When the concentration of the polysaccharide solution used for the adsorption was 10^{-12} g/ml, the average thickness attained was at most 5×10^{-5} A which would be the thickness if all the polysaccharide molecules present in 0.5 ml had been adsorbed. On such a sparsely coated slide, 20 to 30 A of antibodies could be specifically immobilized. What is then the mechanism which permits to have a ratio by weight of antibodies to antigen of roughly 10^6? Whatever the mechanism, the conclusion seems inescapable that a very great number of anti-body molecules must be specifically immobilized by a single mole-cule of antigen. On could assume that the combination antigen antibody adsorbed on the surface could act as a nucleus around which a large number of antibody molecules could congregate es-pecially so under the influence of the electric current. A def-inite arrangement of the chromium atoms of the metalized surface seems important. As mentioned above it was only with certain types of metalized slides that an IEA test could be successfully conducted. Preliminary electron diffraction patterns obtained with the metalized surfaces have indicated that the crystals size of the chromium surface might play a key role in these interactions.

Finally it should not be forgotten that in the case of the polysaccharide, even without current, a very thick layer of anti-bodies close to 600 A could be adsorbed in a few hours on a densely packed layer of polysaccharide 5 A thick anchored to lay-ers of octadecylamine (2). No such thick layer would be formed

if the polysaccharide was not present. It is therefore not too
surprising if with the help of a weak electric current one can
detect the presence of polysaccharides in concentrations as small
as 10^{-13}g/ml.

These findings seem important from the physical as well as
the biological points of view. The forces involved in the building up of thick specifically adsorbed layers are not clearly
understood.

From the biological aspect, the immunological reactions
taking place at a solid-liquid interface might be much more subtle
than it would appear from the present theories based mostly on
experiments carried out in the liquid phase. Reactions at interfaces are of prime importance in biology.

REFERENCES

1.   Rothen, A., and Landsteiner, K. J., Exp. Med., 76, 437, 1962.
2.   Rothen, A., J. Biol. Chem., 168, 75, 1947.
3.   Rothen, A., Rev. Sci. Instr., 28, 283, 1957
     Rothen, A., in Physical Techniques in Biological Research,
        D.H. Moore, Ed., Vol. IIA Academic Press, 1968.
4.   Rothen, A., Bio-Med. Engineering, 2, 177, 1967.
5.   Mathot, C., Rothen, A., and Casals, J., Nature, 202, 1181
        1964.
6.   Mathot, C., Rothen, A., and Scher, S., Nature, 207, 1263,
        1965.
7.   Mathot, C., Rothen, A., and Scher, S., Experientia, 22, 833
        1966.
8.   Rothen, A., and Mathot, C., Immunochemistry, 6, 241, 1969.
9.   Rothen, A., Mathot, C., and Thiele, E., Experientia, 25,
        420, 1969.

ELECTROPHORESIS AND ADSORPTION STUDIES OF PROTEINS AND THEIR

DERIVATIVES ON COLLOIDS AND CELLS

D. J. Wilkins[1] and P. A. Myers
With the technical assistance of C. A. Warren

New England Institute for Medical Research

Ridgefield, Connecticut 06877

INTRODUCTION

The work to be described concerns the adsorption of some proteins and their derivatives at the solid/water interface. This study is derived from previous work on the modification of model particles by adsorption in an attempt to investigate the problem of what constitutes "foreignness" as far as the body is concerned. Specifically, a cell system in mammals, the reticulo-endothelial system identifies foreign particles from native ones and removes them by phagocytosis. The mechanism of this process is but poorly understood and previous work has been directed toward the initial step of recognition. This was studied using a model colloidal system of polystyrene latex (PSL) with its surface properties modified by adsorption of various macromolecules (1). It was found that such surface modification influenced the organ distribution of injected particles.

[1] Present address: Battelle Institute
7 Route de Drize
1227 Carouge/Geneva
Switzerland

The changes in surface properties were monitored by microelectrophoresis and during work on the change of electrophoretic
mobility of the PSL as a function of added macromolecule adsorption
behavior was found which strongly indicated orientation of the
adsorbing species. This communication describes some of this work
in detail in an attempt to understand the interaction of the adsorbing molecules with the substrate. In view of the invariable
presence of protein at the cell surface, indeed it is required by
the Danielli bilipid layer theory (2), it is felt that such a
study may be of interest. Further it is known that many enzymes
function at intracellular interfaces and the fact that they lose
their activity, both quantitative and qualitative, upon cell
destruction inclines one to inquire if this is not perhaps because
the orientation or conformation of the enzyme molecule is not
unfavorably altered.

                              METHODS AND MATERIALS

    Microelectrophoresis was carried out in an apparatus similar
to that described by Bangham et al (3). This consists of a
cylindrical cell with platinized platinum electrodes suspended in
a water bath at 25 ± 0.1°. Particles in the stationary layer were
observed via a microscope linked to a closed circuit television
system. All measurements represent the mean of at least ten particles timed in both directions and are expressed in terms of mobility
in microns/sec/v/cm rather than converting to ζ-potentials. Unless
otherwise mentioned all measurements were made in 0.145 M NaCl,
otherwise referred to as saline. Measurements of mobility as a
function of pH were made as follows: 25.0 mg. of protein or
derivative was dissolved in 4.0 ml. of NaCl solution. To this
stirred solution was added 1.0 ml. of 2% w/v PSL and stirring
continued at room temperature for at least one hour. At that time
0.2 ml. samples were removed and added to 20.0 ml. of saline with
pH adjusted with either 0.145 M HCl or NaOH. The pH of the suspension was then measured and the electrophoretic mobility determined. As was pointed out by Abramson (4), and as will become
clear later, it is very important to carry out determinations by
coating at a high concentration and then diluting for observation.

    The procedure for adding various concentrations of protein or
derivatives to various colloidal particles was as follows: One
ml. of a suspension of particles at a suitable concentration for
electrophoretic mobility determination was added to a series of
beakers containing the adsorbate in a suitable concentration range.
The total volume in each beaker was 20.0 ml. and the final salt
concentration 0.145 M NaCl. After adsorption at 30 minutes the pH
and then the electrophoretic mobility was determined. Except at

high concentrations the pH was constant over a large range of
concentrations with most of the materials used. If necessary
(v. infra) the pH was adjusted with either 0.145 M NaCl or 0.145
M HCl. The following colloidal particles were used as adsorbent
in the protein adsorbtion studies.

1. Polystyrene latex (PSL), diameter 1.099 microns, kindly
supplied by Dr. John W. Vanderhoff, Dow Chemical Co., Midland,
Michigan. This material was extensively washed with distilled
water by Millipore filtration.

2. Methylated polystyrene latex (MPSL) is the above material
methylated as follows: One g. of PSL was filtered, washed with
methanol, and resuspended in 100 ml. methanol and 0.85 ml. concentrated HCl and after stirring overnight the material was filtered
and washed five times with 200 ml. of water and suspended in 50.0
ml. of water.

3. Powdered silica is an ultrapure sample kindly provided by
Dr. Vandergrint of Corning Glass, Corning, New York. The material
was suspended in distilled water and centrifuged. The very fine
particles were removed leaving a sample of average size 1.0 μ
diameter (estimated).

4. Sheep red blood cells (SRBC) were fixed with acetaldehyde
as described by Heard and Seaman (5). It has been shown (5) that
such cells behave electrophoretically like fresh cells and are
useful since no complications can arise from hemoglobin released
from lysed cells under extreme environmental conditions.

5. Mineral Oil (Drakeol 6-VR) kindly supplied by Pennsylvania
Refining Company, Butler, Pennsylvania, was emulsified by adding
0.5 ml. to 50.0 ml. of water and sonifying for 30 secs at position
6 using a Branson Sonifier.

Most of the proteins and their derivatives were prepared by
similar means and hence it is not necessary to describe all the
preparations in detail. The gelatin used was acid extracted por-
cine material (kindly supplied by Ucopco Division, Wilson and Co.,
Chicago, Illinois). Bovine plasma albumin (BPA) powder, was
fraction V material obtained from Armour Pharmaceutical Co., Chicago,
Illinois. Polylysyl gelatin (PLG) was prepared as follows: ε-N-
trifluoro acetyl-L-lysine was prepared (6) and converted to its
carboxy amino anhydride by bubbling phosgene through a solution
in ethyl acetate (7). After recrystallization from ethyl acetate
the anhydride in dioxane solution was reacted with an aqueous

buffered solution of gelatin (8). The protective trifluoro acetyl
groups were removed using aqueous piperidine and the final material
dialyzed exhaustively against distilled water, millipore filtered
(0.45 μ diam.) and lyophilized. Poly-L-ornithyl gelatin (POG) and
poly-L-lysyl bovine plasma albumin (PLBPA) were prepared in a
similar manner. Polylysine (PL) was prepared from the ε-N-trifluoro
acetyl-α-N-carboxyl-L-lysine anhydride by polymerizing with
triethylamine (9). The material was then processed as for PLG
(v. supra). Methylated bovine plasma albumin (Me BPA) was obtained
by esterifying using methanol and HCl (10).

The adsorption isotherm of PLG on PSL was determined using
I^{131}-labelled PLG (11). A PSL suspension containing 5.5 mg. of
PSL was added to a series of cellulose nitrate (0.5 x 2.5 inches)
centrifuge tubes containing PLG solutions in a suitable range of
concentrations. The total volume in each tube was 5.0 ml. and the
final salt concentration 0.0145 M NaCl. After equilibration for
30 minutes at 37° the tubes were centrifuged at 35,000 x g for
15 minutes and washed several times with 0.0145 M NaCl solution
until the PSL pellet showed no further change in counts. Microelectrophoresis was also carried out on suspensions of the washed
PSL in 0.0145 M NaCl.

For injection into rats I^{125}-labelled PSL (1) was suspended
in a range of concentrations of PLG in 0.145 M NaCl. This material
was filtered by membrane filtration (Millipore Corporation, Bedford,
Mass.) to remove excess PLG and resuspended in 0.145 M NaCl with
gentle sonification. Injections were carried out in Nembutal-
anesthetized rats and distributions and rate of clearance from the
blood determined as previously described (1). The results were
expressed as the percentage of the injected dose accumulated in
the spleen after 15 minutes.

RESULTS

Determinations were made of the electrophoretic mobility in
saline of all of the particles used in these studies, as a function of pH. The results are shown in figure 1 and illustrate the
different surface nature of the materials used.

The value for the mobility of acetaldehyde treated SRBC's (-1.40
at pH 7.5) is very similar to that found by Seaman et al (5).
Although there is a certain amount of controversy concerning the
ionogenic groups, at the surface of PSL (12,13) the fact that the
mobility is so markedly reduced by esterification strongly suggests the presence of carboxyl groups.

The electrophoretic mobilities of all proteins and derivatives
used after adsorption on PSL are shown in figures 2 and 3 as a
function of pH.

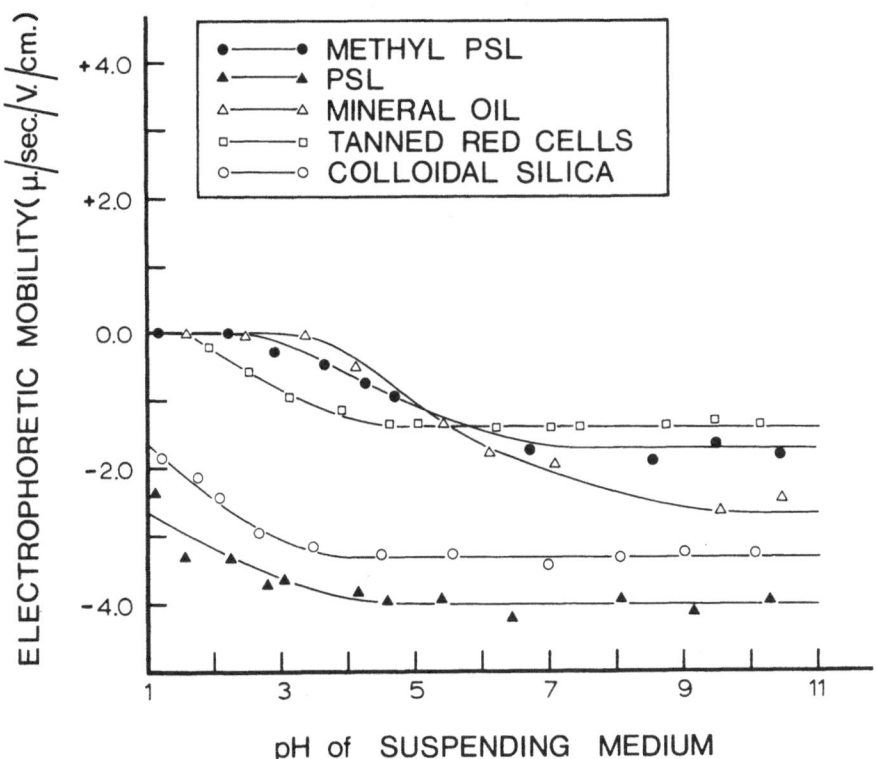

Fig. 1.   Electrophoretic mobility of the particles used as a
function of the pH of the suspending medium (0.145 M NaCl).

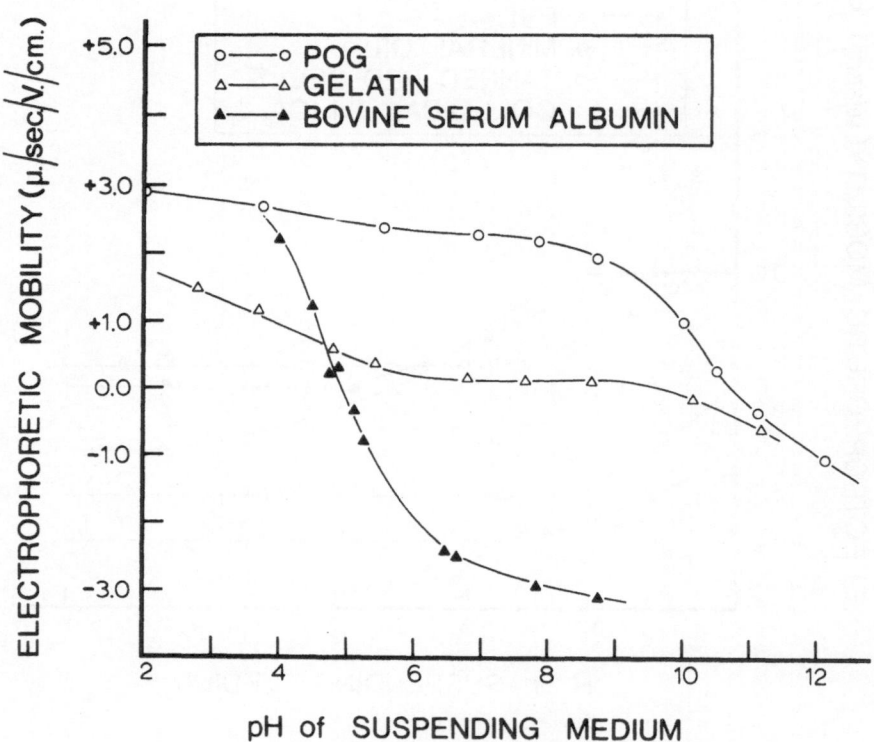

Fig. 2. Electrophoretic mobility of some proteins and derivatives adsorbed on PSL as a function of the pH of the suspending medium (0.0145 M NaCl).

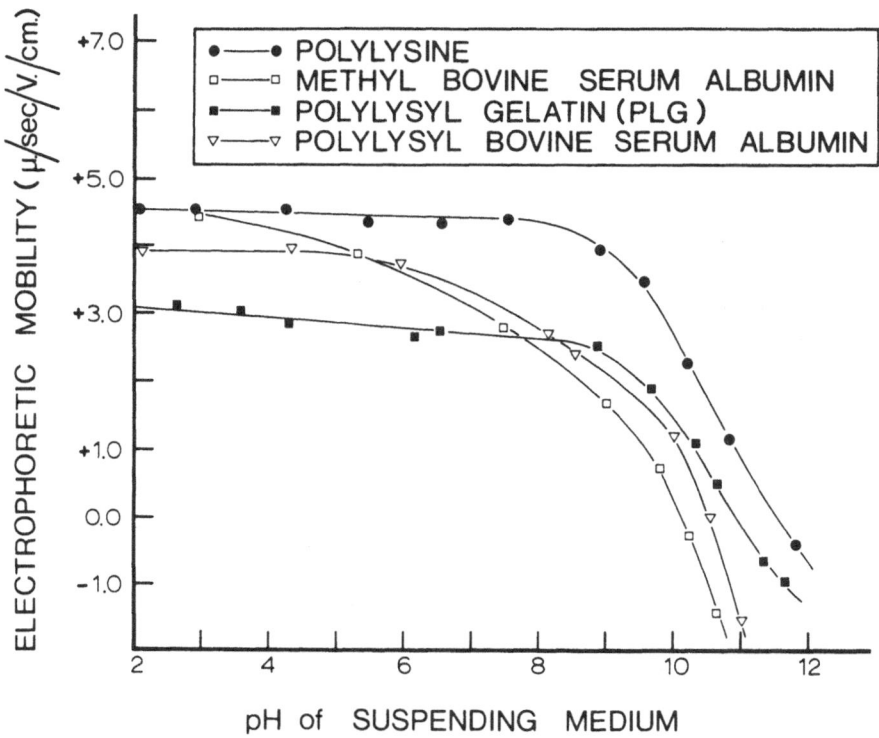

Fig. 3. Electrophoretic mobility of some proteins and derivatives adsorbed on PSL as a function of the pH of the suspending
medium (0.0145 M NaCl).

It can be seen that the particular type of gelatin used is near
isoelectric over the pH range 6-10, a rather unusual behavior for
a protein. Addition of either lysine or ornithine to gelatin can
be seen to increase the positive mobility and the isoelectric
point. The pH-mobility curve for adsorbed BPA is very similar in
shape and isoelectric point to values obtained using the soluble
material and a moving boundary method (4). Addition of lysine or
indeed esterification with methanol again increase the positive
mobility and isoelectric point. Polylysine due to the absence of
any acidic groups has the highest positive mobility and isoelectric
point.

Figure 4 shows a comparison of the adsorption isotherm of PLG
on PSL together with the electrophoretic mobility of the samples.

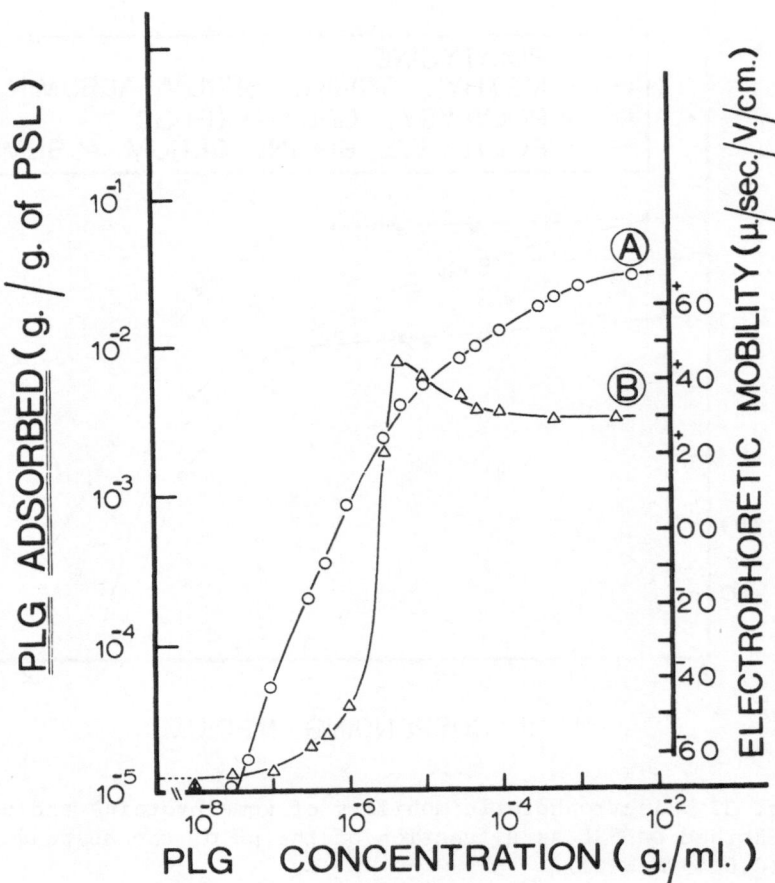

Fig. 4.   Adsorption and electrophoresis of PLG on PSL as a
function of PLG concentration; curve A- amount of PLG bound by PSL
after centrifuging and exhaustive washing; Curve B- electrophoretic
mobility of the washed PSL/PLG from A. Both adsorption and electrophoresis in 0.0145 M NaCl.

It will be noted that the electrophoretic uptake (curve B) shows a
distinct maximum positive mobility and then decreases to a constant
value. It is this maximum in the mobility which is of interest,
since it presumably reflects a particular orientation of the PLG
molecules upon adsorption. In order to see whether the effect was
only obtained by adsorbing PLG on PSL adsorption and electrophoresis
was also carried out on the particles examined in figure 1. The
results are shown in figures 5 and 6, and it is immediately obvious

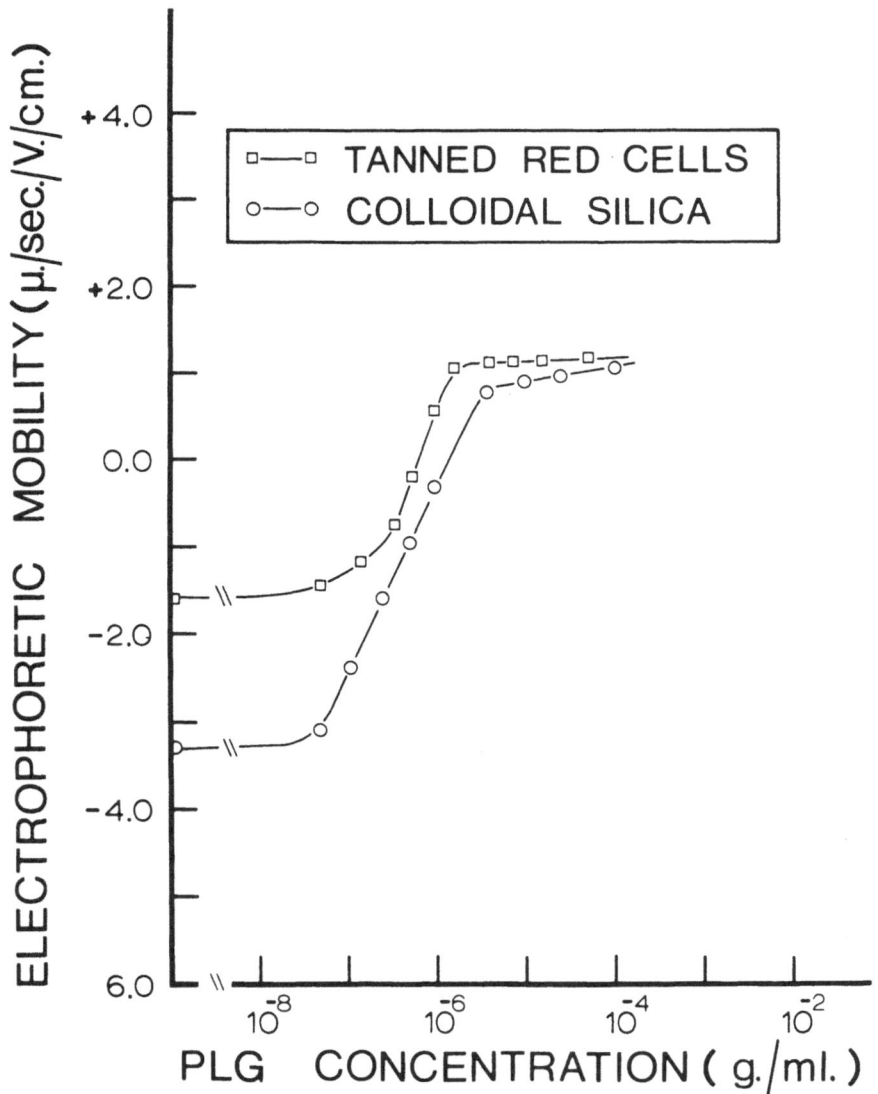

Fig. 5. Electrophoretic mobility of various particles as a function of PLG concentration in the suspension (0.145 M NaCl).

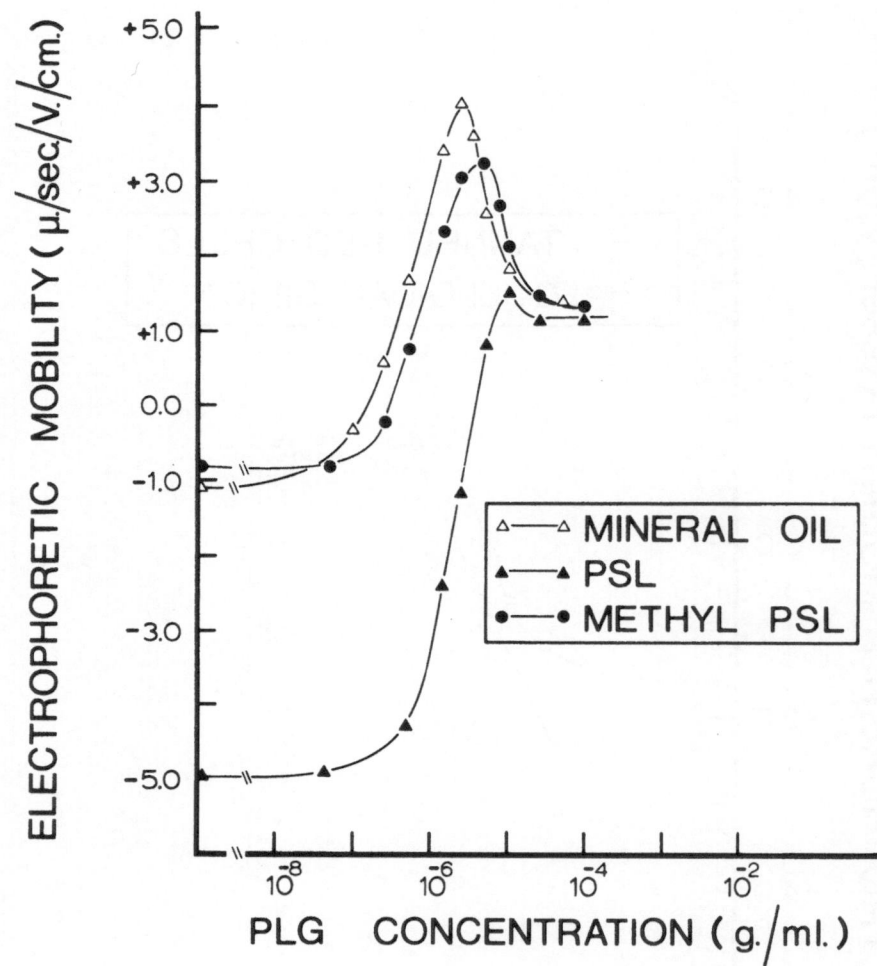

Fig. 6. Electrophoretic mobility of various particles as a
function of PLG concentration in the suspension (0.145 M NaCl).

that a maximum in mobility is not obtained on all substrates.
The electrophoretic adsorption behavior on different particles of
the proteins and derivatives (the pH-mobility characteristics of
which are shown in figures 2 and 3) was also examined and the
results are shown in figures 7 and 8.

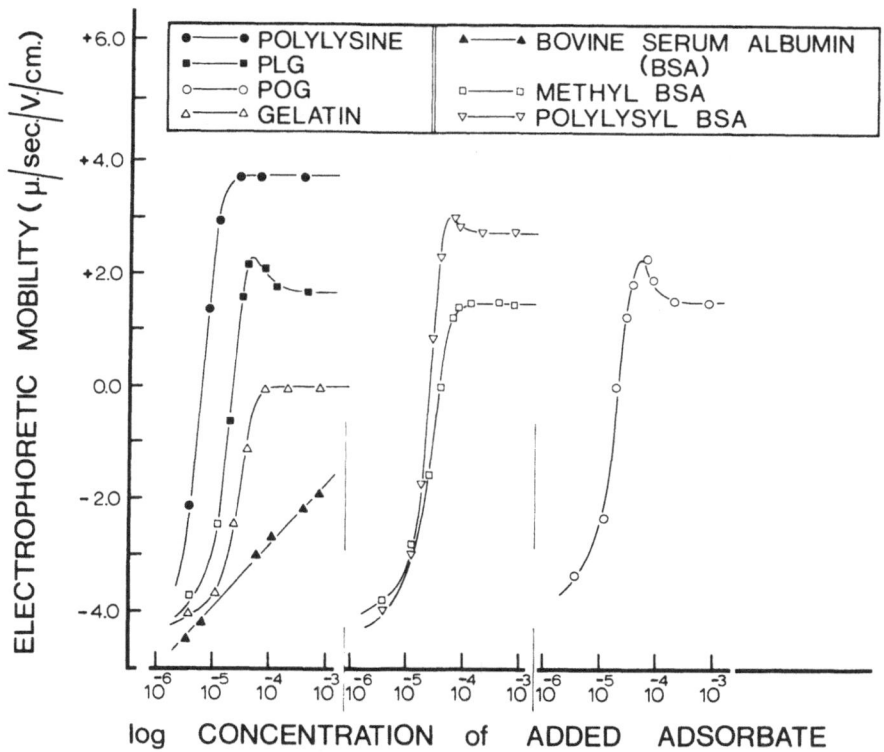

Fig. 7. Electrophoretic mobility of various proteins and their derivatives adsorbed on PSL as a function of the concentration of the adsorbate (0.145 M NaCl).

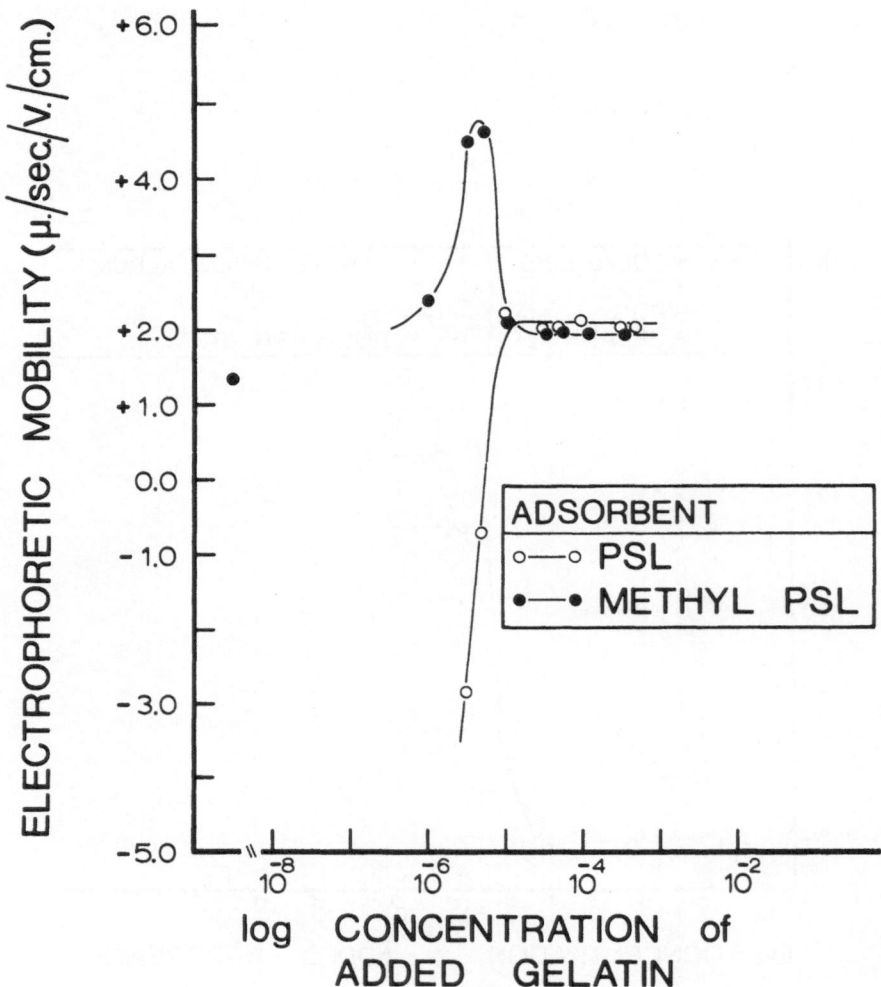

Fig. 8.   Electrophoretic mobility of PSL and Methyl PSL as a function of gelatin concentration at pH 2   (0.145 M NaCl adjusted with 0.145 M HCl).

These results demonstrate that the anomalous high positive mobility found with PLG is not limited to lysyl derivatives of proteins or amino acid derivatives of gelatin. In fact with a protein like gelatin under conditions of low pH the effect is seen on MePSL. Gelatin was chosen for this experiment since it is obviously not denatured at pH 2. Figure 9 shows the results of a typical intra-venous clearance experiment, in which the percentage of the injected

dose accumulated in the spleen 15 minutes after injection is
plotted as a function of the coating concentration of PLG.
It will be observed that the biological behavior is in fact
reflected in the electrophoretic characteristics of the particle.

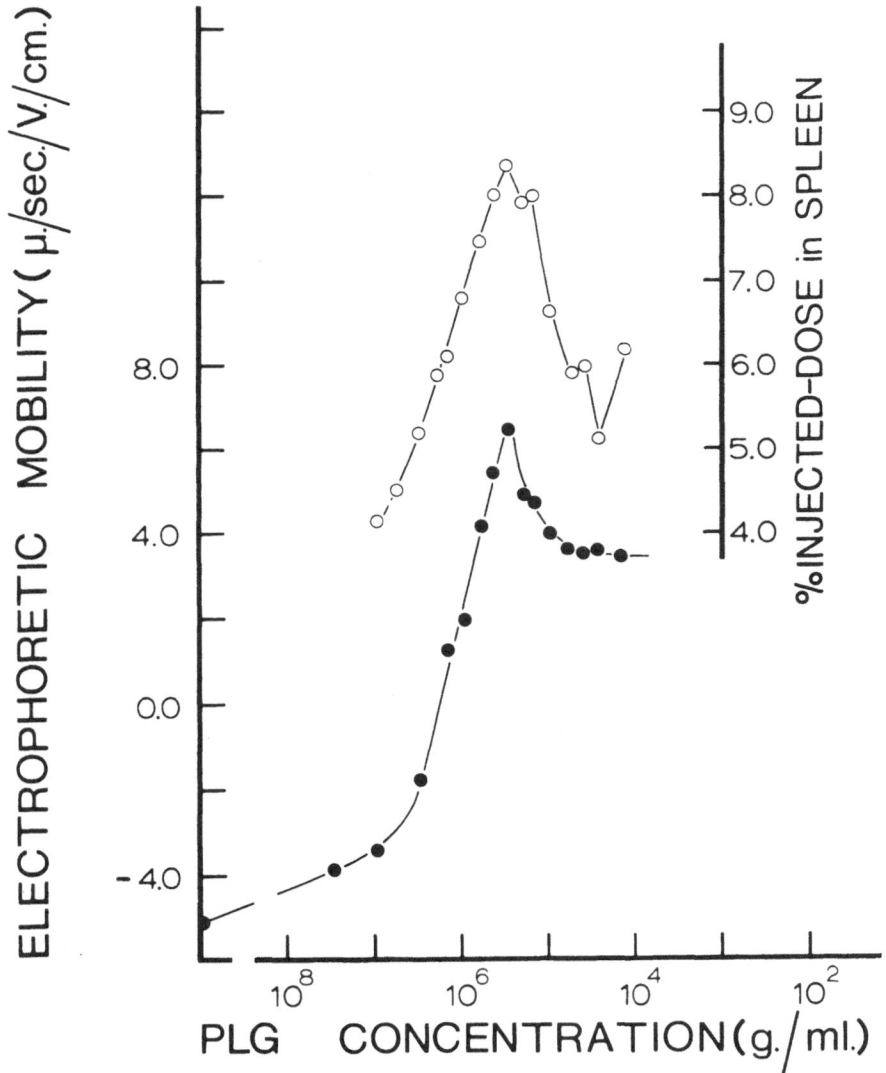

Fig. 9. Electrophoretic mobility and amount of PSL taken up
by the spleen fifteen minutes after injection as a function of PLG
concentration: -o- mobility of PSL/PLG particles; -o- uptake by
the spleen of PSL/PLG particles. Injected dose, 0.73 mg. PSL/
100 g. rat (0.145 M NaCl).

DISCUSSION

It is hoped that this presentation has demonstrated the use-
fulness, and some of the drawbacks, to the technique of microelectrophoresis for protein studies. As with most surface chemistry
of this sort the actual interpretation of what is happening at the
molecular level is not easy. Furthermore as has been noted in the
past (20) and recently by Mathot and Rothen (14) there is not an
extensive literature upon which to draw in the area of the adsorption of proteins onto solids. Loeb (15) and Abramson (4) demonstrated just how useful microelectrophoresis of adsorbed proteins
could be and how well the results compared to those obtained by
other means. Since then the importance of consideration of electrophoretic surface change has been shown to be of great useful-
ness in a variety of biological problems. Bangham (16,17) has
demonstrated the importance of surface charge in blood clotting and
in certain enzyme reactions and Bedford has demonstrated correlations of the maturation of sperm with their surface charge (18)
while the work of Seaman (5), Weiss (19) and others has added
greatly to our knowledge of the nature of cell surfaces, both
normal and neoplastic. Hopefully some of the material presented
here will clarify some of the ideas on the electrophoresis of
adsorbed protein and point to some of the precautions necessary for
correct interpretation of data. It is immediately obvious that the
surface nature of the colloid particle can affect the electrophoretic
results and that the concentration of added protein or polymer is
very important. It was pointed out by Abramson (4) that it is
preferable to coat at a high concentration to achieve adequate
coverage and then dilute to a suitable concentration for observation. It is now obvious that this is also necessary to avoid the
possibility that the protein covered surface will show an abnormal
high mobility since it can be observed (e.g., fig. 6) that at high
enough concentrations the mobilities all reach a similar value.
However, at high enough concentrations the choice of particle is
unimportant since any interaction is damped out. Of course at
high protein concentrations consideration has to be given to
possible changes in viscosity and pH which would lead to mobility
changes. Another useful feature of the method is its ability to
detect changes in protein chemistry or particle surface chemistry.
Thus for example the effects of methylation of PSL were demonstrated by figure 1 and of adding lysine to gelatin by figures 2
and 3.

The high positive mobility of basic protein derivatives on
PSL, for example, at certain concentrations is interpreted as a
configurational change due to interaction with the substrate
surface. If the interaction were purely electrostatic then since
all the surfaces studied are electrophoretically negative one

would expect basic groups from the protein to be as close as possible to negative sites on the surface leaving a preponderance of negative groups in the plane of shear and hence no excessive positive mobility. Since this is in fact the opposite of what occurs, the proposal is made that anionic groups of the proteins adsorb preferentially to the largely hydrophobic PSL surface exposing an excess of basic groups in the plane of shear and a consequent elevated positive mobility. This hypothesis is supported by several facts: (i) anions due to their lower hydration adhere more readily on hydrophobic surfaces (20), (this is also demonstrated by the fact that mineral oil becomes negative at higher pH); (ii) increasing the hydrophobic nature of the PSL surface increases the positive mobility, and using a completely hydrophobic surface such as mineral oil emulsion shows an even higher positive mobility; (iii) polylysine which has no anionic groups does not show an increased positive mobility; (iv) hydrophilic surfaces such as silica or red cells do not show the effect. From figure 4 it can be seen that at the point of highest positive mobility, the coverage is approximately 1.0 mg. PLG/ m^2 of PSL, very similar to the coverage determined by several workers (21) for a protein "monolayer" at the air/water interface. As further PLG is added the mobility decreases again, presumably due to the multi-layer formed "damping" out the effect of the substrate surface. A similar effect was noted by Matijevic and Ottewill (22) who showed a periodic change in electrophoretic mobility of silver halide sols as a function of added cationic detergent. This was ascribed to successive layers of detergent having opposite orientations on a charge basis which was damped out at higher concentrations. It is interesting to note that PLG does not show a Langmuir isotherm on PSL but forms a multilayer, presumably because the PLG-PLG interaction is stronger or more slowly reversible than the PLG-H$_2$0 interaction. It has been noted that although proteins usually show a Langmuir adsorption isotherm, at least on inorganic substrates (23) theoretically at least, they should not, since the adsorption is for all practical purposes irreversible. One final effect of the increased positive mobility should be noted. The elevated positive mobility shown in figure 4 is of a material which after adsorption was filtered, washed and resuspended while those shown, for example, in figure 6 are particles suspended in the adsorbate solution. This suggests that the presumed conformational change is very rigid, and able to withstand fairly harsh physical treatment.

In view of the importance of interfaces in biology, both intracellular and extracellular, it is interesting to speculate on the biological significance of such orientations. It has been shown previously that the distribution of an intravenously administered colloid is influenced by the surface character of the

injected colloid. Electrophoretically positive colloids showed an
increase in splenic accumulation over negative colloids (24). Thus
the fact that the spleen can recognize far more subtle differences
in surface characteristics (see fig. 9) as shown by how precisely
the spleen uptake mirrors the anomalous adsorption behavior takes
on an added interest. Of course, "recognition" of self from non-
self almost certainly is more complex than differentiating between
positive and negative surfaces and only when more is understood
about the complex interactions between surfaces and blood elements
is the problem likely to be completely resolved (25).

It is interesting to speculate on a few further possible
implications of oriented proteins on hydrophobic surfaces. It is
well known that Danielli's (2) model for the cell membrane depends
upon the presence of an exterior coating of globular protein. Yet
it has recently been pointed out by Weiss (26) that electrophoretically it is impossible to detect $-NH_2$ groups at the cell surface.
It is tempting to think of some lipid-protein interaction similar
to that described above possibly accounting for this anomaly,
although preliminary experiments are not very encouraging.
Certainly, as Wallach et al (27) have pointed out, the interaction
at the cell surface between proteins and lipids is almost entirely
hydrophobic. Although doubts have been raised about the validity
of the bilipid layer (28) model it certainly accounts for some of
the properties of cell membranes. What is rather surprising is
that so little attention has been paid to the presence (or absence)
of protein exterior to the lipid layers.

Finally, when cells are destroyed the activity of their enzyme
systems diminishes, both quantitatively and qualitatively (29).
It has been suggested, from this and other arguments, that many
intracellular enzyme reactions take place at interfaces. Certainly
with phospholipases the surface charge is very important in the
enzyme substrate interaction and hence in the rate of hydrolysis (17).
It is possible that enzymes at interfaces are adsorbed and interact with the adsorbent layer in a manner sufficiently similar to
that discussed above that this might be a worthwhile field of
study.

ACKNOWLEDGMENTS

The work was supported in part by grants from The John A.
Hartford Foundation, Inc., the National Institutes of Health
(Grant Number AL 07161-02), and the Office of Naval Research
(Grant Number Nonr-4170 (00, NR 301-740)).

BIBLIOGRAPHY

1. Wilkins, D. J. and Myers, P. A., Brit. J. Exp. Pathol., _47_,
    568 (1966).

2. Danielli, J. F. and Harvey, E. H., J. Franklin Inst., _214_, 1
    (1932).

3. Bangham, A. D., Flemens, R., Heard, D. H. and Seaman, G.V.F.,
    Nature, _182_, 642 (1958).

4. Abramson, H. A., Moyer, L. S. and Gorin, M. H., "The Electro-
    phoresis of Proteins", Reinhold, N. Y. (1942).

5. Heard, D. H. and Seaman, G.V.F., Biochim. Biophys. Acta, _53_,
    336 (1961).

6. Sela, M., Arnon, R. and Jacobson, I., Biopolymers, _1_, 517
    (1963).

7. Fasman, G. D., Idelson, M., and Blout, D. R., J. Amer. Chem.
    Soc., _83_, 709 (1961).

8. Sela, M. and Arnon, R., Biochem. J., _75_, 91 (1960).

9. Daniel, E. and Katchalsky, E., in "Polyamino Acids, Poly-
    peptides and Proteins", Ed. M. A. Stahman, Univ. of Wisconsin
    Press (1962).

10. Fraenkel-Conrat, H. and Olcott, H. S., J. Biol. Chem., _161_,
    259 (1945).

11. Bocci, V., Intern. J. Appl. Radiation Isotopes, _15_, 449 (1964).

12. Ottewill, R. H. and Shaw, J. N., Kolloid-Z.U.Z. Polymere,
    _218_, 34 (1967).

13. Van den Hul, H. J. and Vanderhoff, J. W., J. Coll. Interfac.
    Sci., _28_, 336 (1968).

14. Mathot, C. and Rothen, R., 43rd National Colloid Symposium
    (Preprints), p. 105, Case Western Reserve University,
    Cleveland, Ohio, June 1969.

15. Loeb, J., J. Gen. Physiol., _6_, 116 (1923).

16. Bangham, A. D., Nature, _192_, 1197 (1961).

17. Bangham, A. D. and Dawson, R. M. C., Biochim. Biophys. Acta,
    _59_, 103 (1962).

18. Bedford, J. M., Nature, 200, 1178 (1963).

19. Weiss, L. and Ratcliffe, T. M., J. Nat. Cancer Inst. 41,
     957 (1968).

20. Bull, H. B., "Physical Biochemistry", John Wiley, N. Y. (1943).

21. Cheesman, D. F. and Davies, J. T., Advan. Protein Chem., 9,
     439 (1954).

22. Matijevic, E. and Ottewill, R. H., J. Colloid Sci., 13, 242
     (1958).

23. Neurath, H. and Bull, H. B., Chem. Revs. 23, 391 (1938).

24. Wilkins, D. J., J. Colloid Interfac. Sci., 25, 84 (1967).

25. Wilkins, D. J., "The Reticuloendothelial System and Athero-
     sclerosis", p. 25, Plenum Press, N. Y. (1967).

26. Weiss, L., Bello, J. and Cudney, T. L., Int. J. Cancer,
     795 (1968).

27. Wallach, F. H. and Gordon, A., Federation Proc., 27, 1263
     (1968).

28. Korn, E. D., Science, 153, 1491 (1966).

29. Hendler, R. W., "Protein Biosynthesis and Membrane
     Biochemistry", John Wiley & Sons, Inc. (1968).

SURFACE CHEMICAL FEATURES OF BLOOD VESSEL WALLS AND OF SYNTHETIC MATERIALS EXHIBITING THROMBORESISTANCE

R. E. Baier

Applied Physics Department, Cornell Aeronautical

Laboratory of Cornell University, Buffalo, NY 14221

R. C. Dutton

Laboratory of Technical Development, National Heart

Institute, Bethesda, Maryland 20014

V. L. Gott, Department of Surgery, Johns Hopkins

Hospital, Baltimore, Maryland 21205

INTRODUCTION

Our problem can be stated simply: What surface properties are required for candidate biomedical materials to be as non-thrombogenic as possible? This symposium paper relates recent progress along three converging paths we have taken while attempting to solve this problem. First, we report qualitative results from in vitro and in situ surface chemical measurements on blood vessels which were carried out by Dutton and Baier at the National Heart Institute. A more quantitative treatment of these data will be published separately. Second, we report at some length the results of Gott, Dutton and Baier -- obtained during two years of collaborative effort -- which suggest an explanation for the excellent thromboresistance observed with prosthetic implants of the metallic alloy Stellite. Third, we briefly review the surface chemical investigation of candidate biomaterials now being prosecuted by Gott and Baier with the support of the Artificial Heart Program of the National Institutes of Health.

These studies rely heavily on contact angle measurements and the critical surface tension concept identified with Zisman's laboratory (1), which has recently been extended to bioadhesional problems (2, 3), and are supported by the surface-specific spectroscopy allowed by new internal reflection techniques (4).

BACKGROUND ON THE APPLICABILITY OF SURFACE CHEMICAL CONCEPTS TO BIOMEDICAL PROBLEMS

The maintenance of favorable blood flow depends upon the lack of adhesion between circulating blood elements and vessel walls (or synthetic materials) with which they make intimate contact. Careful examination of the formation, structure and properties of biological interfaces is required to understand these phenomena, eventually. The more urgent task is to discover which known mechanisms assist or impede biological adhesion, since lack of this information is a limiting factor in producing biocompatible prosthetic implants and nonthrombogenic extracorporeal circuits. Current efforts attempt to extend technical data on adhesive behavior to bioadhesional problems by presenting some basic relations between the surface properties of materials and adhesion, and outlining the useful comparative concept of "critical surface tension" (2, 3). There is little doubt of the importance of these relations when considering some initial interactions at biologically interesting surfaces. Many illustrations can be drawn from the large relevant literature on the role surface phenomena play in determining adhesion in biomedical environments.

The extreme localization of surface forces has been emphasized (5, 6, 7). Examples are plentiful that these physical forces alone, without the need for chemical bonding, are sufficient as a basis for good adhesion (8, 9). Wetting and spreading considerations have been reviewed (5, 11), and the point made that careful attention to the relations between contacting phases is required (2, 3). A parameter called "critical surface tension (γ_c)", derived from contact angle data, allows the empirical ranking of the relative surface energies of organic materials (12). "Critical surface tension" is a unique characteristic of the solid surface, and is readily correlated with the actual outermost chemical constitution of most substances. For example, fluorocarbon surfaces are of intrinsically low surface energy, have low γ_c values, and show poor adhesive qualities; surfaces largely composed of oxygen and nitrogen atoms are of relatively high energy, have higher critical surface tensions, and form good adhesive bonds. Organic surfaces are thus conveniently grouped into high-energy and low-energy categories, and the wetting, spreading, and adhesive behavior on each may be differentiated. The major influence of water on biological adhesion has been suggested to be a reflection of its conversion of intrinsically high energy surfaces into

considerably lower energy surfaces (13, 14, 15). As an illustration
of the important role which proteins must play at biological interfaces, the wetting properties of some simpler polyamides have
been given (16, 17, 18). Wetting studies on the nylons, for instance,
using carefully purified diagnostic liquids showed that hydrogen-
bonding across the solid-liquid interface is manifested at polyamide
surfaces (16, 17). Polyacrylamide, a model polymer with great
affinity for water, shows slightly altered wetting properties,
reflecting the presence of adsorbed moisture (18, 19); but this water
does not mask the potential H-bonding contribution of the amide
groups to adhesion. The wetting properties of a model polypeptide,
polymethylglutamate, having various chain conformations show
that the H-bonding functionality can be completely masked when
certain polymer configurations and side chain arrangements occur,
however (18). These concepts are applicable to complex protein
molecules as demonstrated in recent surface chemical investigations
of collagen and its denatured product, gelatin (16, 20).

Recognition of the role of surface forces in biological
adhesion is exemplified in studies of cell-to-cell adhesion (21, 22),
and cell-to-foreign surface adhesion (23, 24). Potential correlations
of biological interactions with critical surface tension (γ_c) are
found in results from tissue culture observations (25), blood
contact studies (26, 27), and cell filtration through prepared bead
columns (23). The major influence of adsorbed films has been
indicated (3, 28). There is growing evidence for "conditioning"
films of adsorbed protein as a prerequisite for cell adhesion to
foreign surfaces (29, 30, 31, 32). The often overlooked role of
adventitious surface-active adhesion modifiers has been brought
out (3, 19), and attention directed to the potential utility of deliberately applied surface-active species as coupling agents for biological
joints (13, 33). One class of such agents might be used to displace
or react with the water already present at biological surfaces (33).
Despite the apparently direct carry-over of current knowledge of
adhesion to biomedical problems, a number of complications still
exist. Among these are disputed theoretical objections, which
suggest stable adhesion even in the absence of intimate molecular
contact (34), and consideration of the auxiliary role of metabolic
processes (35).

Even in the presence of such difficulties, contact angle
studies provide a useful body of required data on the wettability and
reactivity of materials of biological interest. The critical surface
tension concept, as used here, has been amply validated during
two decades of its application to well-defined organic polymers.
Major influences of surface chemical constitution and of the
presence of adsorbed species are admirably reflected in the γ_c
parameter. For example, the results reported here indicate that
fully hydrated vessel interfaces have general surface characteristics
of poor wettability by most organic substances but easy penetrability

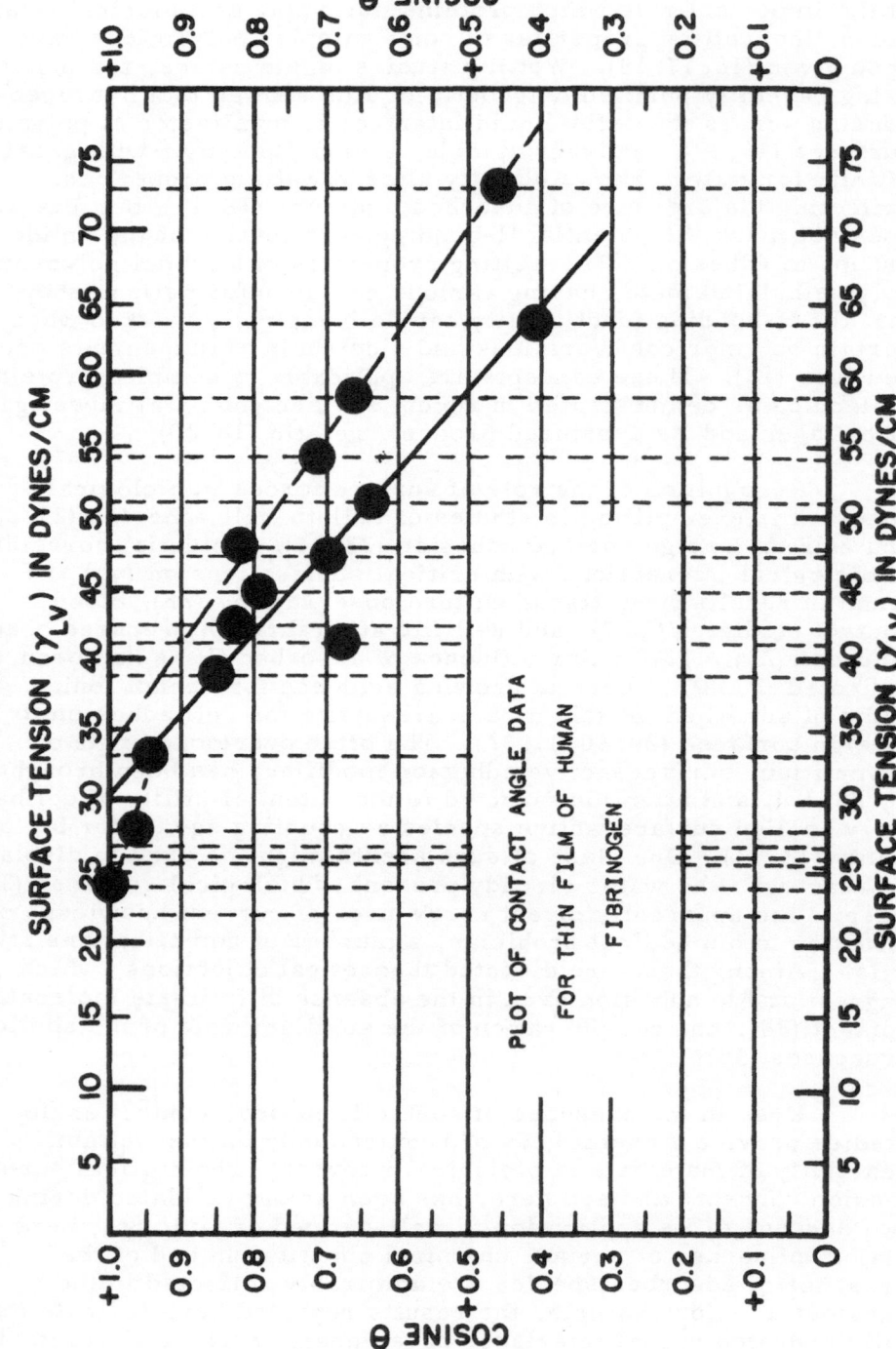

Figure 1

Figure 2

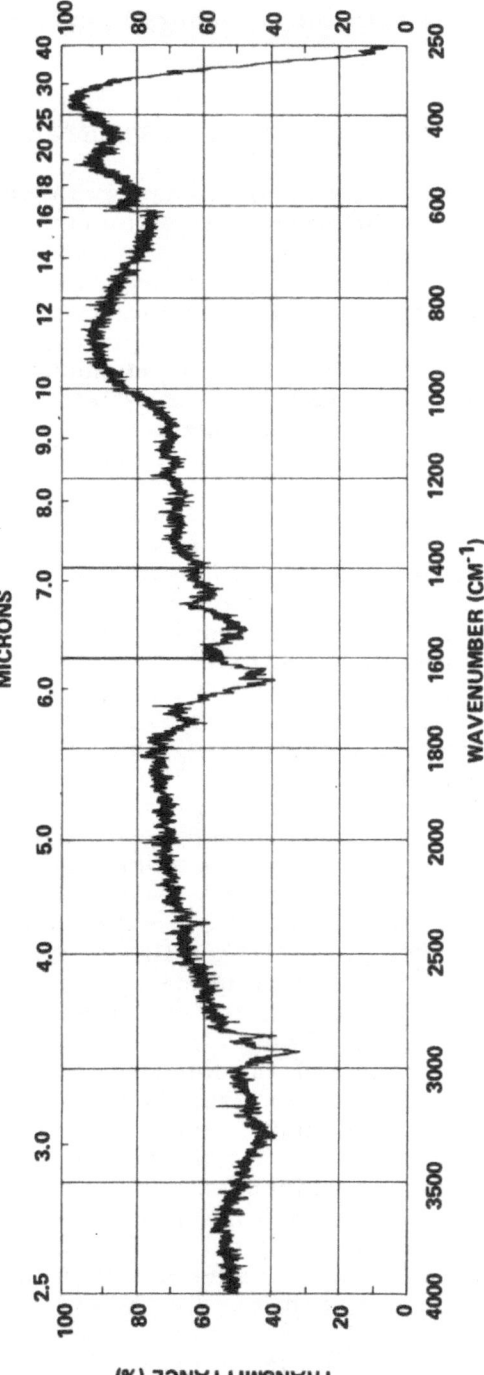

BLOOD VESSEL INTIMA - MAIR

by H-bonding substances. Our results also indicate the importance of adsorbed protein films and of low-energy surfaces in determining thromboresistance.

"WETTABILITY" OF BLOOD VESSEL WALLS

There have been clear disagreements among both Western (36,37,38) and Russian publications (39,40) about the "wettability" of vessel endothelium and its influence on the state of fluidity of blood. There is general agreement, however, that knowledge of the actual surface properties of the inner walls of natural blood conduits would prove valuable in our attempts to mimic nature by providing synthetic nonthrombogenic flow channels. We used contact angle measurements and MAIR (multiple attenuated internal reflection) infrared spectroscopy of excised veins as initial methods to partially reveal their surface chemical makeup.

The type of data produced by these techniques may be judged from Figures 1 and 2. Figure 1 is a plot of contact angle data obtained with a film of human fibrinogen (Fisher Scientific, Purified Fraction F1, Human Fibrinogen Powder), and represents as complex a situation as commonly encountered with biological surfaces. The solid line through the data points (plotted as cosine of the average contact angle Θ versus surface tension of the liquid used γ_{LV}) represents a reasonable trend for the liquids which show stable equilibrium advancing contact angles on the protein surface. These liquids are primarily halogenated organics. The upper dashed line suggests a separate trend for polar, penetrable hydrogen-bonding liquids such as water and glycols. Failure of the H-bonding liquids to cluster more closely about the line drawn through the nonpenetrable, organic liquid data generally indicates the availability of some H-bonding sites in the surface under study (16, 17, 18, 41). The lower dashed line at the upper left of Figure 1 shows the anomalous incomplete spreading for low surface tension hydrocarbon liquids associated with the presence of adsorbed water on the surface being studied (13, 14, 15, 16, 19). In any case, the line drawn through the most reliable data points, when plotted in this format, is used to extrapolate to an intercept with the cos Θ = 1 axis. This intercept is then defined as the approximate critical surface tension (γ_c) for the surface under investigation. This plot for human fibrinogen has been used to illustrate the maximum scatter of data which must be tolerated; it also serves to illustrate that the estimated critical surface tension for purified fibrinogen films is in the range between 30 and 35 dynes/cm. An even higher γ_c value had been found with bovine fibrinogen (29), although the infrared spectra of human fibrinogen films were essentially the same as that published for bovine fibrinogen samples (29). So, in addition to serving an illustrative role, Figure 1 (together with the previous published data (29)) shows that

critical surface tensions reported below for the vessel intima --
which are substantially lower -- are not artifactually influenced
by adsorbed fibrinogen layers masking the actual intimal surfaces.

Figure 2 is the MAIR infrared spectrum typical of the inside
surfaces of excised, thoroughly rinsed (in physiologic saline), and
dried jugular veins. This spectrum is excellent in quality, and
shows that the vessel intimal surface, selectively analyzed to
only a micron or so in depth, is definitely proteinaceous in
composition. Spectral evidence for mucopolysaccharides in this
outermost layer is scant. The strong absorption bands at about
2900 cm^{-1} and 1750 cm^{-1} in Figure 2 are associated with hydrocarbon content and ester components, respectively. These bands
show unambiguously that the vessel intima is abundantly supplied
with lipids, probably admixed and associated with the protein
components. Long chain fatty acids, for instance, are molecules
which show specifically similar infrared spectra.

Knowing that fatty acids and other lipids are surface-active
compounds, and preferentially orient at interfaces to mask higher-
energy protein sites while exposing their own low-energy CH_3 and
CH_2 groups, this spectrum prompts the suggestion that the inside
walls of blood veins are dominated by outermost layers of lipid --
most likely tightly bound to protein -- and have an intrinsically
low surface energy. The critical surface tension range for
variously packed fatty acid films runs from about 22 to 28 dynes/cm.

Figure 3 is a simplified plot of the wetting data obtained
for these excised jugular vein segments, as laid open and dried
into flat sheets on glass slides. The trend line drawn from data
taken at 50% relative humidity is an average through both polar
H-bonding and non-H-bonding liquids. The low critical surface
tension intercept, between 25 and 30 dynes/cm, is certainly consistent with the postulate that hydrocarbons dominate the surface
constitution of the intimal surfaces. In an attempt to approach
more realistic conditions, these flattened veins were allowed to
swell completely by imbibing distilled water from a pool placed
around the periphery of each sample. The line on Figure 3 labeled
"water swollen" represents the trend of only the halogenated
organic wetting liquids, since the water compatible liquids
penetrated the swollen, hydrated surface immediately. The shift
of the apparent critical surface tension intercept to a value near
32 dynes/cm is completely compatible with a conversion of this
surface to one dominated by an adsorbed water film (13, 14, 15).

As a further approach to a realistic assessment of the
"wettability" of blood veins in situ, experiments with living animals
were undertaken. Anesthetized dogs were subjected to a midline
laparotomy, their superior mesenteric veins exposed and cannulated,
and long venous branches were completely flushed free of blood

Figure 3

SURFACE TENSION (γ_{LV}) IN DYNES/CM
COSINE Θ
AT 50% R.H.
WATER SWOLLEN
WETTABILITY OF BLOOD VESSEL INTIMA IN VITRO
5 10 15 20 25 30 35 40 45 50 55 60 65 70 75
+1.0 0.9 0.8 0.7 0.6 +0.5 0.4 0.3 0.2 +0.1 0

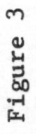

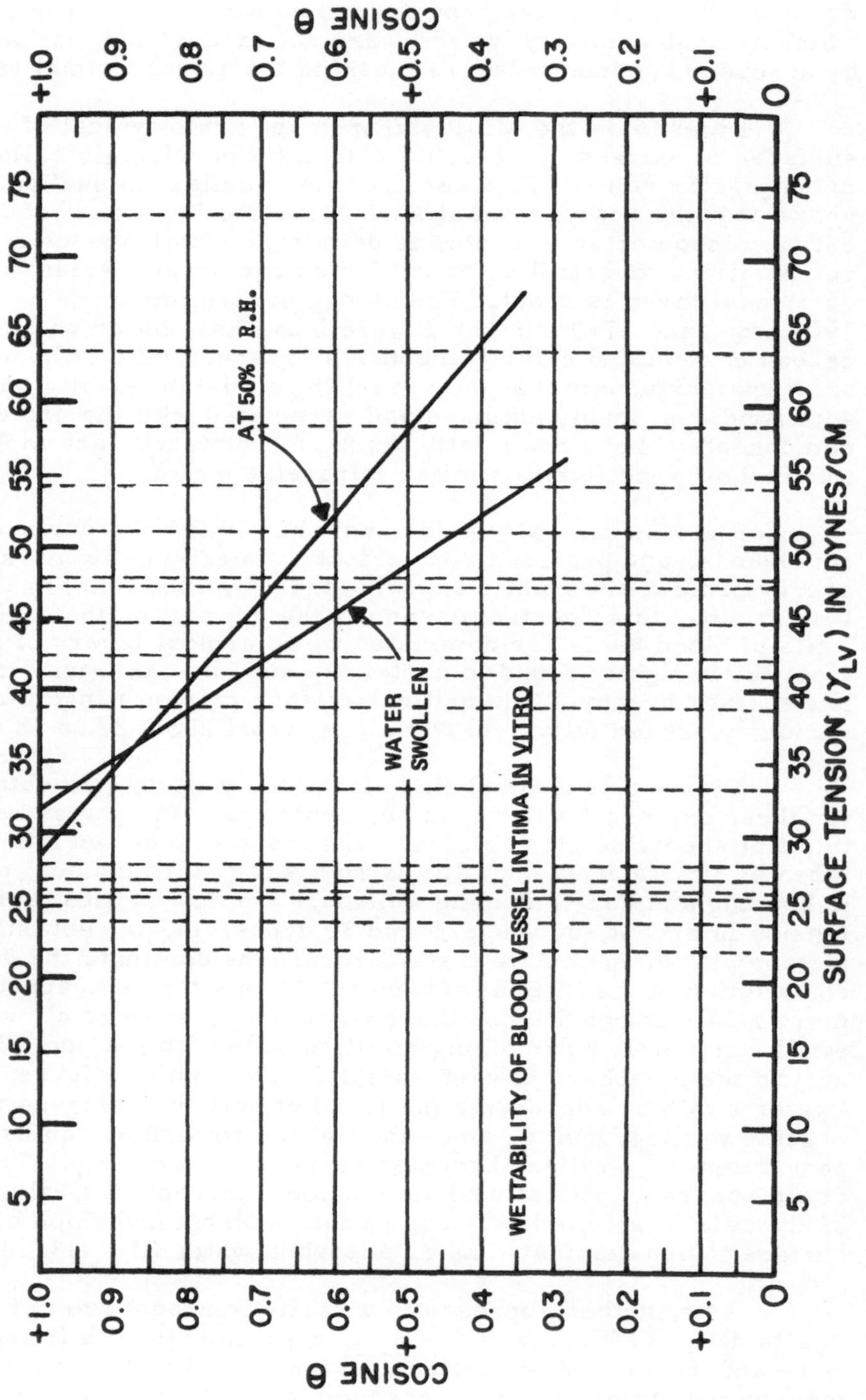

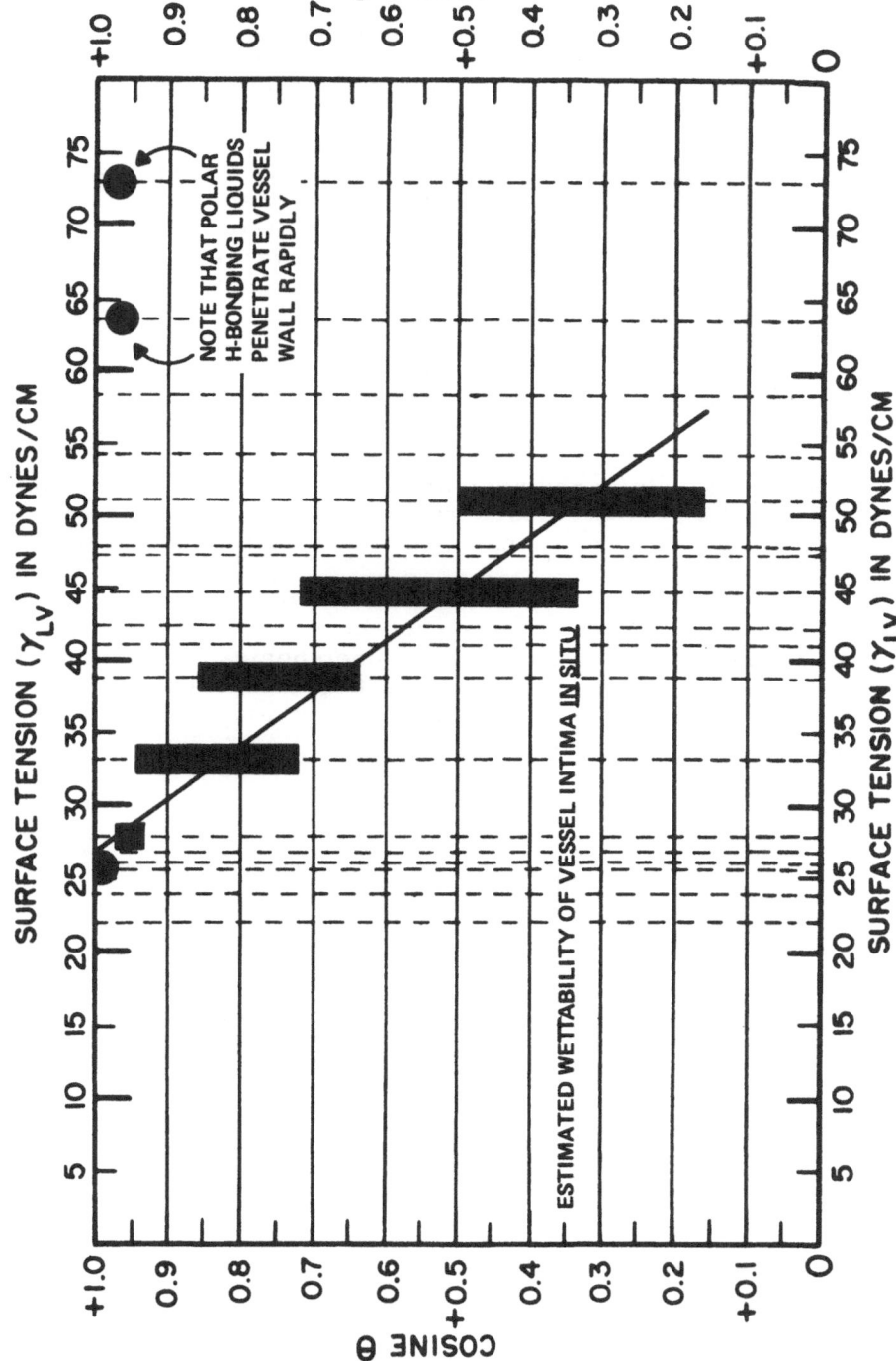

Figure 4

with physiologic saline.   Various wetting liquids representing a
range of structural types and surface tensions were then injected
into the veins.   Purposely formed air bubbles in these liquid trains
were then microphotographed, and the contact angles at the vein/
liquid/air interfaces determined from the photography.   Table 1
lists the liquids used, their approximate surface tensions, the
estimated contact angles observed, and Figure 4 is a graph of
these data.   It should be noted that the results are similar to those
from veins in vitro in the sense that water miscible liquids exhibited
very low contact angles -- correlated with their ready penetrability
of the hydrated vessel walls -- while non-interacting organics
showed a regular progression of decreasing contact angles in
accord with their surface tensions.   Again, a critical surface
tension intercept between 25 and 30 dynes/cm is the best inference
from the data and this range correlates well with surface layers
dominated by hydrocarbon components.

Our conclusion from these contact angle measurements and
from these MAIR infrared analyses is that the blood vessel intima
is an intrinsically low-surface-energy lipoprotein lining with an
outermost layer of predominantly hydrocarbon composition.

A natural corollary is that damage to this surface layer
would expose the higher-energy protein matrix beneath it and
result in spontaneous intravascular thrombosis.   A further
corollary is that adsorption of circulating components to the intima,
resulting in increased surface energy, would also carry an increased
probability for thrombogenesis.

Table 1

APPROXIMATE CONTACT ANGLES FOR VARIOUS
AIR/LIQUID INTERFACES PRODUCED IN VEINS

LIQUIDS AND THEIR APPROXIMATE SURFACE TENSIONS (IN DYNES/CM)		ESTIMATED CONTACT ANGLE RANGE OBSERVED (IN DEGREES)
NORMAL SALINE	73	10 - 15
GLYCEROL	63	10 - 15
METHYLENE IODIDE	50	60 - 80
1-BROMONAPHTHALENE	44	45 - 70
1-METHYLNAPHTHALENE	38	30 - 50
ISOPROPYLBIPHENYL	33	20 - 45
MINERAL OIL	27	15 - 20
DODECANE	25	0 - 10

THE ROLE OF A LOW-SURFACE-ENERGY HYDROCARBON COMPOUND IN DETERMINING THROMBORESISTANCE OF STELLITE 21

There is no obvious reason why Stellite 21 should be intrinsically more thromboresistant than any other metallic alloy, pure metal, or metal oxide. Stellite is simply an alloy of cobalt, chromium, and tungsten -- metals which themselves have shown no thromboresistant character -- which is very hard, brittle, and takes a high polish.

Indeed, metallographically-polished, organic-free Stellite 21 plates and rings show surface properties, as judged from contact angle experiments at high and low relative humidities, identical to those of other high energy materials including many metals and glasses (13, 14, 15). It is no surprise, then, that Stellite prostheses with these surface properties are as adversely thrombogenic as other metallic implants.

Yet, Stellite 21 struts and valve seats used in replacement heart valves admittedly show excellent resistance to thrombus formation, and other metals do not. Our measurements on a Starr-Edwards heart valve, removed after a considerable period of human implantation, were the first to show that the Stellite portions of this prosthetic implant had a remarkably low-energy surface. It was immediately suggested, and later confirmed in discussions with the manufacturer, that this low-surface-energy reflected the attachment to the metal of organic contaminants from a hydrocarbon-based final polishing compound. We have since examined other Stellite surfaces, notably those on the pins and valve seat of the Schimert-Cutter experimental heart valve, and found them to be similarly organic-coated (although not having quite as low critical surface tensions). The common "tallow" additives in commercial polishing agents are simple mixtures of stearic and palmitic acids (long chain fatty compounds).

The surface chemistry of such fatty acids is well-known, and the ready arrangement of these molecules into closepacked surface layers during polishing explains the low-energy character of surfaces polished in their presence. Thus, the outermost surface constitution of thromboresistant Stellite 21 is not metallic at all but, rather, consists of closely packed methyl groups of the organic compounds present during the final polishing steps.

We have extensively investigated the surface properties of Stellite plates and rings to better define the role of this adventitious contaminant in determining the blood compatibility of the material as used in prosthetic devices. These are our conclusions:

Table 2

Contact Angles of Various Pure Liquids on Modified Stellite Surfaces

Wetting Liquid and Surface Tension (γ_{LV}) at 20°C, dynes/cm		Stellite Plate Normal Polish Distilled Water Rinsed (a)	Same After Vigorous Detergent Wash (b)	Stellite Cage Of Starr-Edwards Valve After Vigorous Detergent Wash (c)	Stellite 21 Ring Normal Polish (d)	Same After Implantation And After Vigorous Detergent Wash (e)	Stellite 21 Ring Poorly Organic Coated (f)	Same After Implantation And Adsorption Of Protein (g)	Same After Vigorous Detergent-Wash (h)	Stellite 21 Ring Only Partially Organic Coated (i)	Same After Implantation And Adsorption Of Protein (j)	Same After Vigorous Detergent Wash (k)
Water	72.8	104	101	73	91	43	93	39	5	68	48	5
Glycerol	63.4	102	--	--	--	--	--	--	--	--	--	--
Formamide	58.2	98	--	--	--	--	--	--	--	--	--	--
Thiodiglycol	54.0	81	80	61	79	37	79	32	5	43	32	29
Methylene Iodide	50.8	67	63	54	59	32	59	44	34	32	48	34
Sym-Tetrabromoethane	47.5	66	--	--	--	--	--	--	--	--	--	--
1-Bromonaphthalene	44.6	62	55	41	54	21	53	35	26	28	37	31
0-DiBromobenzene	42.0	57	--	--	--	--	--	--	--	--	--	--
1-Methylnaphtnalene	38.7	55	48	31	46	8	48	30	18	19	26	20
Dicyclohexyl	33.0	45	37	15	29	0	35	16	0	0	15	0
Hexadecane	27.7	36	30	--	18	0	27	8	0	0	12	0
Tetradecane	26.7	32	27	--	--	0	--	--	0	0	--	0
Tridecane	25.9	28	24	--	--	0	--	--	0	0	--	0
Dodecane	25.4	26	23	--	--	0	--	0	0	0	--	0
Decane	23.9	14	12	6	0	0	9	0	0	0	0	0

Figure 5

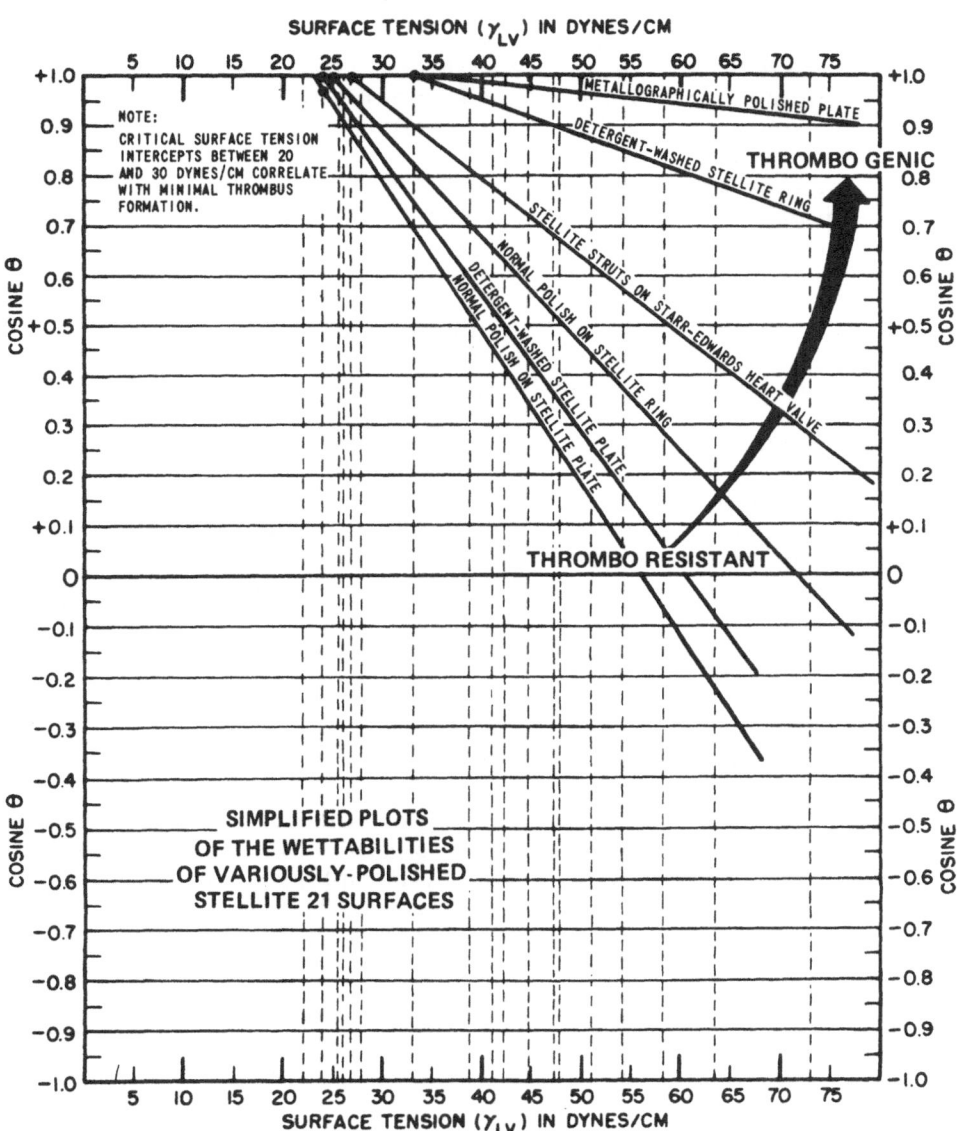

SURFACE TENSION (γ_{LV}) IN DYNES/CM

NOTE:
CRITICAL SURFACE TENSION
INTERCEPTS BETWEEN 20
AND 30 DYNES/CM CORRELATE
WITH MINIMAL THROMBUS
FORMATION.

METALLOGRAPHICALLY POLISHED PLATE

DETERGENT-WASHED STELLITE RING

THROMBO GENIC

STELLITE STRUTS ON STARR-EDWARDS HEART VALVE

NORMAL POLISH ON STELLITE RING

DETERGENT-WASHED STELLITE PLATE

NORMAL POLISH ON STELLITE PLATE

THROMBO RESISTANT

SIMPLIFIED PLOTS
OF THE WETTABILITIES
OF VARIOUSLY-POLISHED
STELLITE 21 SURFACES

SURFACE TENSION (γ_{LV}) IN DYNES/CM

COSINE θ

(1)   Thromboresistant Stellite has an organic, low-surface-
         energy material as its outermost layer, most likely a long-
         chain fatty acid, as revealed by both contact angle
         measurements and MAIR infrared surface spectroscopy.

(2)   On thromboresistant Stellite, and not on other metals tested
         or thrombogenic Stellite, this layer is tenaciously bonded.
         This has been demonstrated by surface quality checks after
         a series of increasingly stringent cleaning techniques
         ranging from distilled water rinsing through detergent
         scrubbing to mechanical abrasion.

(3)   Implantation experiments have shown that only surfaces
         having strongly-bonded low-energy layers are significantly
         thromboresistant. Incompletely coated surfaces, or weakly-
         bound coatings, are less thromboresistant.

(4)   Experiments with polished and unpolished plastic surfaces
         have shown that the influence of surface roughness on the
         above results is relatively minor.

EXPERIMENTAL DATA supporting these conclusions follows.

        Table 2 is a compilation of the average contact angles
measured for a number of highly purified liquids placed on variously
polished and cleaned Stellite parts. Figure 5 illustrates simplified
plots of these data in the Zisman format (as cosines of the contact
angles versus liquid-vapor surface tensions for the liquids (1))
which allows determination of approximate critical surface tensions
for the materials in question.

        Column (a) of Table 2 records the wetting data for a plate
of Stellite, polished exactly as valve cages and vena cava rings --
which showed excellent thromboresistance -- had been polished
routinely. The very high contact angles for all liquids on this
surface show unambiguously that an organic coating, most likely
exposing closepacked methyl groups, completely masks the metallic
surface. Infrared spectra of this surface, obtained by pressing it
tightly against an internal reflection element and using MAIR
(multiple attenuated internal reflection) accessories (4), confirmed
the presence of a hydrocarbon component in the outermost surface
layer. Extensive, and prolonged, distilled water rinsing of this
surface did not change the contact angles obtained.

        Column (b) of the table records the slightly lower contact
angle values obtained after a vigorous detergent washing of the
plate. The observation that the wetting was not significantly
changed illustrates the firmness with which the organic layer was
bound. Further scrubbing of the plate with a hard brush and
detergent further lowered the contact angles, but only mechanical

erosion (using an organic-free abrasive) or direct flaming of the
surface could completely restore the metallic (i. e. high energy)
nature of the surface.

Column (c) records the contact angles on the polished Stellite
cage of a Starr-Edwards heart valve which had been implanted and
then detergent-washed just before the measurements were made.
The fact that these contact angles were lower than those on similarly
washed, non-implanted Stellite plates suggests that during removal
of the adsorbed protein films (formed during implantation) some
of the organic coating was also stripped from the surface. The
fact that the critical surface tension of these struts was still very
low, between 20 and 30 dynes/cm, even after such stripping
indicates that the organic layer was strongly-bound and present in
considerable quantity.

Columns (d) and (e) confirm these suppositions by comparing
the results on a polished Stellite 21 vena cava ring -- which
proved to be significantly thromboresistant when implanted -- and
that same ring after implantation and after vigorous scrubbing
with a strong detergent and hard-bristle test tube brush. The
data in column (e), including a water contact angle greater than
40°, clearly indicate the presence of residual organic material.
Rings with this little organic coverage did not prove thromboresistant when reimplanted, but were not as thrombogenic as rings
made completely organic-free by flaming.

The last two groups of data in Table 2, comprising columns
(f), (g) and (h) and columns (i), (j) and (k), illustrate the surface
modification of Stellite rings by adsorbed protein films in vivo and
the reason for the ineffectiveness of poorly organic-coated Stellite
rings for preventing thrombus formation. Comparison of the
contact angle values obtained after detergent washing of both sets
of rings (see columns (h) and (k)) with values from thromboresistant
rings similarly treated (see column (e)) shows clearly that the
organic coating on both sets of rings was extremely labile. It did
not survive -- certainly not long enough to influence thrombus
formation -- the autoclaving, implantation, and cleaning steps. It
is appropriate to recall here the 1968 study of Dutton and co-workers,
who showed that monolayer organic coatings on glass and silicone
substrates -- although sufficient to completely change the initial
wetting properties of these surfaces -- were not sufficient in
quantity nor strongly enough bound to influence thrombus formation
on these substrates (42).

Nonthromboresistant Stellite surfaces also accumulated
similar protein films by adsorption (compare columns (g) and (j)),
as is evident in the infrared spectra of these protein films, given
as Figure 6.

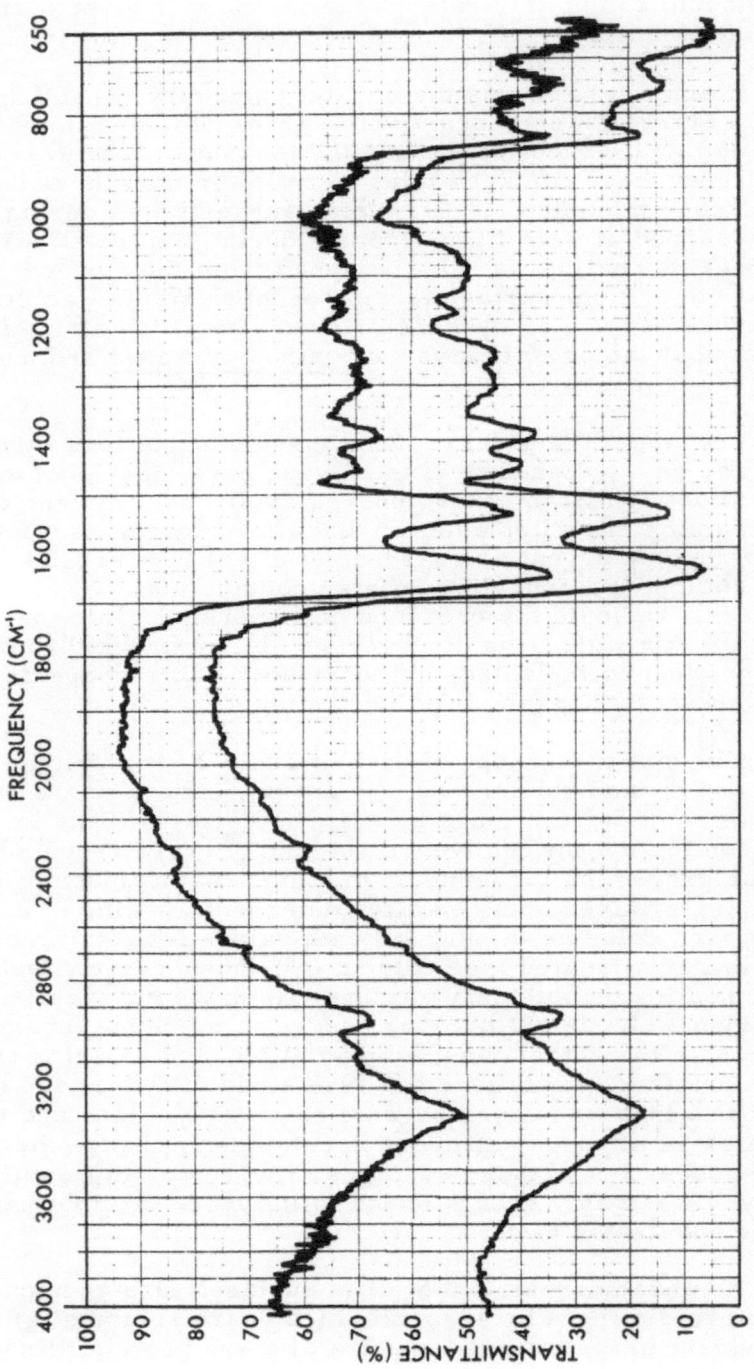

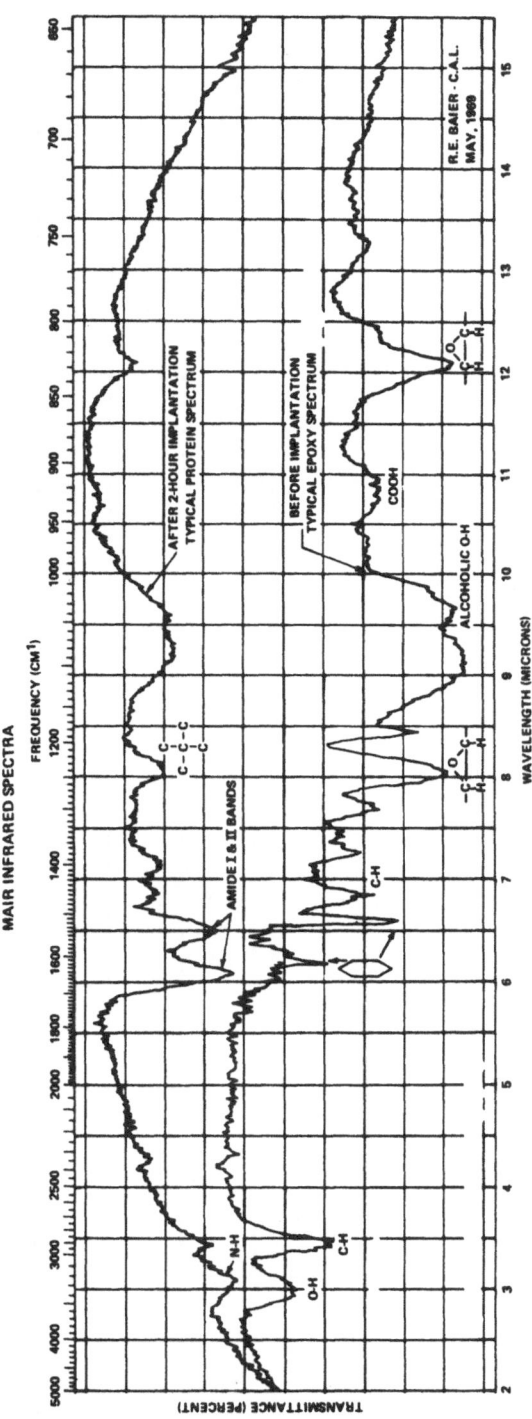

Figure 7

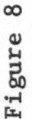

Figure 8

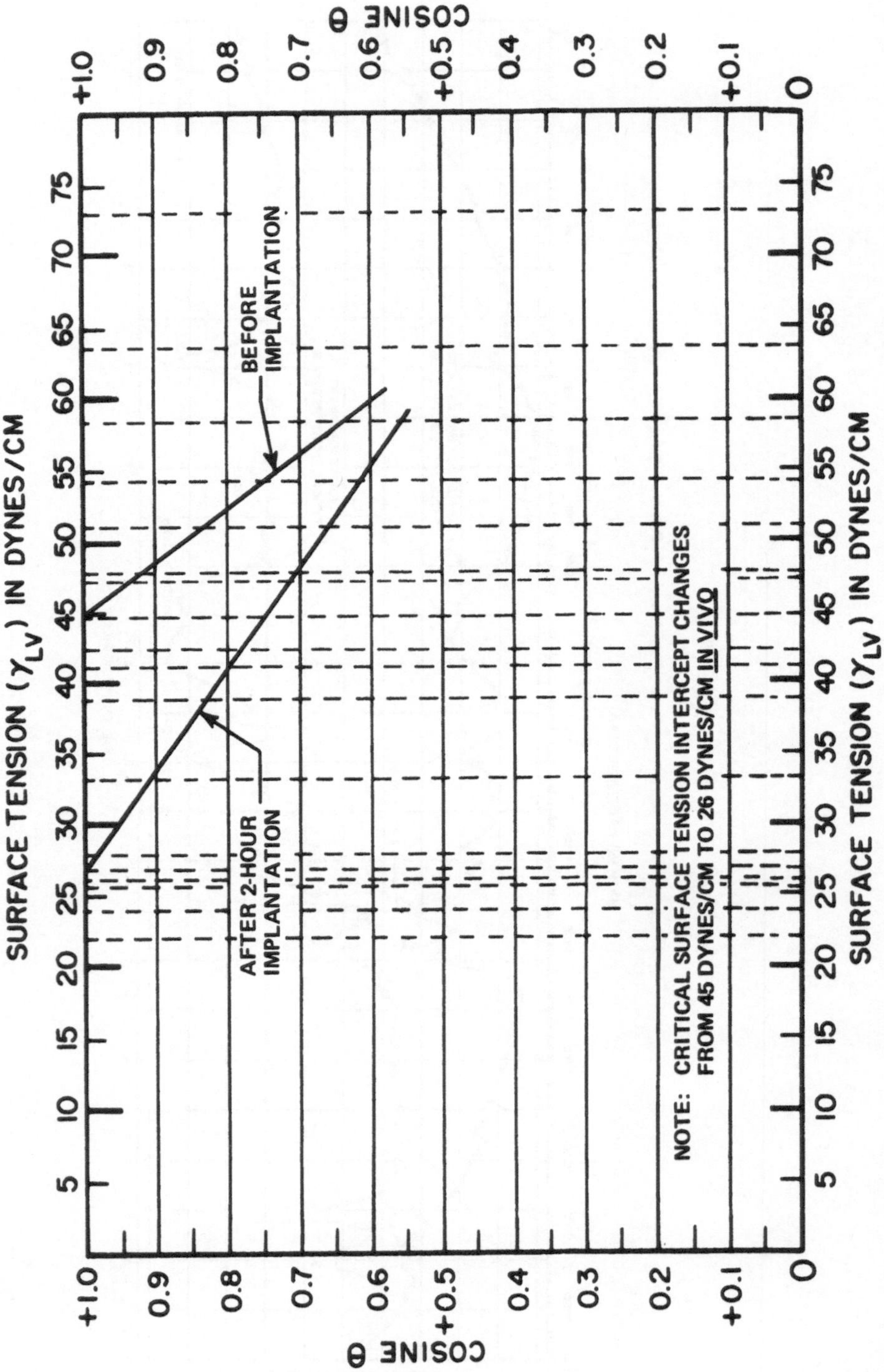
SURFACE TENSION (γ_{LV}) IN DYNES/CM
COSINE θ
BEFORE IMPLANTATION
AFTER 2-HOUR IMPLANTATION
NOTE: CRITICAL SURFACE TENSION INTERCEPT CHANGES FROM 45 DYNES/CM TO 26 DYNES/CM IN VIVO

Concurrently with the above-described experiments, investigations of polished and unpolished vena cava rings of Lexan (polycarbonate) were undertaken. These latter studies produced erratic results but, nonetheless, allowed an assessment of the influence of surface roughness on the properties of Stellite specimens. Our most recent experimental series showed that unpolished, visually rougher-surfaced Lexan rings were not detectably more thrombogenic than polished specimens having the same chemical makeup (as shown by both contact angles and infrared spectra). Therefore, it seems safe to conclude that the thromboresistance which has been empirically attributed to polished Stellite is not simply due to the enhanced surface smoothness of the polished specimens.

OTHER CANDIDATE BIOMEDICAL MATERIALS

We are now conducting an investigation of surface properties of new, potentially thromboresistant biomaterials. This study includes in vivo evaluations of implanted rings whose critical surface tensions (γ_c) have previously been determined from the contact angles of specially calibrated liquids on their surfaces. Surface chemical constitution of each material is verified through sensitive surface spectroscopy which utilizes the MAIR (multiple attenuated internal reflection) infrared technique. The materials are re-analyzed after implantation to determine those surface changes, particularly the adsorption of proteinaceous films, which reflect the relative thrombogenecity of the materials in vivo.

Various plastic vena cava rings have been studied. Enough data has been gathered to allow some cautious generalizations to be drawn about the change in the materials' surface properties which occurs during implantation. We hope to correlate the nature of this change with the biocompatibility of the various materials.

Our initial generalizations are these:

(1) During implantation, all materials rapidly adsorb a tenaciously adherent coating of protein.

(2) On most materials, this adsorbed protein layer substantially changes the critical surface tension, to higher values for initially low-energy surfaces and to lower values for initially high-energy surfaces.

(3) Since there are obvious differences in the thrombogenecity of the various materials, and also differences in the wetting data and infrared spectra for the adsorbed films which form on them, the relation connecting the

structure and properties of these absorbed layers with the in vivo test results requires additional investigation.

(4) Although strongly-bound low-energy surface layers of hydrocarbon composition appear to correlate with the natural thromboresistance of blood vessels and with the observed thromboresistance of commercially polished Stellite, such an initial layer may be neither necessary nor sufficient for imparting thromboresistance to other materials. Relatively high-energy surfaces of some polymers, for example, might be capable of adsorbing components from blood which convert their surfaces to low-energy character and thence become nonthrombogenic.

As an illustration of the type of results being achieved, data for epoxy rings before and after implantation are included as Figures 7 and 8. Figure 7 shows the shift in surface infrared spectrum caused during implantation, and Figure 8 shows the concomitant wettability change accompanying the deposition of the protein film.

Presentation of further results, and specific examples of other new materials, must be postponed until a later time.

DISCUSSION

Our experimental findings are summarized in tabular form as Table 3. We have concluded that the final polish normally given to Stellite, using a polish containing a waxy additive, leaves a hydrocarbon component strongly bound in the outermost metallic layers. This coating imparts a very low wettability to the finished surface. Since Stellite parts polished in the absence of waxy materials, as well as those flamed or lightly abraded after the normal polish, do not resist thrombus formation when in contact with blood, we suggest that the presence of the hydrocarbon materials, when tightly bound or when "lapped" into the surface layers, contributes to the thromboresistance of the normally polished Stellite primarily by providing a semi-permanently decrease critical surface tension. Low critical surface tension is also a feature of naturally thromboresistant blood vessel walls.

The critical surface tension of about 25 dynes/cm for vessel intima and polished Stellite indicates that their surfaces are predominantly composed of methyl groups (43,44). Critical surface tensions much above 30 dynes/cm, for nonheparinized materials, are not compatible with obtaining significant thromboresistance. Clean metals and metal oxides, including Stellite, generally show very small contact angles, approaching zero in most cases, for the

liquids used here, and these surfaces are known to be thrombogenic.
Metals other than Stellite (titanium, for example) do not acquire as
strongly-bound organic layers during polishing and do not show
significant thromboresistance. On polished titanium, very low
contact angles are obtained after detergent-washing, suggesting
that the hydrocarbon component is easily removed. During
polishing then, the Stellite traps more wax in its surface than other
metals do; this hydrocarbon material may be available to replenish
the surface from a considerable depth. Alternatively, the
mechanism of hydrocarbon fixation may be the formation of a
metallic soap between components of the alloy, notably chromium,
and fatty acids from the polishing agent. This is an example of
how the specific composition and bulk properties of materials can
influence their final surface properties. In the case of Stellite,
the alloy's extreme hardness dictates a rigorous polishing process,
and thus favors strong binding of hydrocarbon.

    The Zisman plot (see Figure 5) of data obtained from contact
angle measurements on Stellite struts of a Heart Valve cage is
most interesting. This valve had actually been implanted, washed
clean, handled by many people as a laboratory specimen, and
washed with concentrated detergent solution and a soft brush,
followed by copious rinsing with distilled water and air-drying,
before contact angles were measured. The critical surface tension
intercept of about 25 dynes/cm and the generally large contact
angles for the liquids used, even after this rather lengthy handling
and cleaning, guarantee that the surface of the valve cage as
implanted was of very low wettability.

    As a demonstration of the longevity of the hydrocarbon
embedded in the Stellite surface layers, experiments with Stellite
plates are cited in Table 2. The wetting results on a freshly
polished plate, rinsed only with distilled water before measurements
were made, and of the same plate after a vigorous wash with
concentrated detergent and a soft brush are comparable. Gentle
hand-stroking of the plates over a Fisher Gamal-B polishing cloth
coated with Linde gamma-alumina (0.3 micron), following by
copious rinsing with distilled water, did not visually modify the
surface finish but did abrade enough material from the plate to
make it wettable (i.e. have contact angles of about 5° or less) with
most liquids used in this investigation. This indicates that removal
of the first few layers of material from the plate surface is
sufficient to change that surface from one of low surface energy to
one of high surface energy.

    When a freshly polished Stellite plate was briefly exposed
to a stream of flowing blood followed by copious rinsing first with
physiologic saline and then distilled water, it was obvious that a
very thin surface film was deposited on the surface, as could be
seen by viewing the plate in glancing reflected light. The film was

Table 3

SUMMARY OF FINDINGS

MATERIAL	BLOOD VESSEL INTIMA	"TALLOW-POLISHED" STELLITE	"TALLOW-COATED" STELLITE	ORGANIC-FREE STELLITE	OTHER ORGANIC-FREE METALS	LEXAN POLYCARBONATE
THROMBORESISTANCE	GOOD	GOOD	POOR	POOR	POOR	POOR
RELATIVE SURFACE ENERGY	LOW	LOW	LOW	HIGH	HIGH	HIGH
RELATIVE RESISTANCE OF SURFACE LAYER TO EROSION	HIGH	HIGH	LOW	HIGH	HIGH	HIGH

identified as predominantly proteinaceous in character from infrared spectra obtained using multiple internal reflection techniques. Similarly, protein films were formed spontaneously on the surfaces of implanted Stellite vena cava rings. The initially adsorbed protein films adhered rapidly and tenaciously. After vigorous washing of the filmed surfaces with hot concentrated detergent and a hard brush, all trace of the film was removed, as judged by visual inspection, contact angle measurements, and by infrared spectroscopy.

Since the hydrocarbon component is also eventually removed by washing and slight wear, exposure of the high-energy metal surface gradually occurs. This limitation should also apply to the thromboresistance of the surface.

The most plausible explanation for the favorable biocompatibility of polished Stellite, then, is its low surface energy, as reflected in a low critical surface tension. Although it is known that blood proteins and formed elements (platelets, in the main) do adhere to low energy materials (27,29,45), the work of adhesion to distract them is substantially less than that necessary to detach them from higher-energy surfaces (2). Rapid, and turbulent, flow regimes (such as are indeed present with an implanted heart valve) would provide the best conditions for detaching small aggregates from low energy surfaces, and allow them to flow downstream where they could be eliminated by enzymatic and other digestive processes. Thus, it is by no means certain, that synthetic low energy surfaces are intrinsically nonthrombogenic. Rather, they may in fact induce thrombus formation but -- because of the diminished strength of adhesion to low energy surfaces -- allow easier removal of growing thrombic masses before they can impair the action of the prosthetic implant. But the apparently intrinsic low-surface-energy coating of natural blood conduits favors the hypothesis that such surface features are closely correlated with thromboresistance.

Additional study of the mechanism of this action is deserved. Substantive low energy coatings can be provided on other materials which must contact the blood. If such surfaces prove to be as thromboresistant as polished Stellite, other plastic and metal components can be deliberately fabricated to include a surface layer of low critical surface tension, preferably renewable by diffusion from the bulk. Some polymeric materials with diffusable low-energy surfactants have already been created (19).

Table 3 indicates the guidelines for future work. Good thromboresistance correlates best with low-energy surfaces having a high resistance to erosion.

REFERENCES

1. Zisman, W. A. (1964). Advances in Chemistry Series 43, pp 1-51, American Chemical Society, Wash. D. C.

2. Baier, R. E., Shafrin, E. G., and Zisman, W. A. (1968). Science 162:1360.

3. Baier, R. E., "Surface Properties Influencing Biological Adhesion", Chapter 2 in Adhesion in Biological Systems (R. L. Manly, Ed.), Academic Press, in press.

4. Harrick, N. J. (1967). "Internal Reflection Spectroscopy", Interscience Publishers, New York.

5. Zisman, W. A. (1962). In "Symposium on Adhesion and Cohesion", (P. Weiss, Ed.), Elsevier, New York.

6. Langmuir, I. (1916). J. Am. Chem. Soc. 38:2286.

7. Langmuir, I. (1925). In "Third Colloid Symposium Monograph", pp 48-75, New York, Chem. Catalogue Co., Inc

8. Budgett, H. M. (1911). Proc. Roy. Soc. A86:25.

9. Henniker, J. C. (1949). Rev. Mod. Phys. 21:322.

10. Zisman, W. A. (1961). "Constitution Effects on Adhesion and Abhesion", Naval Research Laboratory Report No. 5699.

11. Weiss, L. (1960). Int. Rev. Cytology 9:187.

12. Shafrin, E. G. (1967). In "Polymer Handbook", (J. Brandrup and E. H. Immergut, Eds.) pp 111-113, Interscience, N.Y.

13. Shafrin, E. G., and Zisman, W. A. (1967). Journal of the American Ceramic Society 50:478.

14. Bernett, M. K., and Zisman, W. A. (1968). J. Colloid Interface Sci. 28:243.

15. Bernett, M. K., and Zisman, W. A. (1969). J. Colloid Interface Sci. 29:413.

16. Baier, R. E., and Zisman, W. A. (1967). "Critical Surface Tension of Wetting for Protein Analogues", Presented at American Chemical Society, 153rd National Meeting, Miami Beach.

17.   Baier, R. E., and Zisman, W. A. (1968). "Wettability and MAIR Infrared Spectroscopy of Solvent-Cast Thin Films of Polyamides and Polypeptides", Naval Research Laboratory Report No. 6755.

18.   Baier, R. E. and Zisman, W. A., "The Influence of Polymer Conformation on the Surface Properties of Polymethylglutamate", Submitted for publication.

19.   Jarvis, N. L., Fox, R. B., and Zisman, W. A. (1964). Adv. In Chem. 43:317, American Chem. Soc., Wash., D.C.

20.   Baier, R. E. and Zisman, W. A., "Wetting Properties of Collagen and Gelatin", In preparation.

21.   Weiss, L. (1967). "The Cell Periphery, Metastasis, and Other Contact Phenomena", Amsterdam:North-Holland Publishing Co.

22.   Weiss, L. (1968). Exp. Cell. Res. 51:609.

23.   Weiss, L., and Blumenson, L. E. (1967). J. Cell. Physiol. 70:23.

24.   Taylor, A. C. (1962). In "Biological Interactions in Normal and Neoplastic Growth", (Ed. by M. J. Brennan and W. L. Simpson), pp 169-182, Little, Brown, and Co., Boston, Mass.

25.   Taylor, A. C. (1961). Exp. Cell Res. Suppl. 8:154.

26.   Brash, J. L. and Lyman, D. J. (1969). J. Biomedical Materials Res. 3:175.

27.   Lyman, D. J., Muir, W. M., and Lee, I. J. (1965). Trans. Amer. Soc. Artif. Int. Organs 11:301.

28.   Eirich, F. R. (1969). In "Interface Conversion for Polymer Coatings", pp 350-373, (Ed. by P. Weiss and G. D. Cheever), American Elsevier Publishing Company, New York.

29.   Baier, R. E., and Dutton, R. C. (1969). J. Biomedical Materials Res. 3:191.

30.   Dutton, R. C., Webber, T. J., Johnson, S. A., and Baier, R. E. (1969). J. Biomedical Materials Res. 3:13.

31.   Scarborough, D. E., Mason, R. G., Dalldorf, R. G., and Brinkhous, K. M. (1969). Lab. Invest. 20:164.

32.   Vroman, L., and Adams. A. L. (1969). J. Biomedical
        Materials Res. 3:43.

33.   Zisman, W. A. (1963). Industrial and Engineering Chemistry
        55:19.

34.   Curtis, A. S. G. (1962). Biol. Rev. 37:82.

35.   Weiss, P. (1958). Int. Rev. Cytology 7:391.

36.   Copley, A. L. (1960). In "Flow Properties of Blood and
        Other Biological Systems", (Ed. by A. L. Copley and
        G. Stainsby) pp 97-117, Pergamon Press, New York.

37.   Moolten, S. F., Vroman, L., Vroman, G. M. S., and
        Goodman, B. (1949). Arch. Internal Med. 84:667.

38.   Ratnoff, O. (1959). In "Connective Tissue, Thrombosis,
        and Atherosclerosis", (Ed., I. Page), Academic Press, N. Y.

39.   Perel'man, I. B. (1965). Colloid J. (USSR) 27:422.

40.   Zubairov, D. M., Repeikov, A. V., and Timberlaev, V. N.
        (1963). Fiziologicheskii Zhurnal SSSR 49:85.

41.   Ellison, A. H., and Zisman, W. A. (1954). J. Phys. Chem.
        58:503.

42.   Dutton, R. C., Baier, R. F., Dedrick, R. L., and Bowman,
        R. L. (1968). Trans. Amer. Soc. Artif. Int. Organs 14:57.

43.   Shafrin, E. G. and Zisman, W. A. (1957). J. Phys.
        Chem. 61:1046.

44.   Shafrin, E. G. and Zisman, W. A. (1960). J. Phys.
        Chem. 64:519.

45.   Lyman, D. J., Brash, J. L., Chaikin, S. W., Klein, K. G.,
        and Carini, M. (1968). Trans. Amer. Soc. Artif. Int.
        Organs 14:250.

LIPID-PROTEIN ASSOCIATION IN LUNG SURFACTANT [1,2]

Morton Galdston, M.D.

Associate Professor of Medicine
New York University School of Medicine
and
Dinesh O. Shah, Ph.D.

Surface Chemistry Laboratory, Marine Biology Division
Lamont-Doherty Geological Observatory of Columbia
University, Palisades, New York 10964

ABSTRACT

We investigated the nature of the lipid-protein association
and the role of the protein moiety in a highly surface active
lipoprotein fraction we isolated from cell-free rabbit lung wash
and also compared the surface properties of dipalmitoyl lecithin,
generally believed chiefly responsible for the surface properties
of lung surfactant, with those of the lipoprotein fraction.

An increase in surface potential of monolayers of the
lipoprotein fraction on calcium binding and a decrease in surface
potential on enzymatic hydrolysis with phospholipase A suggest
the polar groups of phospholipid are free to interact with Ca^{++}
and phospholipase A and the lipid-protein association in the
lipoprotein fraction is of the van der Waals type. Measurements
of relative surface viscosity and of film state indicate the
protein moiety presumably increases intermolecular spacing between
lipid molecules in the lipoprotein fraction, helping to maintain
films in the liquid state under high surface pressure. We
observed distinct differences in surface properties of films of
dipalmitoyl lecithin and of the lipoprotein fraction.

[1]These investigations were supported by U.S.P.H.S.Grant No.
HE 05256-10 (CVA) and by the National Science Foundation Grant No.
GB-5273 and by the Sea Grants Program Administration Grant GH-16.

[2]Lamont-Doherty Geological Observatory Contribution No. 1405

INTRODUCTION

The surfactant lining of the alveolar air-liquid interface
helps to stabilize the alveolar spaces, to prevent alveolar
collapse at low lung volumes and to modulate the retractive force
of lung (1). Lung surfactant is thought to consist of an insoluble
lipoprotein layer, "lining film", and beneath this an extracellular
saline dispersable lipoprotein, the "lining complex" or"alveolar
surfactant", which acts as a reserve from which the "lining film"
is formed (2). An extracellular layer corresponding to surfactant
fixed in situ and independent of the external plasma membrane of
the alveolar epithelial cells was recently identified by
electronmicroscopy (3). Surfactant probably originates in the
large alveolar (Type II)cells of the epithelial layer (4).

Though the chemical composition of lung surfactant is unknown
the concensus is its high content of disaturated phosphatidyl
choline chiefly accounts for its unique surface properties (5).
Highly surface active lipoproteins (6, 7, 8), mixtures of phospholipids, neutral lipids and polysaccharides (9); and of lipids,
predominantly phospholipids together with cholesterol and triglycerides (10) have been isolated from lung. The lipoprotein we
isolated from rabbit lung wash was shown by immunologic techniques
to come from lung and not from blood (8).

Our purpose herein is to present the current status of our
investigations on the nature of the lipid-protein association and
the role of the protein moiety in the lipoprotein fraction we
isolated and to compare the surface properties of dipalmitoyl
lecithin and of the lipoprotein fraction. The association between
lipid and protein appears to be of the van der Waals type; the
protein moiety helps maintain films of lung surfactant in the
liquid state at high surface pressure; and the surface properties
of dipalmitoyl lecithin and the lipoprotein fraction differ in
many important respects.

METHODS

A highly surface active lipoprotein fraction, designated
fraction B, was isolated from cell-free rabbit lung wash, designated
stock O, as previously described (8). The surface properties of
these preparations, of the lipid extract (Folch method, 11) of
the lipoprotein fraction and of chromatographically pure L-α-
dipalmitoyl lecithin (Dajac Laboratories, Philadelphia, Pa.) were
investigated by measurements of surface pressure and surface
potential related to M^2/mg of material spread, using intermittent
manual film compression.

To learn about the nature of the lipid-protein association in the lipoprotein fraction enzymatic hydrolysis by snake venom Naja naja phospholipase A and calcium binding were investigated. To learn about the role of the protein moiety in the lipoprotein fraction relative surface viscosity and physical state of films were investigated.

Surface pressure, the difference between the surface tension of a clean surface and a film-covered surface, was measured in a lucite trough, 400 cm^2 surface area, with paraffin-coated glass barriers, by the modified Wilhelmy plate method (12) with a sand-blasted rectangular platinum plate (26 guage and 5 cm perimeter), suspended from a torsion balance (500 mg maximum) mounted on an adjustable elevating stand (Fig. 1). Surface potential, which is related to surface molecular concentration and orientation, was measured with an ionizing (α-radiation air) electrode and a Ag-AgCl electrode connected to an electrometer (13) simultaneously with surface pressure (Fig. 1).

Relative surface viscosity, an index of film fluidity was measured by the canal method (ref. 12, p. 253) with a teflon bar containing a canal (0.1 cm wide, 2 cms. long) with a flood gate attached to its front, placed over the middle of the trough (Fig. 1). A monolayer was spread on the interface between the teflon bar and a paraffin-coated glass barrier used to compress the monolayer. Five minutes were allowed for the monolayer to spread, then it was compressed to arbitrarily selected levels of 8 and 30 dynes/cm.

The flood gate was then opened, allowing the monolayer to pass through the canal with surface pressure held constant by movement of the barrier toward the teflon bar. The time required for 1 cm^2 of a monolayer to pass through the canal on subsolutions of 0.02 M NaCl or 0.01 M CaCl$_2$ is designated t_{Na^+} or $t_{Ca^{++}}$, respectively. The average of at least five such measurements was taken at each surface pressure level. The ratio

$$\left(\frac{t_{Ca^{++}} - t_{Na^+}}{t_{Na^+}} \right) \times 100$$

is the percent increase in time due to binding of calcium to monolayers (13) and indicates the increase in relative surface viscosity.

The state of a film was studied by gently blowing air with a bellows on talc previously sprinkled on the film (14). If the talc moves freely the film is in the liquid state, if very little or not at all, in the gel or solid state, respectively.

Films of stock O, the cell-free preparation of rabbit lung wash, and of its lipoprotein fraction were spread from solutions

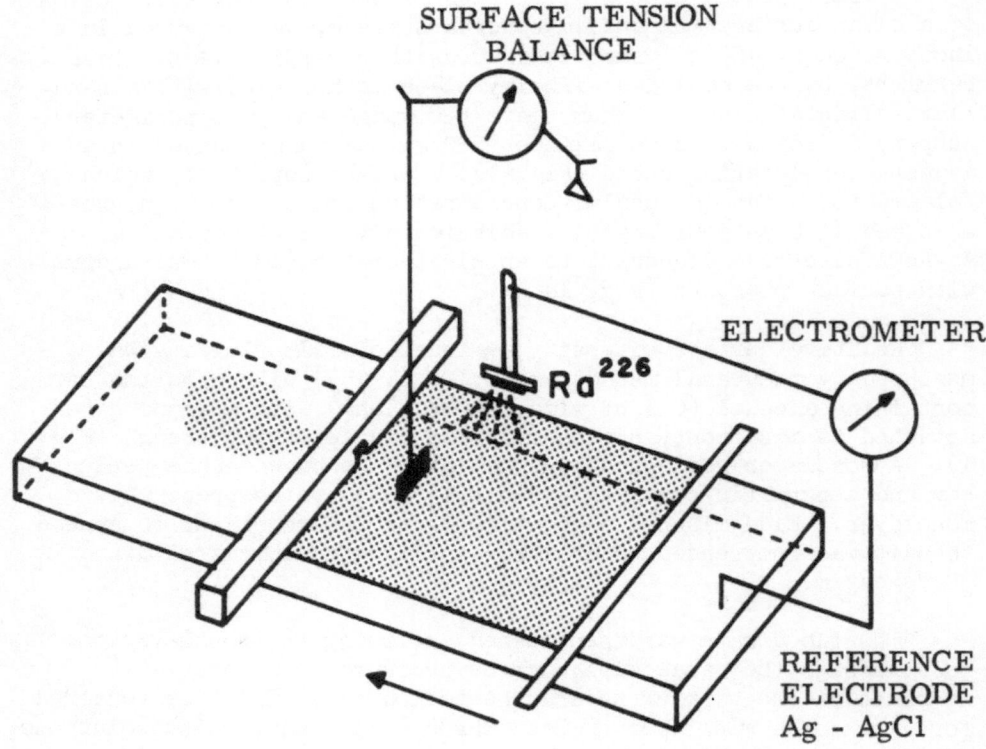

<u>Fig. 1.</u> Schematic presentation of methods used to measure
pressure by the modified Wilhelmy plate method, surface
potential by radioactive and reference electrodes and
surface viscosity by the canal method.

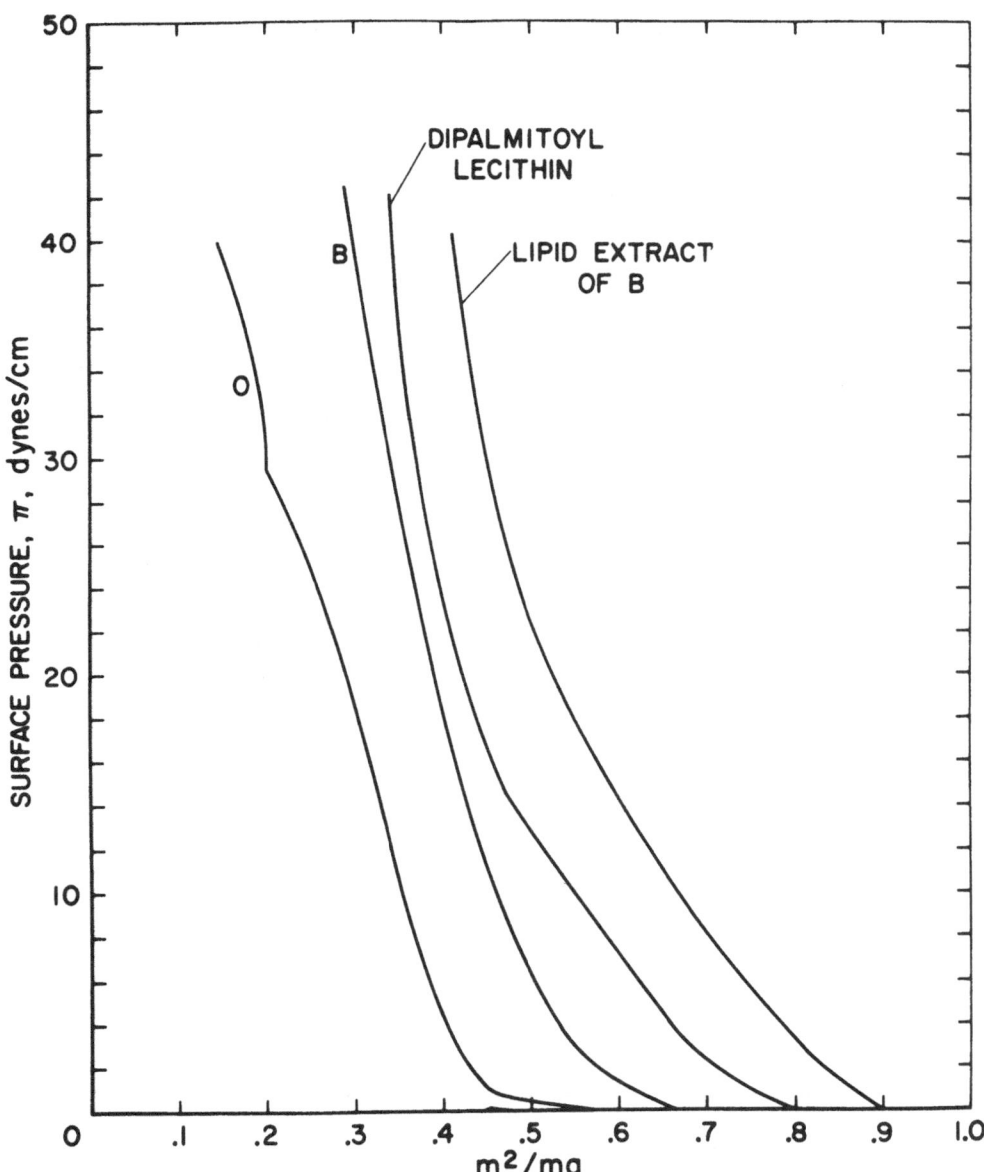

Fig. 2. Surface pressure-area curves of cell-free lung wash,
         stock O, of its lipoprotein fraction, B, of the lipid
         extract of the lipoprotein fraction, B, and of dipalmi-
         toyl lecithin on subsolutions of 0.02 M NaCl or 0.01 M
         CaCl$_2$, pH 6.0 and 25 C. The curves were identical on
         these subsolutions; therefore, only one set is shown.
         Similar curves were also obtained on 0.155 N NaCl.

TABLE I

SUMMARY OF SURFACE PROPERTIES

Preparation	Film Area M²/mg		Collapse Pressure Dynes/cm Subsolution		Maximum Surface Potential mv Subsolution	
	At onset of surface pressure	At maximum surface pressure	0.02 M NaCl	0.01 M CaCl₂	0.02 M NaCl	0.01 M CaCl₂
Cell-Free Lung Wash (Stock O)	0.45	0.15	40	40	400	465
Lipoprotein Fraction of Lung Wash (Fraction B)	0.66	0.30	43	43	400	470
Lipid Extract of Lipoprotein Fraction	0.90	0.42	41	41	410	445
Dipalmitoyl Lecithin	0.80	0.35	42	42	550	600

TABLE II

RELATIVE SURFACE VISCOSITY* CHANGES OF DIPALMITOYL LECITHIN AND LIPOPROTEIN FRACTION

MONOLAYERS IN THE PRESENCE OF Ca^{++} IN SUBSOLUTION, pH 6.0, 25C.

Difference in surface pressure $(\Delta \pi)$ dynes/cm	substance	t_{Na^+} sec 0.02 M NaCl	$t_{Ca^{++}}$ sec 0.01 M $CaCl_2$	$\left(\dfrac{t_{Ca^{++}} - t_{Na^+}}{t_{Na^+}}\right)$ X 100%
8	Dipalmitoyl lecithin	1.49	1.49	0
8	Lipoprotein fraction	1.50	1.60	7
30	Dipalmitoyl lecithin	0.53	0.73	40
30	Lipoprotein Fraction	0.50	0.59	18

* $\eta_{monolayer} = \left(\dfrac{\Delta \pi w^3}{12\ l}\right) t$

where $\Delta \pi$ = pressure difference across canal
w = width of canal = 0.1 cm
l = length of canal = 2 cm
t = time required for 1 cm^2 of film
to flow through canal $\left(\dfrac{sec}{cm^2}\right)$

of known quantities of lyophilized preparations in 0.1% amyl
alcohol in distilled water (15) by means of an Agla micrometer
syringe. Amyl alcohol facilitates spreading and increased the
reproducibility of the surface pressure-area and surface potential-
area curves. Films of lipid extract of the lipoprotein fraction
and of dipalmitoyl lecithin were spread from solutions of
chloroform-methanol-hexane (1:1:3). Hexane was added to assist
spreading of lipid.

   Enzymatic hydrolysis of monolayers of the lipoprotein
fraction and of dipalmitoyl lecithin by phospholipase A was
investigated by injecting 20 μg of boiled snake venom Naja naja
containing phospholipase A under a monolayer kept at fixed surface
pressure levels. The subsolution consisted of 400 ml tris buffer
in 0.02 M NaCl and 0.01 M $CaCl_2$ at pH 7.2. Phospholipase A
hydrolyzes lecithin to lysolecithin and free fatty acid, causing
a change of surface potential (16). A decrease in surface
potential in the first two minutes was assumed to be proportional
to the initial rate of hydrolysis.

RESULTS

   Consistent force-area curves were obtained with films of all
of the preparations studied (Fig. 2). Films of the lipid extract
of the lipoprotein fraction are more expanded than films of
dipalmitoyl lecithin and these in turn exceed the expansion of
films of the lipoprotein fraction and also of stock O. The steep
part of curves of the lipoprotein fraction and of dipalmitoyl
lecithin are nearly equally incompressible, somewhat more so
than the curve of the lipid extract of the lipoprotein. Though
all of the preparations attain about the same maximum surface
pressure (40 - 42 dynes/cm) on the NaCl and $CaCl_2$ subsolutions,
surface potential differs: 550 mv for dipalmitoyl lecithin,
410 mv for the lipoprotein fraction and its lipid extract and
400 mv for stock O, and is consistently about 35 to 70 mv higher
with Ca^{++} than with Na^+ of equal normality in the subsolution
(Fig. 3 and Table I).

   Films of all of the preparations remain in the liquid state
on both subsolutions, except those of dipalmitoyl lecithin. On
subsolutions of NaCl dipalmitoyl lecithin films are in the liquid
state below a surface pressure of 35 dynes/cm, in the gel state
between 35 to 40 dynes/cm and in the solid state above 40 dynes/cm
(17). When calcium ions are in the subsolution, lecithin films
change to the solid state at a lower surface pressure level
(33 to 35 dynes/cm).

   The relative surface viscosity of films of dipalmitoyl
lecithin and of the lipoportein fraction is about the same at low
surface pressure, 8 dynes/cm, but at high surface pressure,

30 dynes/cm, dipalmitoyl lecithin is about twice as viscous as the lipoprotein fraction in the presence of Ca^{++} (Table II).

Hydrolysis of films of dipalmitoyl lecithin upon injection of phospholipase A into the subsolution occurs until a film surface pressure of 20 dynes/cm, is attained, but with the lipoprotein fraction hydrolysis continues to a film surface pressure of 30 dynes/cm   (Fig. 4).

DISCUSSION

We studied rabbit lung washings freed of cells and cellular debris to minimize the likelihood of contamination with blood and tissue elements and separated into fractions having distinct chemical composition (8). One of these fractions consistently gives two bands on disc gel electrophoresis, one staining for lipid and the other for protein, suggesting the lipid and protein are associated. The distribution of these bands is distinctly different from those present in rabbit serum.

The lipd-protein association does not occur during centrifugation nor during lyophilization, since there is similar band mobility in the starting material, cell-free stock 0, and in lyophilized and also in non-lyophilized preparations of the lipoprotein fraction (8).

Immunologic studies indicate the liprotein fraction we isolated is indigenous to lung and is not a blood contaminant (8).

Lipid comprises about 63% of the lipoprotein fraction (57% phospholipid and 6% non-phospholipid); the remainder is protein. Of the total lipid, about 90% is phospholipid and 10% non-phospholipid. Lecithin is the predominant phospholipid on thin layer chromatography, having a fatty acid composition similar to egg lecithin, but without polyunsaturated fatty acids normally present. There are also trace amounts of lysolecithin and of phosphatidic acid (8).

The cell-free lung wash, stock 0, and its lipoprotein fraction are normally surface active according to criteria employed by respiratory physiologists, attaining in initially highly compressed films a minimum surface tension of zero or nearly zero dynes/cm upon compression (5 cm/min) from 100% to 20% of the trough area and exhibiting hysteresis on re-expansion to 100% of the trough area (18). For these studies a specially designed trough with a submerged compression bar, avoiding spillage of material over the rim of the trough when compression is continued beyond film collapse area, was used.

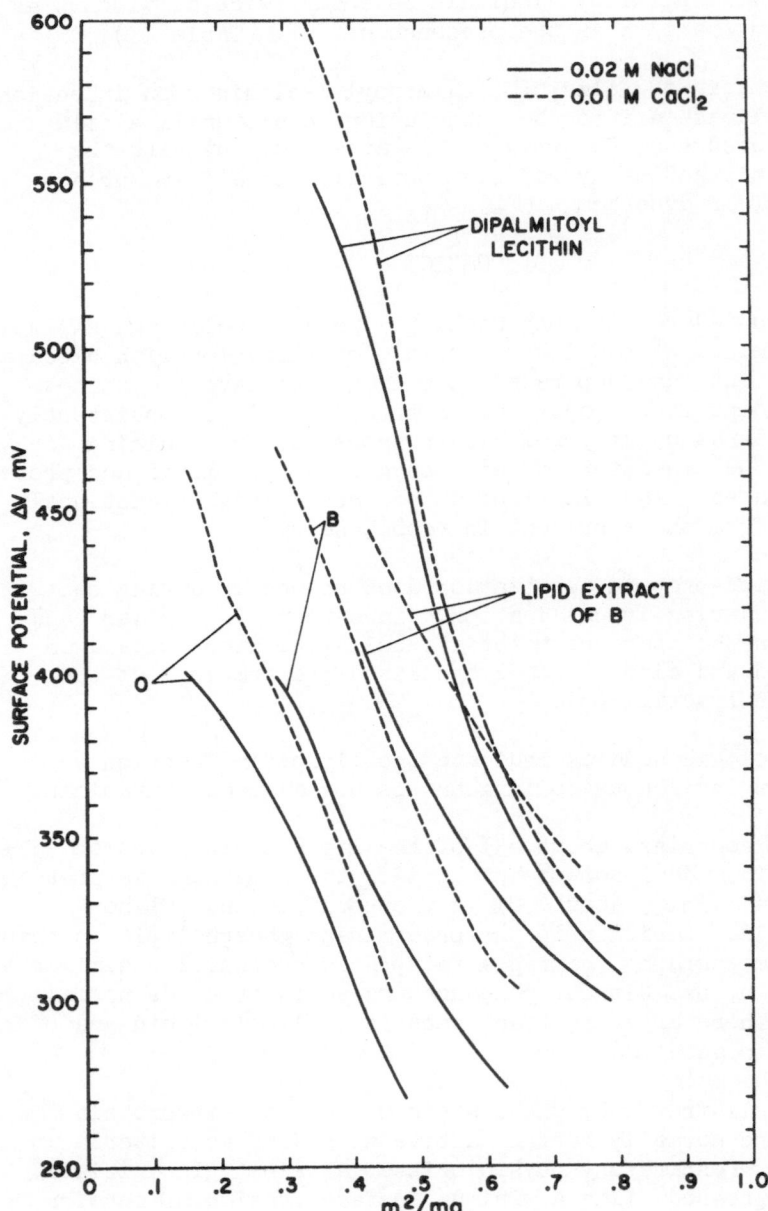

<u>Fig. 3</u>.   Surface potential-area curves of cell-free lung wash,
          stock O, of its lipoprotein fraction, B, of the lipid
          extract of the lipoprotein fraction, B, and of dipalmitoyl
          lecithin on subsolutions of 0.02 M NaCl and 0.01 M CaCl$_2$,
          pH 6.0 and 25 C.

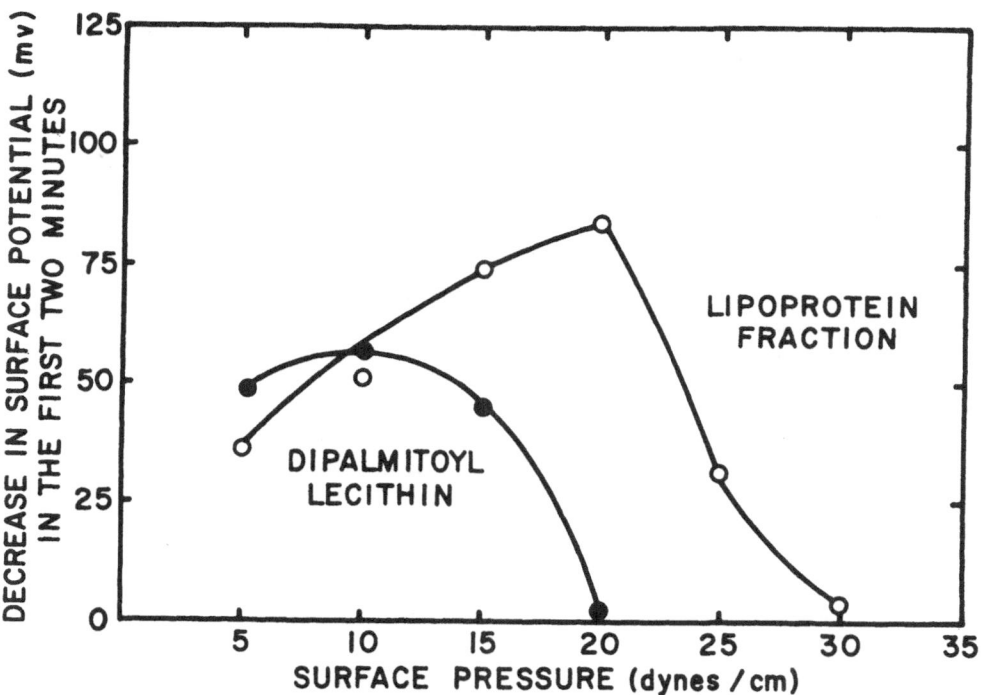

<u>Fig. 4.</u>   Comparison of the rate of enzymatic hydrolysis of
           monolayers of the lipoprotein fraction, B, and of
           dipalmitoyl lecithin at different surface pressure levels
           by snake venom Naja Naja phospholipase A, based on
           decrease in surface potential two minutes after addition
           of the enzyme to the substrate, tris buffer, pH 7.2,
           0.05 M and 0.01 M CaCl$_2$. Each point represents a newly
           spread monolayer.

To investigate the surface properties of the several preparations, we applied parameters employed by surface chemists to investigate protein monolayers, i.e., surface pressure and surface potential related to M^2/mg of material spread and physical state of film.

Though films of stock O, of its lipoprotein fraction and of the lipid extract of the lipoprotein fraction occupy different areas, they all attain about the same collapse pressure (Fig. 2). They also have about the same maximum surface potential on a subsolution of 0.02 M NaCl, (Fig. 3 and Table 1) suggesting at high surface pressure they contain the same chemical species. The kink in the force-area curve of stock O (Fig. 2) suggests the less surface-active material present in the original material is squeezed out of the monolayer at high surface pressure.

To learn about the nature of the lipid-protein association in the lipoprotein fraction, we investigated the interaction of calcium ions with monolayers of this fraction and with its lipid extract, and also enzymatic hydrolysis of the lipoprotein fraction by phospholipase A.

Though the presence of Ca^{++} does not affect the collapse pressures of these monolayers (Fig. 2), surface potential is consistently higher in the presence of Ca^{++} as compared with Na^+ (Fig. 3 and Table I). The increase in surface potential is a measure of the magnitude of the interaction. The lipoprotein fraction interacts somewhat more strongly with Ca^{++} than the starting material, stock O, and the lipid extract of the lipoprotein fraction least of all.

The direction of change in level of surface potential with calcium binding indicates the probable site of interaction and probably orientation of the polar groups of lipid in films of the lipoprotein fraction. Calcium ions could interact with phosphate groups of lipid or carboxyl groups of protein in the lipoprotein. Surface potential decreases with binding of Ca^{++} to carboxyl groups (19) and increases with binding to phosphate groups (13). Since the surface potential of the lipoprotein monolayers increases in the presence of Ca^{++} (Fig. 3 and Table I), calcium binding most likely occurs with phosphate groups in the lipid component of the lipoprotein. This suggests the polar groups of the lipoprotein fraction are exposed to the aqueous phase and are thus available to interact with calcium in the subsolution.

The results of the studies of hydrolysis of monolayers of the lipoprotein fraction by phospholipase A (Fig. 4) tend to support this conclusion. If protein were bound to lipid in this fraction by polar groups or by ionic interaction, the rate of hydrolysis would have been impeded (20). This did not occur, suggesting the polar groups of phospholipid in the lipoprotein are exposed to the water phase. It, therefore, appears that the interaction between protein in the phospholipid in the lipoprotein fraction is likely of the van der Waals type. Additional indirect evidence comes from the occurrence of about the same maximum surface potential in films of the lipoprotein fraction and in those of the lipid extract of the lipoprotein on a subsolution of Na^{++} (Fig. 3 and Table I). This suggests the lipid-protein association in the lipoprotein does not influence the dipole moment of lipid molecules and is not the result of ionic interaction but probably of the van der Waals type.

Since dipalmitoyl lecithin is commonly believed responsible for the surface properties of lung surfactant we compared the surface properties of the lipoprotein fraction and of dipalmitoyl lecithin. Both attain the same collapse pressure, but films of the lipoprotein fraction are much more condensed (Fig. 2). In the presence of Ca^{++} the surface potential of the lipoprotein fraction increases somewhat more, 70 as compared with 50 mv for dipalmitoyl lecithin (Fig. 3 and Table I).

At low surface pressure (8 dynes/cm) films of dipalmitoyl lecithin and of the lipoprotein fraction have the same relative surface viscosity in the presence of Ca^{++} as compared with Na^+ but at a high surface pressure (30 dynes/cm) dipalmitoyl lecithin is about twice as viscous as the lipoprotein fraction due to interaction with Ca^{++} ions (Table II). This is an important difference between the behavior of films of dipalmitoyl lecithin and the lipoprotein fraction.

The properties of films of dipalmitoyl lecithin differ from those of the lipoprotein fraction in other respects, attaining the gel state on a substrate of NaCl at a surface pressure of 35 to 40 dynes/cm, and the solid state above 40 dynes/cm. With Ca^{++} in the subsolution, lecithin films change to the solid state at a lower surface pressure 33 to 35 dynes/cm. Films of the lipoprotein fraction remain in the liquid state at all levels of surface pressure on the NaCl and $CaCl_2$ subsolutions used in these investigations.

The studies of the rate of enzymatic hydrolysis of films of the lipoprotein fraction and dipalmitoyl lecithin by phospholipase A (Fig. 4) provide information about the relative intermolecular spacing in these monolayers. Since hydrolysis ceases in films of dipalmitoyl lecithin at a surface pressure of 20 dynes/cm and in films of the lipoprotein fraction at 30 dynes/cm, it appears the intermolecular spacing between lipid molecules is much smaller in dipalmitoyl lecithin than in the lipoprotein fraction. Therefore,

the active site of the enzyme cannot penetrate dipalmitoyl
lecithin monolayers above 20 dynes/cm and monolayers of the
lipoprotein fraction above 30 dynes/cm.

BIBLIOGRAPHY

1. Radford, E.P., Jr., In "Handbook of Physiology, Vol. 1, p.
   447 Amer. Physiol. Soc., Washington, D.C., 1964, Eds. Fenn,
   W.O. and H. Rahn
2. Pattle, R.E., Advances in Pulmonary Physiology. Ed. C.G.
   Caro, The Williams and Wilkins Co., Baltimore, 1966.
3. Weibel, E.R. and J. Gil, Resp. Physiol. 4 (1968) 42.
4. Buckingham, S., Heinemann, H.O., S.C. Sommers and W.F. McNary,
   Amer. J. Pathol., 48 (1966) 1027.
5. Clements, J.A. and D.F. Tierney. 1965. Vol. II, p. 1573
   Amer. Phsyiol. Soc., Washington, D.C., Eds. Fenn, W.O. and
   H. Rahn.
6. Abrams, M.E., J. Appl. Physiol., 21 (1966) 718.
7. Klein, R.M. and S. Margolis, J. Appl. Physiol., 25 (1968) 654.
8. Galdston, M., D.O. Shah and G.Y. Shinowara, J. Colloid and
   Interface Sci., 29 (1969) 319.
9. Scarpelli, E.M., B.C. Clutario and F.A. Taylor, J. Appl.
   Physiol., 23 (1967) 880.
10. Redding, R.A., C.T. Hauck, J.M. Steem and M. Stein, New
    Engl. J. Med., 280 (1969) 1298.
11. Folch, J., M. Lees and G.H. Sloane Stanley, J. Biol. Chem.
    226 (1957) 497.
12. Davies, J.T. and E.K. Rideal, Interfacial Phenomena,
    Academic Press, 2nd ed., p. 46.
13. Shah, D.O. and J.H. Schulman, J. Lipid Res. 6 (1965) 341.
14. Adam, N.K. 1941, The Physics and Chemistry of Surfaces,
    Oxford Press, London, 3rd., p. 55.
15. Dervichian, D., Nature, 144 (1939) 629.
16. Shah, D.O. and J.H. Schulman, J. Colloid Interface Sci. 25,
    (1967) 107.
17. Galdston, M. and D.O. Shah, Biochim. Biophys. Acta, 137
    (1967) 255.
18. Tierney, F., Dis. Chest, 47 (1965) 247.
19. Spink, J.A. and J.V. Sanders, Nature, 175 (1955) 644.
20. Hughes, A., Biochem. J., 29 (1935) 437.

ABSENCE OF LIPOPROTEIN IN PULMONARY SURFACTANTS

Emile M. Scarpelli and Giuseppe Colacicco

Department of Pediatrics, Albert Einstein College of

Medicine, New York, N.Y. 10461 and York College,

City University of New York, Flushing, N.Y. 11365

The "alveolar-capillary barrier", which separates the air of
the pulmonary alveoli from the blood of the pulmonary capillaries,
has been rather well defined by electron microscopy as consisting
of the capillary endothelial cell, interstices, alveolar epithelial
cell, and acellular lining layer. The latter covers the plasma
membrane of the epithelial cell and is the external limiting surface that makes contact with alveolar air (1,2)(Fig. 1). The
chemical components of the alveolar lining layer, which have been
termed "the surfactant system of the lung" (3), probably include
proteins, carbohydrates, phospholipids and other lipids (2,3,4).

The discovery that saline extracts of pulmonary tissue and
foam obtained from normal lungs are highly surface-active led to
the hypotheses that pulmonary surfactants are essential determi-
nants of alveolar stability (5) and that they modulate alveolar
liquid balance to prevent pulmonary edema (6). Indeed, experimental alteration of the surface properties of the lung results in
alveolar instability, alveolar collapse, and pulmonary edema (7,8).
In addition, several pathological conditions of the lung have been
related to disruption of normal surface properties of the alveoli (3).

The pulmonary surfactants reside in the alveolar lining layer
and are components of the surfactant system (3). As such they
probably form a surface film at the interface between the lining
layer and alveolar air, and thus impart to the lung the surface
properties that are essential for normal function. The idea that
the physiologically important surfactants of the alveolar lining
layer are phospholipids was established by Klaus and co-workers (9).

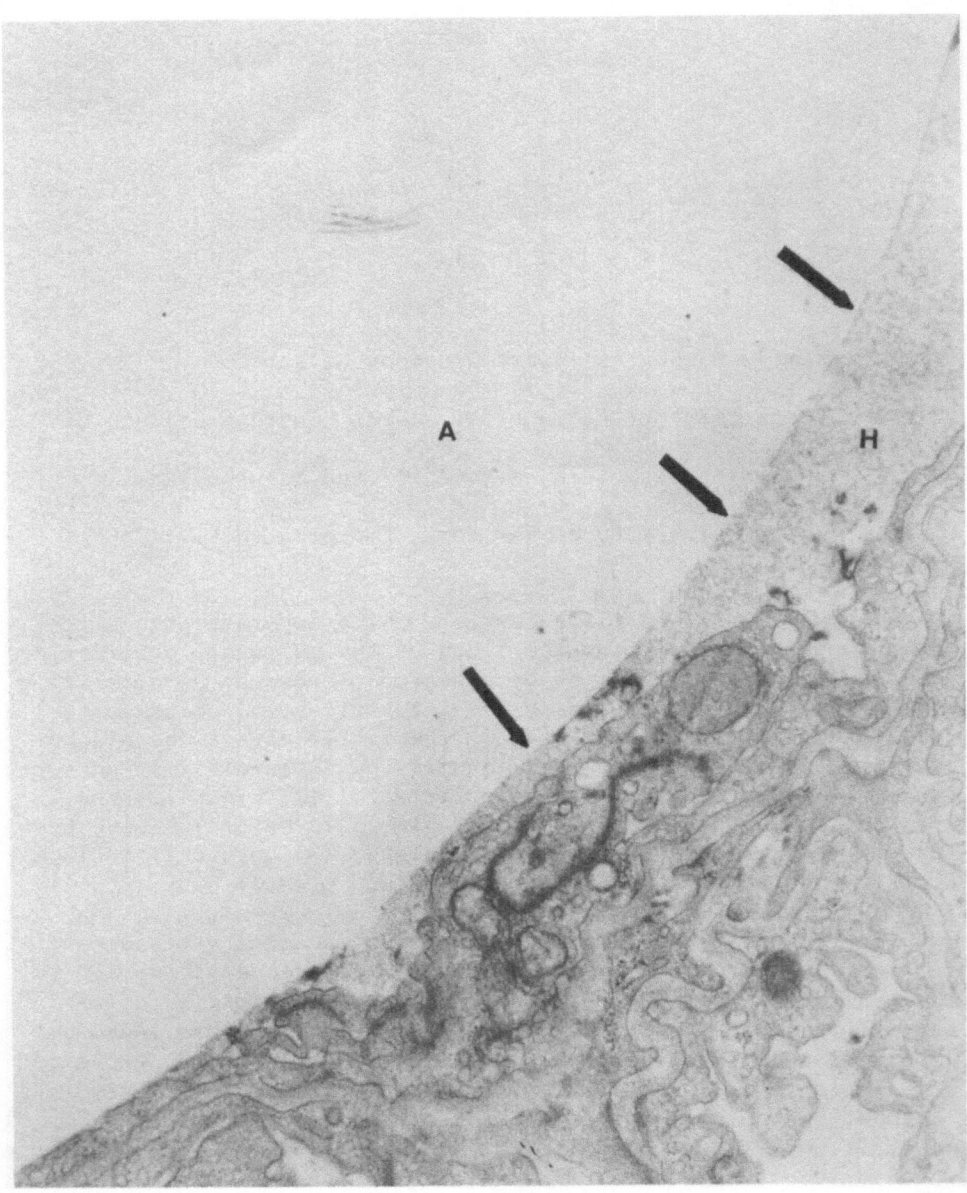

Fig. 1. Electron micrograph of normal rat lung. Magnification 40,000 x. A: alveolus. Arrow: surface film. H: hypophase. (Courtesy of Dr. Yutaka Kikkawa).

This has been confirmed by other investigators (10-13) and it is now apparent that the following phospholipids constitute a "family of pulmonary surfactants": phosphatidyl inositol, lysolecithin, sphingomyelin, phosphatidyl dimethylethanolamine (PDME), and phosphatidyl choline. Since the latter, lecithin, is recovered in highest concentration in pulmonary extracts and the other phospholipids are present in small to trace amounts, it is assumed that lecithin is the essential surfactant.

The natural state of the phospholipid surfactants is not known. It has been generally accepted that they are part of a lipoprotein, notwithstanding the fact that the evidence for this does not exist. Since knowledge of the state of the phospholipids is essential to considerations of the physical and chemical properties of the alveolar lining, we have searched for the lipoprotein in pulmonary extracts. We have found that the phospholipids are not bound to protein, that they cannot be classified as lipoprotein, and that determination of their relationship to other components of the surfactant system, including other lipids, proteins, and carbohydrates, must await further investigation.

PRESUMPTIVE EVIDENCE FOR LIPOPROTEIN SURFACTANTS

Pattle (14,15) and Clements (9,16) and their co-workers were the earliest proponents of the lipoprotein nature of pulmonary surfactants. The evidence upon which the first conslusions were based included the similarity between infrared absorption spectra of pulmonary foam and lecithin-gelatin mixtures (14) and the failure of surface films of total lipid (protein-free) extracts of pulmonary tissue to reduce surface tension below 20 dynes/cm when compressed in a modified Wilhelmy balance (9). Other evidence that was used to support the "lipoprotein theory" included a) the relatively high concentration of nitrogen in films harvested from pulmonary extracts (17), b) the recovery of surface-active precipitates after treatment of pulmonary extracts with trichloroacetic acid (11), and c) the reduced surface-activity of pulmonary foam following incubation with trypsin (14).

More recently three groups of investigators have reported the isolation of a surface-active pulmonary "lipoprotein" (18-20). Abrams (18) working with homogenized and centrifuged pulmonary tissue, isolated a pellicle which floated on NaCl solution of density 1.15 following low speed centrifugation. The pellicle was highly surface-active and contained a protein with the electrophoretic mobility of "α-globulin", in addition to lecithin, phosphatidyl ethanolamine, and other lipids. We repeated these experiments (21) and analyzed the various fractions for protein and lipoprotein content by disc electrophoresis and specific lipid

and protein stains. We have demonstrated that a) the centrifugation method of Abrams neither qualitatively nor quantitatively
isolated the phospholipids in the crude extract, b) the pellicle
contained up to eight different proteins, c) albumin (whose electrophoretic mobility is near that of α-globulin) was present in
highest concentration, and d) there was no lipoprotein as demonstrated by the absence of Sudan Black B staining of the separated
proteins. Klein and Margolis (19) obtained a surface-active
pellicle of phospholipid and protein after ultracentrifugation of
pulmonary extracts in KBr solutions of densities 1.15 and 1.21 at
100,000 g, 0 to 4°C, for up to 18 hours. Although the proteins
were not characterized by any qualitative method, it was assumed
that a lipoprotein had been isolated. We have demonstrated that
Klein and Margolis did not use a sufficiently stringent ultracentrifugation procedure to isolate the surfactants (22,23). The
phospholipid surfactants may be purified by 48 hour centrifugation
at density 1.080, 115,000 g, 15°C; the purified lipid fraction
contains less than 2% protein, which has the electrophoretic mobility of albumin (22,23).

Galdston, Shah and Shinowara (20) reported isolation of pulmonary lipoprotein, which was identified by its reaction with Oil
red O stain following disc electrophoresis. We repeated their
studies (21,24) and found that a) the Oil red O positive band did
not contain phospholipid and was identical with the albumin band
from pulmonary extracts, and b) the color reaction of Oil red O was
quite different from that obtained with several purified serum
lipoproteins. In addition, with the aid of the Sudan Black B
prestaining which gave positive reaction with several serum lipoproteins, we demonstrated that none of the protein bands contained
lipoprotein.

THE NON-LIPOPROTEIN NATURE OF PULMONARY SURFACTANTS

We have conducted several experiments in search of the pulmonary lipoprotein. These have included standard techniques that are
used for the isolation and characterization of serum lipoproteins.

Sephadex G-200 Chromatography

Pulmonary washings were obtained by irrigating excised lungs
from dogs with 0.15M NaCl. Samples of the pulmonary washing were
applied to Sephadex G-200 columns (4). Five fractions were obtained
as indicated by spectrophotometry (280 mμ) of the effluents (Fig. 2).
These were monitored for a) lipid phosphorus, b) phospholipid by
thin layer chromatography, and c) protein and lipoprotein by disc
electrophoresis. These methods have been described elsewhere (4,21,2⁴

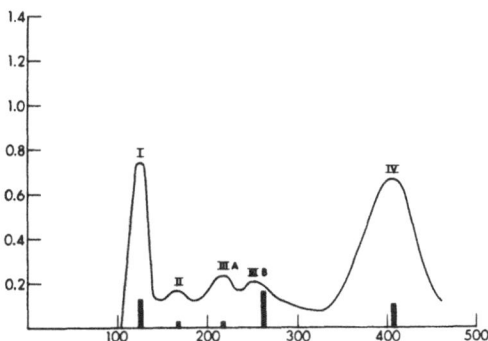

Fig. 2. Absorbance (280 mμ, vertical axis) of effluent from
Sephadex G-200 fractionation of dog pulmonary washing. Horizontal
axis = effluent volume. Fraction I was recovered in the "void
volume". (See reference 4).

A fact of significance is that phospholipid and protein,
which together were supposed to make up the pulmonary "lipoprotein",
were recovered in separate fractions: The phospholipids in fraction
I and the proteins in the other fractions. Relevant to our discussion is fraction I: it was very turbid and contained all the phospholipids of the original extract, including phosphatidyl ehtanolamine, sphingomyelin, lysolecithin (traces), PDME, and lecithin in
highest concentration. Its high absorbance was due to the turbidity
which was eliminated by the addition of sodium dodecylsulfate.
Protein analysis by the quantitative Lowry method, which were performed on concentrated samples of fraction I and on the aqueous
layer after extraction of lipid with chloroform:methanol, revealed
a weight ratio of protein to phospholipid of less than 1/500. Disc
electrophoresis showed no lipoprotein and, in addition, there was
no lipoprotein in the other fractions.

When high density lipoprotein from rat serum was applied to
the same Sephadex G-200 column, only one fraction was obtained in
a sharp peak which was the intact lipoprotein. Thus lipoprotein
could not be demonstrated in pulmonary washings by a method that
has been used successfully by others (25) and us to isolate and
characterize serum lipoproteins.

Preparative Polyacrylamide Electrophoresis

Pulmonary washings of dogs were applied to preparative
polyacrylamide gel columns (7.5% gel concentration) and subjected
to electrophoresis in the Buchler Instruments "Poly-Prep" apparatus (24).

Up to six peaks were recorded spectrophotometrically (280 mμ) in
the effluent (Fig. 3). Each fraction was examined for protein by
disc electrophoresis and for phospholipid by lipid phosphorus
determination and thin layer chromatography. Material was removed
from the top of the separation gel after each run and studied in
the same way. This included components of pulmonary washings that
did not enter the separation gel. Ninety one to 100% of the phospholipids applied were recovered in the top of the gel. Virtually
all the proteins were recovered in the various fractions. The
phospholipid fraction from the top of the gel, which contained
phosphatidyl ethanolamine, PDME, sphingomyelin, lysolecithin,
lecithin, and other lipids, contained a trace of protein with the
electrophoretic mobility of albumin.

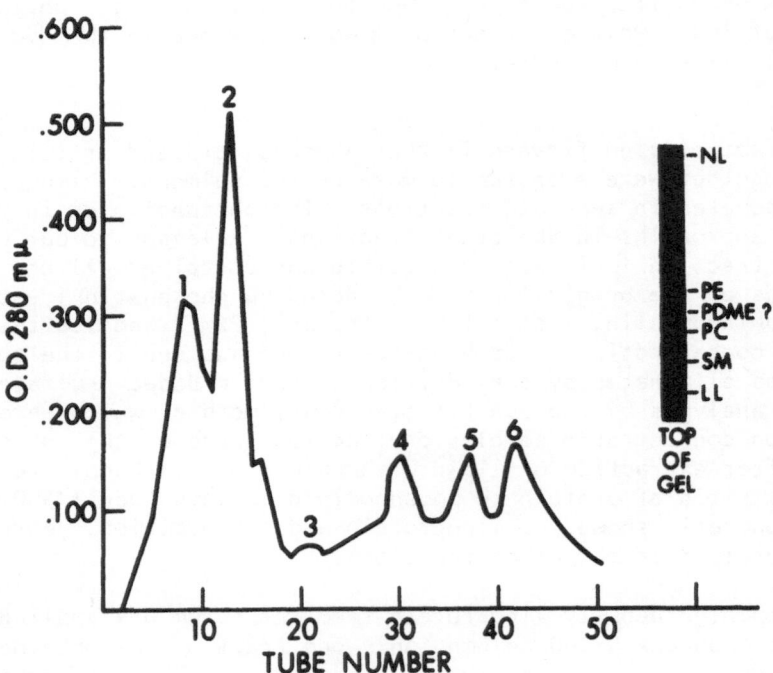

Fig. 3. Preparative electrophoresis of dog pulmonary washings.
Vertical axis: Absorbance at 280 mμ for protein. Unlike with the
lipid Fraction I of Sephadex Column (Fig. 2) the eluates were clear.
All the lipids present in the crude pulmonary washings were found
on top of gel after electrophoresis: NL, neutral lipids (cholesterol, triglycerides, cholesterol esters, and fatty acids); PE,
phosphatidyl ehtanolamine; PDME, phosphatidyl dimethylethanolamine;
PC, phosphatidyl choline; SM, sphingomyelin; LL, lysolecithin.

Selective Staining and Disc Electrophoresis

Two extraction methods, viz., pulmonary washing and mincing
of pulmonary tissue, were used to obtain samples from several
mammalian species including sheep, dog, cat, rabbit, and monkey.
Some samples were mixed with Sudan Black B stain for lipoprotein
prior to disc electrophoresis: other samples were separated by
disc electrophoresis (both 3.75% and 7.5% gels) and the gels were
stained with amido black, a non-specific protein stain; a third
set of extracts was applied to Sephadex G-200 columns and the
effluent fractions were analyzed by the same staining and electrophoretic methods (24). Serum was analyzed in parallel experiments.
None of the proteins in the pulmonary washings reacted with Sudan
Black B, whereas three to eight lipoprotein bands were observed in
the serum samples. The pulmonary minces of dog, sheep, cat, and
rabbit contained no proteins that reacted with Sudan Black B, i.e.,
no lipoprotein. Similarly, each Sephadex fraction from the separation of dog and sheep extracts contained no Sudan Black B-positive
protein. Mince extracts from the lungs of squirrel monkeys, which
were heavily contaminated with blood, contained a slow-moving
lipoprotein band that was recovered in fraction III of the Sephadex
column. This lipoprotein was not associated with the pulmonary
phospholipid surfactants since the latter were recovered in fraction I. A similar lipoprotein band was present in monkey serum.
The results of these experiments are summarized below.

Table 1. Disc electrophoresis of lung surfactant preparations of
normal lung of various species

SUBJECT (NO. OF ANIMALS)	MINCE 3.75% gel	7.5% gel	WASHING 3.75% gel	7.5% gel	SEPHADEX FRACTIONS 1	2	3	SERUM* 3.75% gel	7.5% gel
DOG (5)	0	0	0	0	0	0	0	6	6
SHEEP (3)	0	0	0	0	0	0	0	4-6	3
CAT (2)	0	0	0	0	-	-	-	4	3
RABBIT (2)	0	0	0	0	-	-	-	4-6	4
MONKEY (5)	+	+	0	0	0	0	+	7-8	6-7

* = Number of lipoprotein bands observed.
0 = No lipoprotein.
+ = Probably serum lipoprotein.
- = Not studied.

Thus surface-active pulmonary washings from several species contained no lipoprotein. Mince extracts contained no lipoprotein except for the samples that were highly contaminated with blood. The lipoprotein found in the latter samples was probably a serum contaminant.

Ultracentrifugation

Pulmonary washings were prepared from the lungs of rabbits with 0.15M NaCl, the density was adjusted to 1.080 with solid KBr, and the samples were centrifuged at 115,000 g, 15°C for 48 hours. All the lipids floated as a single fraction into a pellicle. After dilution and purification by two additional centrifugations, the third pellicle contained less than 2 µg protein for 100 µg phospholipid; on disc electrophoresis the protein was albumin, but there was no lipoprotein (22,23). Each of the three pellicles contained all the lipids and the surface activity of the original pulmonary washing.

Conclusions

The several analytical methods that were used to study pulmonary extracts in our laboratory lend no support to the thesis that pulmonary phospholipid surfactants are part of a lipoprotein. These experiments demonstrate that the physical characteristics of the phospholipids are sufficiently different from those of the proteins to permit the separation of the two on the basis of molecular or aggregate size (Sephadex chromatography), electrophoretic mobility (polyacrylamide gel), and density (ultracentrifugation). It should be noted also that each of these methods has been used successfully by us and others to isolate and identify serum lipoproteins. We have also demonstrated that lipoproteins may find their way into pulmonary extracts when the latter are heavily contaminated with blood, which complicates the interpretation of results.

Corroborative Evidence

Several corroborative reports have appeared since our initial characterization of the phospholipid surfactants as non-lipoprotein (4). According to the electron microscopy studies of Kikkawa (1) (Fig. 1) and Weibel and Gil (2), the alveolar lining layer is composed of two phases: a superficial layer of phospholipids and a base aqueous layer probably containing proteins, lipids and mucopolysaccharides. These layers have been referred to as surface film and hypophase (26,27). In the electron micrographs the surface phospholipids appeared as lamellae, rather than as a monomolecular film, probably because of the tissue fixation method that was used; the small repeating distances of the lamellae indicated that proteins were excluded (2). Schematic reconstruction of the electron density pattern from the electron microscope permitted an approximation of molecular orientation in the alveolar lining layer; the picture that emerged supports the thesis that phospholipids are not bound to protein. McClenahan and Ohlsen (28) isolated a surface-active fraction from washings of human lung. The proteins were separated and injected into rabbits. The resulting antiserum reacted with human serum albumin only; a water soluble protein other than normal serum albumin was not found in the pulmonary extracts. It should be noted that albumin is the protein present in highest concentration in pulmonary washings, but that other proteins may also be recovered in the washings (4,21,24). Steim and co-workers (29) purified the surface-active phospholipids from an acellular precipitate of pulmonary washings. The precipitate was layered over a linear sucrose gradient and ultracentrifuged for 20 hours at 4°C. A band centering at density 1.035 contained the surface-active phospholipids and less than 3% protein and carbohydrate. This is in agreement with our own ultracentrifugation studies (22, 23). Finally, it is interesting to note that in the disease of humans "alveolar proteinosis", in which there is a large intra-alveolar accumulation of pulmonary phospholipids, the lipids are not bound to protein and are in fact associated with only very small amounts of soluble protein, as determined by electrophoresis of pulmonary washings from patients with the disease (30).

SUMMARY

The evidence that the surface-active pulmonary phospholipids are not part of a lipoprotein is summarized as follows:

Sephadex separation of PL and Proteins	Scarpelli, et al, (4)
Electrophoretic separation of PL and Proteins	Scarpelli, et al, (21)
Absence of specific LP staining reaction	Scarpelli & Taylor, (31)
Purification of surface-active PL by UC	Steim, et al, (29) Colacicco & Scarpelli, (22,23)
Pulmonary ultrastructure compatible with PL but not with LP	Kikkawa, et al, (1) Weibel & Gil, (2)
Absence of antigenic pulmonary LP	McClenahan & Ohlsen, (28)
No recoverable LP in pulmonary washings	Scarpelli, et al,(4,21-24)
Absence of LP in alveolar proteinosis	Ramirez-R. & Harlan, (30)

(PL = phospholipid, LP = lipoprotein, UC = ultracentrifugation)

The most relevant conclusion is that lung surfactants are phospholipids and not lipoprotein. Any specific pulmonary lipoprotein surfactant must be still demonstrated.

REFERENCES

1. Kikkawa, Y., E.K. Motoyama and C.D. Cook. The ultrastructure
    of the lungs of lambs. Amer. J. Pathol. 47:877-903 (1965).

2. Weibel, E.R. and J. Gil. Electron microscopic demonstration
    of an extra-cellular duplex lining layer of alveoli. Resp.
    Physiol. 4:42-57 (1968).

3. Scarpelli, E.M. The Surfactant System of the Lung. Lea and
    Febiger Publishers, Philadelphia (1968).

4. Scarpelli, E.M., B.C. Clutario and F.A. Taylor. Preliminary
    identification of the lung surfactant system. J. Appl.
    Physiol. 23:880-886 (1967).

5. Clements, J.A. Surface phenomena in relation to pulmonary
    fraction. Physiologist 5:11-28 (1962).

6. Pattle, R.E. Properties, function and origin of the alveolar
    lining layer. Proc. Roy. Soc. (London) B148:217-240 (1958).

7. Pattle, R.E. and F. Burgess. The lung lining film in some
    pathological conditions. J. Pathol. Bacteriol. 82:315-331,
    (1961).

8. Yoshida, S. Effects of surfactants or fat solvent on static
    pressure-volume hysteresis of excised dog lung. Amer. J.
    Physiol. 203:725-730 (1962).

9. Klaus, M.H., J.A. Clements and R.J. Havel. Composition of
    surface-active material isolated from beef lung. Proc. Natl.
    Acad. Sci. 47:1858-1859 (1961).

10. Fujiwara, T. and F.H. Adams. Isolation and assay of surface-
    active phospholipid components from lung extracts by thin-
    layer chromatography. Tohoku J. Exptl. Med. 84:46-54 (1964).

11. Brown, E.S. Isolation and assay of dipalmityl lecithin in
    lung extracts. Amer. J. Physiol. 207:402-406 (1964).

12. Morgan, T.E., T.N. Finley and H. Fialkow. Comparison of the
    composition and surface activity of "alveolar" and whole lung
    lipids in the dog. Biochim. Biophys. Acta 106:403-413 (1965).

13. Gluck, L., E.K. Motoyama, H.L. Smits and M.V. Kulovich. The
    biochemical development of surface activity in mammalian
    lung. I. Pediat. Res. 1:237-246 (1967).

14. Pattle, R.E. and L.C. Thomas. Lipoprotein composition of the film lining the lung. Nature 189:844 (1961).

15. Pattle, R.E. Surface tension and the lining of the lung alveoli, In Advances in Respiratory Physiology, C.G. Caro (Ed.), Williams and Wilkins Publishers, Baltimore (1966), pp. 83-105.

16. Clements, J.A. The alveolar lining layer, In Development of the Lung, A.V.S. deReuck and R. Porter (Eds.), Little, Brown and Co., Publishers, Boston (1967), pp. 202-221.

17. Buckinghsm, S. Studies on the identification of an anti-atelectasis factor in normal sheep lung. Amer. J. Dis. Child. 102:521-522 (1961).

18. Abrams, M.E. Isolation and quantitative estimation of pulmonary surface-active lipoprotein. J. Appl. Physiol. 21:718-720 (1966).

19. Klein, R.M. and S. Margolis. Purification of pulmonary surfactants by ultracentrifugation. J. Appl. Physiol. 25:654-658 (1968).

20. Galdston, M., D.O. Shah and G.Y. Shinowara. Isolation and characterization of a lung lipoprotein surfactant. J. Colloid Interface Sci. 29:319-334 (1969).

21. Scarpelli, E.M., S.J. Chang and G. Colacicco. Evaluation of methods for isolation and identification of pulmonary surfactants. (Submitted for publication).

22. Colacicco, G. and E.M. Scarpelli. Purification of pulmonary surfactants by ultracentrifugation. (Manuscript in preparation).

23. Colacicco, G. and E.M. Scarpelli. Physico-chemical properties of lung surfactants, this symposium.

24. Scarpelli, E.M., S.J. Chang and G. Colacicco. A search for the surface-active pulmonary lipoprotein. (Submitted for publication).

25. Fleischer, B. and S. Fleischer. Interaction of the protein moiety of bovine serum α-lipoprotein with phospholipid micelles. II. Isolation and characterization of the complexes. Biochim. Biophys. Acta 147:566-576 (1967).

26. Heinenmann, H.O. and A.P. Fishman. Nonrespiratory functions
     of mammalian lung. Physiol. Rev. 49:1-47 (1969).

27. Scarpelli, E.M., K.H. Gabbay and J.A. Kochen. Lung surfac-
     tants, counterions and hysteresis. Science 148:1607-1609
     (1965)

28. McClenahan, J.B. and J.D. Ohlsen. Protein components of human
     surfactant. Clin. Res. 16:134 (1968).

29. Steim, J.M., R.A. Redding, C.T. Hauck and M. Stein. Isolation
     and characterization of lung surfactant. Biochem. Biophys.
     Res. Comm. 34:434-440 (1969).

30. Ramirez-R, J. and W.R. Harlan, Jr. Pulmonary alveolar
     proteinosis. Amer. J. Med. 45:502-512 (1968).

31. Scarpelli, E.M. and Taylor, F.A. Unpublished observation.

RELATION OF WATER TRANSPORT TO WATER CONTENT IN SWELLING

BIOLOGICAL MEMBRANES

Joel L. Bert and Irving Fatt

Department of Mechanical Engineering, University of

California, Berkeley, California 94720

ABSTRACT

In biological and some synthetic polymer materials, which
swell on imbibing water (corneal stroma, costal and articular
cartilage, skin, cellulose acetate, and polymethyl methacrylate),
the phenomenological parameter, k/η, relating rate of water flow
to pressure gradient in Darcy's law [$q = -(k/\eta)(dP/dx)$] has been
found to be a function of water content. Log k/η appears to be
linearly related to log H, where H is grams of water/grams of
dry swelling material. A pore model predicts an increase in "pore"
radius with increasing hydration. The "pore" radius of the corneal
stroma predicted by the model is in agreement with the radius
calculated from light scattering and diffusion data.

INTRODUCTION

Most biological materials have a fixed water content in a
living system. If these materials are dehydrated by artificial
means they will reimbibe water to restore the normal water content. Some biological materials have a smaller water content
in situ than they can hold if simply immersed in water. The
corneal stroma is a well-studied example of such a system. An
"active" process is supposedly operating in the living system to
keep the water content of the corneal stroma to its normal level.[1]

Fatt and Goldstick[2] have shown that the transient water
imbibition process in swelling membranes can be described by an
equation analogous to the transient diffusion equation. The Fatt

and Goldstick equation is

$$\frac{\partial H}{\partial t} = D(H) \frac{\partial^2 H}{\partial \psi^2} \tag{1}$$

where the terms of this and all subsequent equations are defined in the nomenclature section. Fatt[3] has shown that the water transport coefficient $D(H)$, in corneal stroma, is given by

$$D(H) = \frac{k}{\eta} \frac{\varepsilon^2}{(\varepsilon+H)} \frac{dP}{dH} \tag{2}$$

The flow conductivity term, k/η, in equation (2) is the same term that appears in Darcy's law,

$$q = -\frac{k}{\eta} \frac{dP}{dx} \tag{3}$$

where equation (3) is for steady state flow of a liquid of viscosity η in a one-dimensional system.

Both transient and steady state studies of water flow in swelling materials (Mishima and Hedbys,[4] Fatt,[3] Bert and Fatt[5]) show that k/η is a function of hydration of the material.

If the swelling material is assumed to consist of a matrix of solid material through which the water flows, then a pore model can be hypothesized and tested. In such a model, the pore radius will be a function of hydration.

PORE MODEL

If in an area A_T of a membrane there are n pores, all of radius r and length ℓ, then Poisueille's law gives the flow rate as,

$$Q = n \pi r^4 \Delta P/8 \eta \ell \tag{4}$$

For the same membrane Darcy's law gives

$$Q = k A_T \Delta P/\eta L \tag{5}$$

Equating (4) and (5) and letting the tortuosity term be τ, where $\tau = \ell/L$, we obtain

$$k A_T = n \pi r^4/8 \tau \tag{6}$$

The volume of water in the membrane is

$$V_w = n \pi r^2 \ell \tag{7}$$

Combining equations (6) and (7) gives

$$k A_T = V_w r^2 / 8 \tau \ell \tag{8}$$

The volume of dry membrane material is assumed to be generated by moving a slice with area of solid material A_d through a distance L. Then

$$V_d = A_d L \tag{9}$$

Equation (9) assumes that the fractional area of solid material in any plane is equal to the fractional volume of solid in the total membrane volume. In membranes with high solids content this is only a gross approximation, but it is sufficient for our purposes. By the same argument the area of water is given by

$$A_w = V_w / L \tag{10}$$

Combining equations (7), (9), and (10) gives

$$A_T = V_d / L + n \pi r^2 \tau \tag{11}$$

Combining equations (7), (8), and (11) gives

$$k = (V_w r^2) / 8 V_d \tau^2 (1 + V_w / V_d) \tag{12}$$

If γ is the density of water and σ is the density of the dry material, then the definition of H leads to

$$H = \gamma V_w / \sigma V_d \tag{13}$$

If $\gamma / \sigma = \varepsilon$ then

$$H = \varepsilon V_w / V_d \tag{14}$$

Combining equations (12) and (14) and dividing both sides by η gives

$$\frac{k}{\eta} = H r^2 / 8 \tau^2 \eta (\varepsilon + H) \tag{15}$$

Equation (15) shows that if there is a unique relation between k/η and H then r^2/τ^2 must also be unique for that combination of k/η and H.

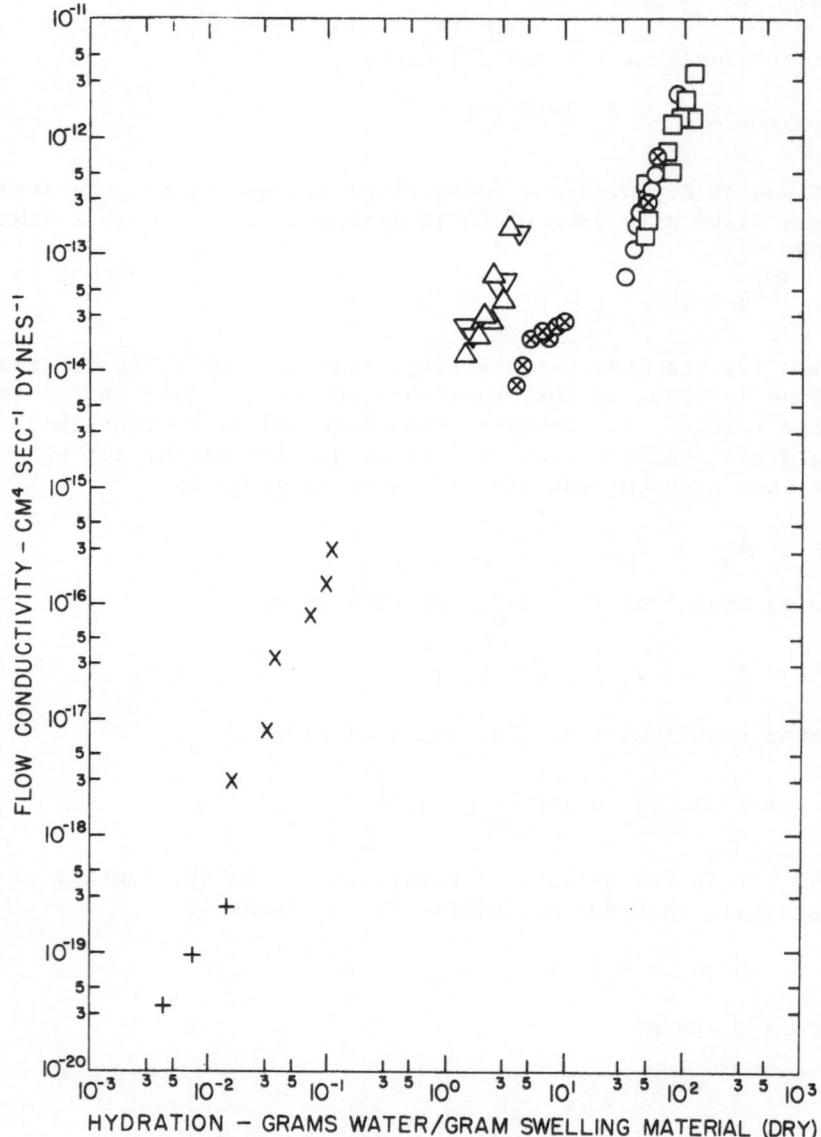

Figure 1.   Flow conductivity, k/η, as a function of
hydration, H. The materials represented
are: Steer corneal stroma (□), rabbit
corneal stroma (○), skin (stratum corneum)
(⊗), steer costal cartilage (△), steer
articular cartilage (▽), cellulose acetate
(×), and polymethyl methacrylate (+).

TEST OF PORE MODEL

Figure 1 shows experimentally determined values of k/η as a function of H for various biological and synthetic membranes. In this work hydration is considered to be the ratio of water to dry swelling material. This hydration differs from the ratio of water to all dry material for biological tissues, but is the same for homogeneous synthetic polymers shown on the graph. The hydration used here is not the same as the hydration described by previous authors,[3,5] who considered the ratio to include all dry material. It is assumed in our treatment of water flow in swelling membranes given here, that water flows only in the swelling matrix. The non-swelling material in the corneal stroma, for example, is so widely spaced as to offer no resistance to water flow. Consequently, when dealing with the swelling phenomenon it appears that the non-swelling material is present only for structural reasons and should not, therefore, be included when describing the hydration of the material.

The data of Figure 1 appear to group along a line whose equation is

$$k/\eta = a \, H^b \qquad (16)$$

Combining equations (15) and (16) and solving for r gives

$$r = [8 \, a \, H^{b-1} \, \tau^2 \, \eta \, (\varepsilon + H)]^{1/2} \qquad (17)$$

Figure 2 shows a plot of equation (17) for r/τ (or $\tau = 1$) and for $\tau = 5$.

Two sets of experimental data are available for testing Figure 2. Maurice[1] measured the "pore" size in corneal stroma by observing the maximum size of a molecule that will diffuse in this tissue. He estimated the pore diameter to be 120 Å. Hart and Farrell;[6] and Hart, Farrell, and Langham,[7] using light scattering data, estimated the distance between polymer chains in the ground substance of the stroma, equivalent to the pore diameter of our model, to be 210 Å. Before Figure 2 can be used to estimate an equivalent pore radius, a proper tortuosity must be determined. It is well known that porous materials with large pores have tortuosities approaching unity; and materials with small pores have higher tortuosity factors. In the case of the corneal stroma, a tortuosity approaching unity is an appropriate choice. The hydration of the normal, *in vivo*, corneal stroma is 76. At this hydration then, Figure 2 would estimate a pore radius of approximately 75 Å, or a diameter of 150 Å, for the corneal stroma. This result seems reasonable when compared with the two pore radius estimations that have already been mentioned.

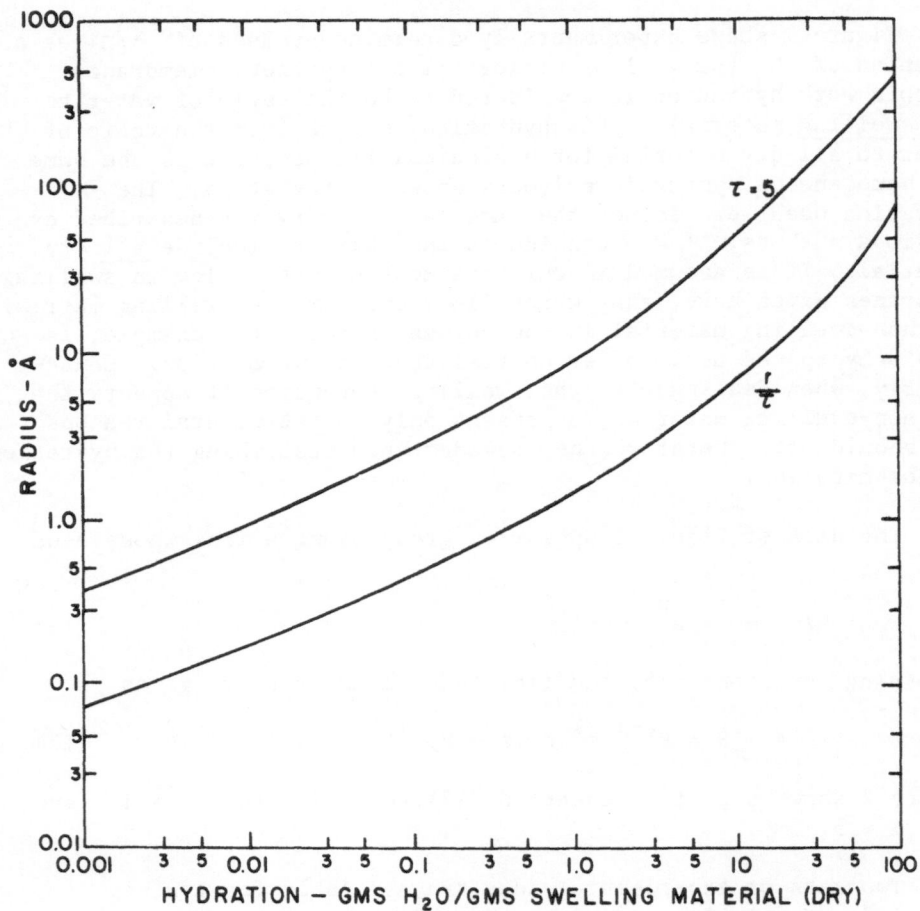

Figure 2.   Equivalent pore radius versus hydration for
materials of Figure 1, according to Equation
(17).

Moreover, Longuet-Higgins and Austin[8] point out that in
membranes with pore diameters greater than 4.5 Å most of the water
transport will be by a convective process, whereas in membranes
with pore size less than 4.5 Å a diffusive process will predominate.
Bert, Fatt, and Saraf[9] have shown that in cellulose acetate, at
hydrations between 0.02 and 0.08, the diffusive and convective
fluxes are comparable. Below this range the convective flux falls
off rapidly, allowing most of the transport to be by diffusion.
For cellulose acetate one would expect the tortuosity to be much
greater than one. Carman[10] has suggested that tortuosity is five

for packs of granulated materials. This value is probably applicable for polymeric membranes. For cellulose acetate then, using Figure 2, one would predict that from a diameter of about 3 Å and smaller diffusion is the primary water transport mechanism. On the other hand, for the biological membranes, whose minimum pore diameter from Figure 2 would be about 4 Å, it is known that convection is the significant water transport mechanism. These values of pore diameter seem to agree fairly well with the values theoretically predicted by Longuet-Higgins and Austin.[8]

Finally, although none of the agreement is brilliant, the results of a pore model as given by equation (17) and as shown in Figure 2, do seem reasonable when tested against the results of Maurice;[1] Hart and Farrell;[6] Hart, Farrell, and Langham;[7] and Longuet-Higgins and Austin.[8]

This work was supported in part by National Institute of Health grants HE-06796 and GM-1418, and by the Petroleum Research Fund administered by the American Chemical Society.

The authors wish to express their sincere appreciation to Pru Mann and Dr. Helen Fatt, for all they have done for us.

NOMENCLATURE

A_d	=	effective surface area of membrane occupied by dry material
A_T	=	surface area of membrane
A_w	=	effective surface of membrane occupied by water
$D(H)$	=	water transport coefficient
H	=	Hydration, grams of water/grams dry swelling material
k/η	=	flow conductivity
ℓ	=	length of pore in membrane
L	=	thickness of membrane
n	=	number of pores per A_T surface area
P	=	swelling pressure in equation (2), and total pressure (i.e.--swelling plus hydrostatic, etc.) in other equations
q	=	volumetric flow rate across area A_T
Q	=	volumetric flow rate
r	=	pore radius
t	=	time
V_d	=	volume of dry membrane material
V_w	=	volume of water in membrane
x	=	distance variable
γ	=	density of water
ε	=	ratio of densities of water to dry membrane material
η	=	viscosity of transported material, water
σ	=	density of dry membrane material
τ	=	tortuosity, ℓ/L

BIBLIOGRAPHY

1. D. Maurice in *The Eye*, ed. by H. Davson, Vol. 1, 2nd ed., 489-601, Academic Press, New York (1969).
2. I. Fatt and T. K. Goldstick, J. Colloid Science, 20, 962-989 (1965)
3. I. Fatt, Exptl. Eye Research, 7, 413 (1968).
4. B. O. Hedbys and S. Mishima, Exptl. Eye Research, 1, 262-275 (1962).
5. J. L. Bert and I. Fatt, Life Sciences, 8, Part 1, 343-346 (1969).
6. R. W. Hart and R. A. Farrell, J. Optical Soc. of America, 59 (6), 766-774 (1969).
7. R. W. Hart, R. A. Farrell, and M. E. Langham, APL Tech. Digest, Applied Physics Lab., Johns Hopkins Univ., Baltimore, 8 (3), 2-11 (1969).
8. H. C. Longuet-Higgins and G. Austin, Biophysics J., 6 (2), 217-224 (1966).
9. J. L. Bert, I. Fatt, and D. Saraf, J. Polymer Sci., in press.
10. P. C. Carman, *Flow of Gases Through Porous Media*, Academic Press, New York (1956).

KINETIC AND EQUILIBRIUM BEHAVIOR OF SIMPLE SUGARS IN A WATER-BUTANOL-LIPID SYSTEM.

Thomas Jenner Moore, M.D., Ph.D.

St. Luke's Hospital Center, New York, New York

College of Physicians and Surgeons, New York, New York

A word of explanation as to why a pediatrician becomes interested in the plasma membrane seems appropriate. Premature human infants are often found to have plasma glucose levels so low that convulsions would ensue were such levels detected in an older child.[1] Yet clinically these babies generally flourish. To further complicate matters, these prematures compared to older children, when given a glucose load, show a prolongation in the rate of disappearance of glucose from the plasma of their extremities.[2]

Certain newly born mammals, rats for example, show a difference in toxicity of morphine with age. This difference has been shown to depend on factors that alter drug uptake in the brain.[3]

These examples suggest that the rate of movement of certain small molecules across plasma membranes differs in the very young when compared to his elder.

In addition, a score of diseases have been described in children where the primary defect is thought to be a disorder in the active or passive movement of small molecules across a cell membrane.[4]

An understanding of any of these related processes is lacking because there exists no satisfactory description of the biophysical and biochemical mechanisms involved in mammalian transport. It is held that a detailed knowledge of simpler forms of biologic transport will provide insights towards solving these clinical problems.

In the nineteenth century, Overton[5] called attention to the correlation between the rate of penetration of a solute into a cell

and its olive oil-water partition coefficient. This lipoid solubility theory was re-examined quantitatively twenty years ago by
Collander and Barlund.[6]   Striking correlations between the rate
of penetration of a variety of small hydrophilic molecules into
the cell and their oil/water partition were demonstrated. Jacobs
and others showed this relationship holds for the human erythrocyte.
Many physiologically important solutes, including sugars, were not
examined. Collander believed that their distribution coefficient
in the lipoid-like organic solvents he employed would be of the
magnitude of 1/100,000,[6] too low for meaningful direct measurements.

If the distribution coefficients of sugars are of this magnitude, penetration into the cell should occur very slowly indeed.
Yet many simple sugars rapidly enter the red blood cell. It is
believed that a specialized mechanism of penetration resides in
the plasma membrane facilitating the diffusion of these solutes
into the cell. This mechanism is termed mediated transport.

There exist certain well defined criteria for identifying
such a mediated transport system.[7]   It operates exclusively with
the existing electrochemical gradient of the permeant and leads to
the disappearance of this gradient.

As stated, the rate of penetration exceeds the rate predicted
by the lipoid solubility theory.

The rate of penetration is not proportional to concentration
but reaches a limiting or saturation value as the concentration is
increased.

The entry rates for particular sugars are species-dependent.
The conformational properties of the penetrating sugar molecule
are thought to be the determinant of this selectivity. As the
number of bulky groups in the axial position increases, the rate
of penetration decreases. In addition, optical enantiomorphs have
been shown to possess different rates of penetration.[8]

A difference in penetration rate exists depending on whether
net transfer or unidirectional flux is measured.

Finally the penetration rate can be reduced by the presence
of small amounts of inhibitors.

The membrane constituents participating in this process have
not been identified with certainty.

Membrane phospholipid might be directly involved in the sugar
transport system of the human erythrocyte. Maudsley and Widdas[9]

suspending erythrocytes in an environment of ^{14}C labelled glucose, recovered the label combined to a lipid fraction having the chromatographic properties of triphosphoinositide. Labelling of this fraction could largely be eliminated by pretreating the ghosts with an inhibitor of transport dinitrofluorobenzene.

Lefevre[10] has recently demonstrated that erythrocyte membrane phospholipid will markedly accelerate the transport of d-glucose from an aqueous phase through a chloroform phase into a second aqueous phase. The output rate into the second aqueous phase was found to be directly proportional to the input glucose level without limit. In addition, he found that the comparative transfer of glucose, ribose, and inositol in this system did not mimic their penetration rate in the intact erythrocyte.

This laboratory has been attempting to construct qualitative analogies between the penetration of certain sugars into the human erythrocyte and the behaviour of these sugars in a water-butanol and lipid system.

MATERIALS AND METHODS

The principle of the Schulman chamber[11], the model membrane used in this laboratory, is that two immiscible liquids, having a stable interface are stirred at the same rate by two blades mounted on a single stirrer shaft. An amphiphilic solute, lipid, introduced into the less polar butanol phase forms a film at the interface. Lipid can also be detected in the bulk aqueous and butanol phases. Radioactive, hydrophilic molecules (sugars in this case) are introduced into the more polar aqueous phase and their rate of appearance in the butanol is measured. Different variables of the system can be changed to measure their influence on the translocation process.

The apparatus consists of battery jars with glass covers and double-bladed stirrers on each stirring shaft. The stirring blades were rotated at a constant speed of 15 rpm. A six unit Phipps & Bird electric stirrer allowed four simultaneous experiments to be conducted. For all experiments 10 ml of 1 M sodium and potassium propionate (included to provide a counter-ion for the interfacial lipid), 200 ml of 1-butanol, and sufficient distilled water to bring the final volume of water--after addition of the sugar solution--to 280 ml, the two layers being mutually saturated, were placed in the jars and allowed to equilibrate. In some experiments, 4 mg of total lipid from erythrocytes was introduced into the butanol in each cell. Then a solution of D-glucose-U-^{14}C or D-galactose-U-^{14}C was injected into the aqueous phase, so that the sugar concentration of the aqueous phase at time zero was either

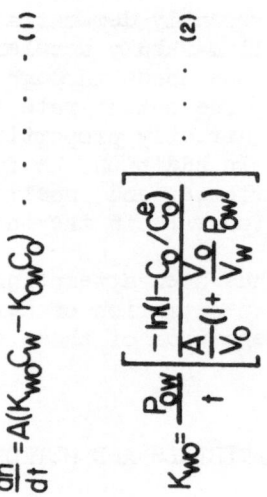

$$\frac{dn}{dt} = A(K_{WO}C_W - K_{OW}C_O) \quad \cdots \cdots \cdots (1)$$

$$K_{WO} = \frac{P_{OW}}{t} \left[\frac{-\ln(1 - C_O/C_O^e)}{\frac{A}{V_O}(1 + \frac{V_O}{V_W}P_{OW})} \right] \quad \cdots (2)$$

THE SCHULMAN CHAMBER

Fig. 1.   The Schulman Chamber and the equations describing
          solute translocation in the system.

0.015 M or 1.3 M. Eight 1.0 ml samples were taken from the butanol phase at 10 min. intervals. After the samples had been collected, the whole system was vigorously stirred to equilibrate the sugar between both phases, and allowed to clear for several hours before samples were drawn from each phase for the determination of the butanol-water distribution coefficient (P_{OW}). Samples were transferred to scintillation vials containing 15 ml of Bray's solution and counted in a Packard model 4000 liquid scintillation spectrometer. Counts per minute were converted to disintegrations per minute by means of an external standard correlation curve. The temperature in the Schulman chamber was maintained at 34°C in a 120 liter circulating water bath.

THEORETICAL CONSIDERATIONS[12]

The kinetic equation for measuring the interfacial transfer coefficient in the Schulman chamber has already been derived. It states that

$$dn_O/dt = A(K_{WO}C_W - K_{OW}C_O) \qquad \text{Equation 1}$$

$$K_{WO} = (P_{OW}) - \ln\left[1 - (C_O/C_O^e)\right] \quad At(aP_{OW} + 1)/V_O^{-1}$$
$$\text{Equation 2}$$

where C_O = concentration of the permeator in the butanol phase at time t; C_O^e = concentration of the permeator in the alcohol phase at equilibrium, $t = \infty$; C_W = concentration of the permeator in the aqueous phase at time t; n_O = number of moles of permeator in the butanol phase; P_{OW} = distribution coefficient for the volumes of butanol and water used = C_O^e/C_W^e; A = area of the interface; K_{WO} = interfacial transfer coefficient from water to butanol (cm-hr^{-1}; V_O = volume of the alcohol phase; and $a = V_O/$(Volume of the aqueous phase.)

RESULTS

The kinetic and equilibrium experiments were conducted at 34°C in every experiment. Figure 2, a representative experiment, is a plot of $K_{WO}t$ (Calculated from Equation 2) against time in minutes. The slope represents K_{WO}, the interfacial transfer coefficient from water into oil. The figure shows that the K_{WO} for D-galactose is greater when erythrocyte phospholipid is added to the system already containing cholesterol.

Table 1 summarizes the results of numerous experiments where the introduction of the phospholipid species into the chamber was varied. In these experiments d-galactose is the translocating sugar. It is seen that both human erythrocyte phospholipid

A TYPICAL PLOT OF $(P_{ow})\left[-\ln\left(1-\dfrac{C_0}{C_0^e}\right)\right]\Big/\left(\dfrac{A}{V_0}\right)(aP_{ow}-1)$

VS. t AT 1.3 M SUGAR. THE SLOPE IS K_{wo}, THE INTER-
FACIAL TRANSFER COEFFICIENT. D-GALACTOSE IS THE
PERMEATOR.

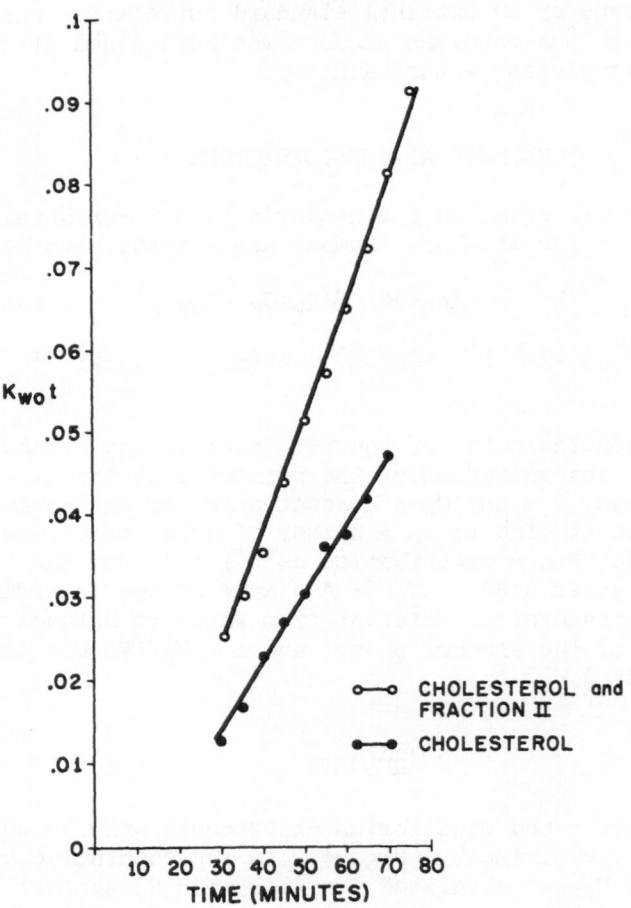

Fig. 2

fractions eluted from the DEAE cellulose column accelerate the
translocation process relative to cholesterol alone. Commercially
available (Applied Science Laboratories) individual phospholipids
known to reside in the human erythrocyte membrane are seen to have
a similar effect.

Table 1

INTERFACIAL TRANSFER COEFFICIENT (K_{wo}) WATER INTO OIL
FOR CHOLESTEROL AND PHOSPHOLIPID IN THE CHAMBER

Lipid Species*	n	K_{wo} D-Galactose	S.D.
Cholesterol 0.5 mgm**	20	0.836	.196
Fraction II 0.5 mgm***	3	1.570	.089
Fraction III 0.5 mgm****	6	1.548	.384
P. Ethanolamine 0.5 mgm**	9	1.071	.106
Lecithin 0.5 mgm**	6	1.522	.277
Sphingomyelin 0.5 mgm**	6	1.915	.321
P. Serine 0.5 mgm**	6	1.568	.560
P. Inositol 0.5 mgm**	9	1.358	.467

*3.5 mgm Cholesterol is in every chamber.
**Applied Science Laboratories
***Human erythrocyte Lecithin, Lysolecithin, P. Ethanolamine,
    Sphingomyelin
****Human erythrocyte P. Serine, P. Inositol

To determine whether the rate of movement of sugar into the
butanol phase reaches a limiting value as the initial concentration
of sugar in the aqueous phase is increased, the initial flux of
glucose into the oil phase with 4 mgm of human erythrocyte lipid
was studied as a function of initial glucose concentration. The
flux dn_o/dt was calculated by solving Equation 2 after experimentally determining K_{wo}. Each point represents the mean of three to
five experiments. Figure 3 indicates that a limiting value is
achieved when the initial aqueous glucose concentration exceeds
0.6 M, the translocation process appears to be saturated.

The previous experiments indicated that the partition coefficient, measured to calculate the interfacial transfer coefficient, varied depending upon the sugar and lipid species in
the chamber. Therefore we decided to ask what affect complex lipid,
known to reside in the human erythrocyte membrane had on the
partition coefficients of a variety of simple sugars. All constituents introduced into the Schulman chamber were reduced in
volume by a factor of fourteen and introduced into small vials.
The vials were vigorously shaken, equilibrated at 34°C for 24 hours

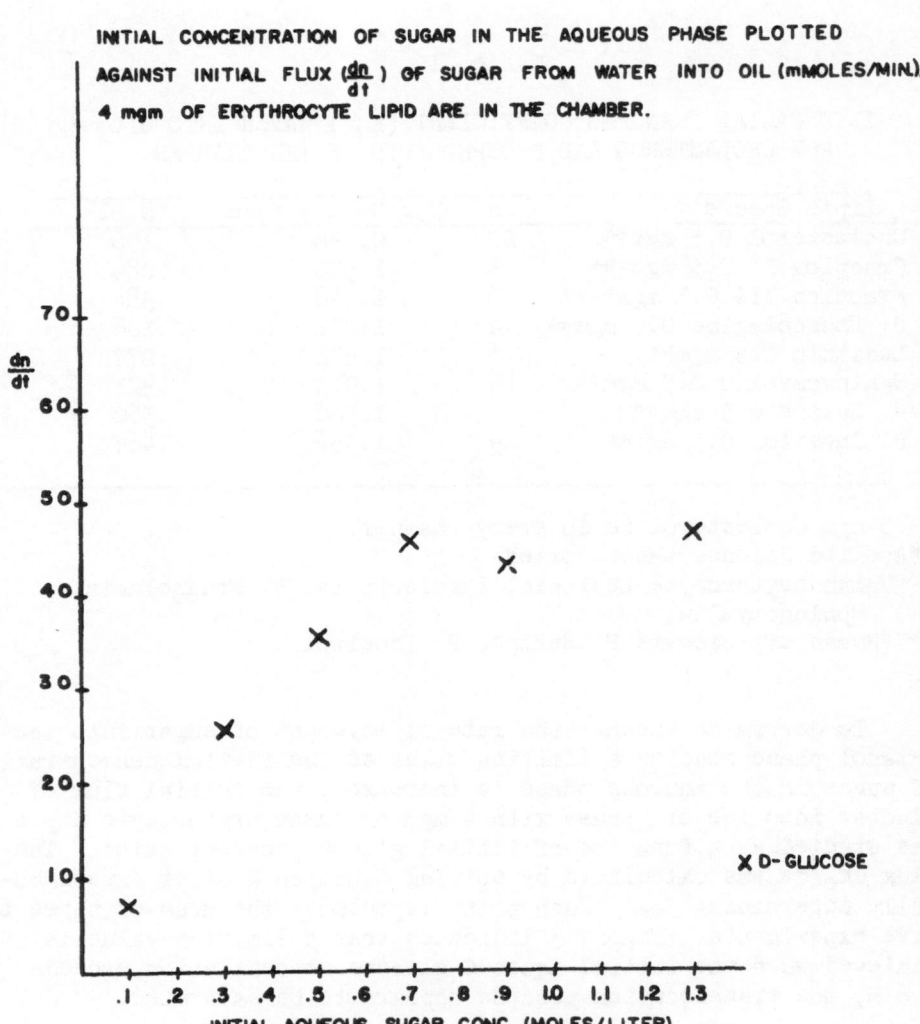

INITIAL CONCENTRATION OF SUGAR IN THE AQUEOUS PHASE PLOTTED
AGAINST INITIAL FLUX ($\frac{dn}{dt}$) OF SUGAR FROM WATER INTO OIL (mMOLES/MIN).
4 mgm OF ERYTHROCYTE LIPID ARE IN THE CHAMBER.

Fig. 3

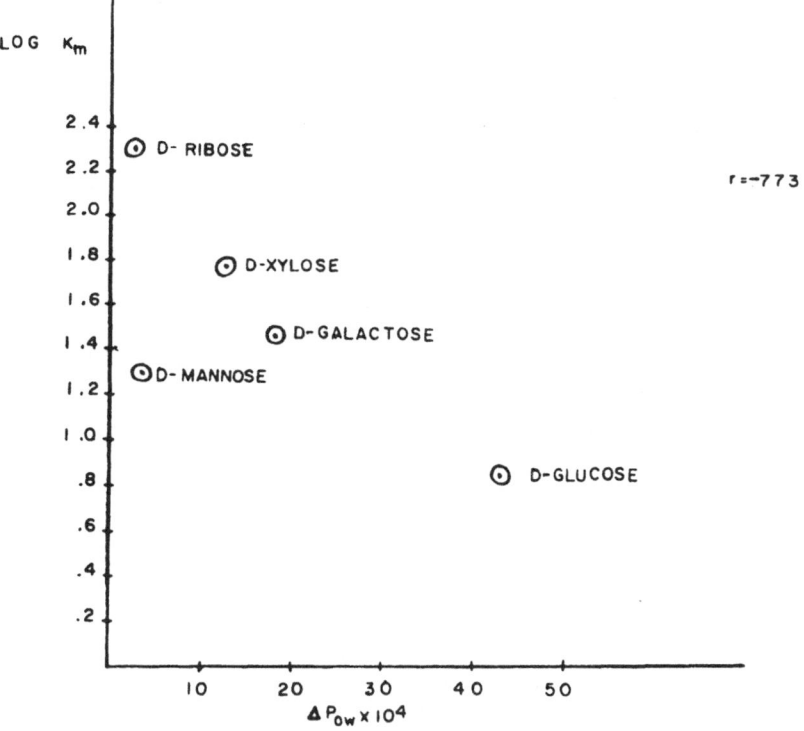

THE CHANGE IN PARTITION COEFFICIENT IN A WATER—BUTANOL SYSTEM FOLLOWING THE ADDITION OF 0.16 mgm TOTAL HUMAN ERYTHROCYTE LIPID PLOTTED AGAINST LOG K_m (AFFINITY FOR THE ERYTHROCYTE CARRIER APPARATUS). INITIAL AQUEOUS PHASE SUGAR CONCENTRATION IS 0.01 M.

Fig. 4

and sampled. Lipid was then added, the vial re-equilibrated, and
sampled. The differences in partition coefficients were then
calculated. The results represent the mean of from ten to twenty
individual experiments. These differences were then plotted
against the log of the affinity for human erythrocyte carrier,
measured by Lefevre and Marshall[8] and correlation coefficients
calculated. The next figure illustrates the effect of total
erythrocyte lipid on the change in partition coefficient for five
sugars at 0.01 M. It is seen that d-glucose is driven into the
oil phase, but sugars that penetrate the erythrocyte membrane
slowly are not. There is a fairly good correlation between the
lipid induced solubilization and the affinity for carrier, and
hence the rate of penetration into the red cell.

To determine the effect of individual phospholipids known to
reside in the erythrocyte membrane, chromatographically pure phospholipids were tested at the same sugar concentration. Figure 4
illustrates the behaviour of phosphatidyl ethanolamine. Again
lipid drives d-glucose into the oil phase and a significant
correlation holds. Table 2 summarizes the behaviour of five
phospholipids individually examined. It is seen that a significant
correlation holds between the penetration rate into the erythrocyte
of the five sugars studied and the change in partition coefficient
following addition of phosphatidyl ethanolamine. Suggestive but
nonsignificant correlations hold for lecithin and phosphatidyl
inositol, while there is no correlation for sphingomyelin or
phosphatidyl serine. Because the biologic measurements are subject
to some uncertainty[13], correlation coefficients were calculated
using penetration rates calculated by a third laboratory.[14] The
resulting r values were in fairly good agreement with those presented, phosphatidyl ethanolamine again being the phospholipid
with the greatest r value.

D-glucose enters the erythrocyte very much faster than its
optical enantiomorph l-glucose. If the correlation holds for
optical enantiomorphs, phospholipids added to the system should
drive d-glucose into the oil phase and l-glucose out of the oil
phase. Table 3 indicates that this did not take place. Each of
the five phospholipids introduced drove both d and l glucose into
the oil phase. The discriminatory property possessed by certain
phospholipids for the d sugars studied does not exist when these
optical enantiomorphs are compared, suggesting non-lipid membrane
constituents also participate in the selectivity process of the
sugar transport apparatus.

SUMMARY

In summary, certain analogies between sugar transport in the
human erythrocyte and the translocation of sugars from an aqueous
to butanol-lipid phase have been presented. The addition of a

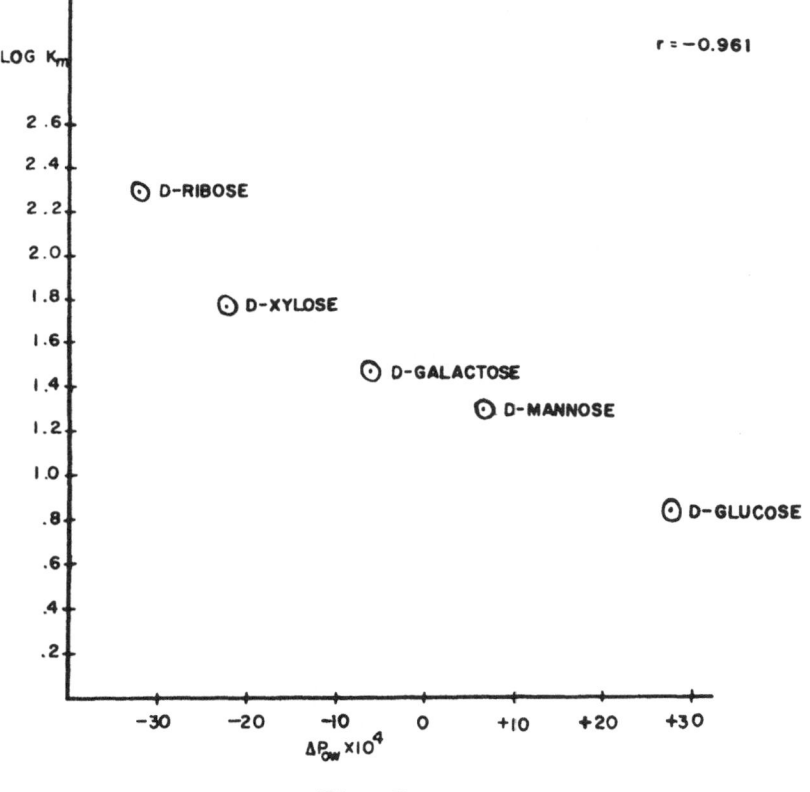

THE CHANGE IN PARTITION COEFFICIENT IN A WATER—BUTANOL SYSTEM FOLLOWING THE ADDITION OF 0.10mgm PHOSPHOTIDYL ETHANOLAMINE PLOTTED AGAINST LOG K_m (AFFINITY FOR THE ERYTHROCYTE CARRIER APPARATUS). INITIAL AQUEOUS PHASE SUGAR CONCENTRATION IS 0.01 M.

Fig. 5

Table 2

THE EFFECT OF INDIVIDUAL PHOSPHOLIPIDS ON THE
CHANGE IN PARTITION COEFFICIENT OF FIVE SUGARS*
IN A WATER-BUTANOL SYSTEM

Phospholipid	r Value of Log K_m vs ΔP_{OW} for 5 Sugars* at 0.01 M.	P 3 d.f.
P. Ethanolamine	-0.961	0.01
P. Inositol	-0.764	N.S.
Lecithin	-0.653	N.S.
Sphingomyelin	-0.394	N.S.
P. Serine	-0.162	N.S.

*D-Mannose, D-Glucose, D-Galactose, D-Ribose, D-Xylose.

Table 3

THE EFFECT OF INDIVIDUAL PHOSPHOLIPIDS ON THE CHANGE IN
PARTITION COEFFICIENT OF THE OPTICAL ENANTIOMORPHS,
D AND L GLUCOSE, IN A WATER-BUTANOL SYSTEM

Phospholipid	D Glucose	L Glucose
P. Ethanolamine	+	+
P. Inositol	+	+
Lecithin	±	+
Sphingomyelin	+	+
P. Serine	-	+
Predicted from erythrocyte penetration rate.	+	-

+ indicates lipid drives sugar into oil phase
- indicates lipid drives sugar out of oil phase

variety of erythrocyte or commercial phospholipids to the system
will accelerate the process when compared to a water-butanol and
cholesterol system. As the initial aqueous sugar concentration is
increased the initial translocation rate does not increase without
limit but achieves a maximum value. Examining the effect of complex lipid on the partition coefficient of sugars illustrates
certain properties these lipids possess that parallels to a
remarkable degree the hypothetical sugar transport apparatus. They

are capable of moving certain sugars into the oil phase, one phospholipid is capable of distinguishing the five d sugars, but not the l sugar. The d sugars are distinguished in a manner highly reminiscent of the red cell carrier.

In recent years membranes have been considered in terms of their dynamic function and more attention has been focused on the protein components of membranes.[15] The present experiments indicate that understanding the physical interaction between complex membrane lipid and permeating solute is a promising avenue for investigating certain of these dynamic functions.

REFERENCES

1. Cornblath, M. and Schwartz, R., 1966, Disorders of Carbohydrate Metabolism in Infancy, W. B. Saunders Co., Phila.

2. Tiernan, R., 1967, Personal Communication.

3. Kupferberg, H. J., and Way, E. L., 1963, J. Pharm. Exp. Therapy, 141:105.

4. Rosenberg, L. E., 1969, in Biological Membranes, Dowben, ed. Chap. 8, Little, Brown & Co., Inc., Boston.

5. Overton, E., quoted in Davson, H., 1964, A Textbook of General Physiology, 3rd ed., p 282, Little, Brown & Co., Inc., Boston.

6. Collander, R., 1947, Acta Physiol. Scand. 13:363.

7. Stein, W. D., 1967, The Movement of Molecules Across Cell Membranes, Academic Press, Inc., New York, Chap. 4.

8. LeFevre, P. G. and Marshall, J. K., 1958, Am. J. Physiol., 194:333.

9. Maudsley, D. A. and Widdas, W. F., 1967, J. Physiol., London, 189:75.

10. LeFevre, P. G., Jung, C. Y., and Chaney, J. E., 1968, Arch. Biochem. and Biophys., 126:677.

11. Schulman, J. H., 1966, Ann. N. Y. Acad. Sci., 137:860.

12. Ting, H. P., Bertrand, G. L., and Sears, D. F., 1966, Biophys. J., 6:813

13. LeFevre, P. G., 1969, Personal Communication.

14. Stein, W. D., 1967, The Movement of Molecules Across Cell
     Membranes, Academic Press, Inc., New York, p 164. (Widdas'
     results at 37°C were used for comparison.

15. Korn, E. D., 1969, Annual Review of Biochemistry, Annual
     Reviews, Inc., p 263.

USE OF SYNTHETIC MEMBRANE MODELS IN THE STUDY OF GASTRIC SECRETORY PROCESSES

Jesse M. Berkowitz and Melvin Praissman

Division of Gastroenterology, Department of Medicine, Meadowbrook Hospital, East Meadow, N.Y. and The Department of Physiology, Mount Sinai School of Medicine, New York, N.Y.

Synthetic membrane models have been used in the study of gastric secretory problems for over thirty years. These techniques were employed to elucidate the mechanisms by which ions were secreted in gastric juice. Modern studies of gastric secretory processes began in the 1930's with the studies of the ionic composition of gastric juice by the late Franklin Hollander (1), a physical-organic chemist turned physiologist, and the membrane model experiments of Torsten Teorell, a physical chemist interested in physiologic problems.

Hollander (1,2) was primarily concerned with the origin of the inorganic constituents of gastric juice and he systematically studied its composition as a function of the volume rate of secretion. These studies were performed in dogs and after an extensive series of experiments, Hollander concluded that the primary secretory product of the gastric glands was a slightly hypertonic (with respect to the interstitial fluid) solution of HCl. The original studies were carried out at a time when flame photometry techniques for Na and K assay were not available. We now know that the primary inorganic secretory product of the gastric glands is a solution of HCl and KCl (the ratio of H to K is approximately 15:1). At low rates of secretion Hollander found that the "basic" chloride, which is now known to be primarily sodium, appeared in the gastric juice at a concentration higher than hydrogen. The sodium-rich solution found at low rates of secretion was hypotonic. By a regression analysis, Hollander concluded that the sodium and the hypotonicity could be accounted for by the leakage of sodium-rich interstitial fluid into the gastric lumen. The combination of the bicarbonate from the alkaline interstitial fluid and the secreted hydrogen ions would result in the evolution of CO_2 producing a decline in solute concentration. The hypothetical entry of the alkaline interstitial fluid was utilized to account for the decreased hydrogen ion concentration, the sodium content and the hypotonicity of basal gastric juice.

STUDIES ON ION AND WATER FLOW ACROSS THE RESTING GASTRIC MUCOSA

Hollander's treatment of the problem was based on a mathematical analysis. In contrast, Teorell approached the problem experimentally: he placed 5 ml. of a slightly hypertonic solution of pure HCl into the resting stomach of anesthetized cat and followed the changes in its ionic composition with time (3). The volume of the instilled fluid remained constant, the hydrogen concentration fell and the "basic" fraction of the chloride concentration increased. Concomitantly, there was a fall in total chloride concentration. As chloride accounted for almost the total concentration of anions, a decline in its concentration reflected a decline in tonicity. Teorell extended these studies with a synthetic membrane system (4). Isosmotic HCl was placed on one side of an uncharged cellophane membrane and isosmotic NaCl on the other side. As the exchange progressed the hydrogen concentration declined, the sodium concentration rose and the volume remained constant on the acid side of the system. Simultaneously, the chloride concentration declined reflecting a fall in solute concentration. There was no Donnan effect because the membrane was uncharged; Teorell reasoned therefore, that HCl, as an ion pair, left the acid side of the system faster than NaCl entered as an ion pair. In both the synthetic membrane studies and the cat stomach, the unequal exchange rates of the ion pairs with no change in volume was assumed to indicate a net loss of solute from the acid solution. However, changes in the volume of the instilled fluid may not have been detected because of the small volumes used (5.0 ml). If there had been a change in the volume of the instilled acid solutions in Teorell's cat experiments, then the model he proposed would be invalid.

To minimize the problems of Teorell's experiments, we instilled large volumes of isosmotic HCl solutions containing a non-absorbable dilution indicator into unstimulated canine gastric pouches (5). The non-absorbable indicator (phenol red) enabled us to assess two things: (1) Whether the instilled material was quantitatively recovered and (2) Any changes in the volume of the fluid placed in the pouch. The change in fluid volume was determined by measuring the indicator concentration. In Table I is a typical instillation experiment, one of twelve. With time, the hydrogen ion concentration declined, the sodium ion concentration increased, and the measures of solute concentration, chloride and osmolality fell. The decline in phenol red concentration (Table I) reflecting an

entry of fluid was confirmed by the recovery of 96.3 per cent
(S.E. ± 0.6) of the indicator in twelve experiments (5). In other
words, the high recovery of phenol red demonstrated that the change
in indicator concentration was not the result of diffusion of the
compound across the gastric mucosa but the consequence of fluid
entry. The results seem paradoxical because the fluid continued
to enter the pouch from the mucosa even though the pouch solute
concentration was less than that of the interstitial fluid.

TABLE I

CHANGES IN IONIC CONCENTRATIONS
Typical Canine Instillation Experiment

TIME hours	Na	H mEq./L.	Cl*	OSMOLALITY mOsm./Kg.H$_2$O	PHENOL RED mg./L.
0	0	155	155	295	40.0
1	2	146	151	290	35.4
2	12	125	144	267	32.4
3	31	98	135	252	28.9
4	52	69	129	238	26.1
5	68	52	130	237	24.3

*The sum of the cation concentrations, Na and H, is less than the
concentration of chloride. This difference is primarily accounted
for by K which has not been listed in the table.

We calculated the net changes in solute and fluid in each
hourly period for each experiment (5). In Table II are the net
ionic and fluid changes from a typical experiment. The net gain
in sodium always exceeded the net hydrogen loss and the net gain
in chloride correlated almost quantitatively with the differences
between the sodium and hydrogen changes indicating a net gain in
solute, i.e., there was not a simple 1:1 exchange of sodium for
hydrogen. The changes in fluid volume were always positive indicating a gain in solution (Table II). The solute concentration
of the solution transported across the gastric mucosa was calculated
and is expressed in mEq./L. In the middle three hours there was
a net flow of hypotonic solution into the gastric pouch (concentration of 75 mEq./L). During these periods the greatest observed
fall in osmolality occurred (see Table I).

TABLE II

NET IONIC AND VOLUME CHANGES
Typical Canine Instillation Experiment

| Hour | mEq. | | | ml. | mEq./L |
	Δ Na	Δ H	Δ Cl	ΔH_2O	conc. of solute
1	1.40	-0.62	1.04	6.5	160
2	0.57	-0.44	0.44	5.6	75
3	1.25	-0.80	0.50	7.3	68
4	1.54	-1.18	0.56	7.3	76
5	1.51	-0.95	0.78	6.4	122

The observations in the canine pouches contradict Teorell's findings in the cat stomach. Teorell found no change in volume and an unequal exchange of NaCl and HCl ion pairs with a decline in solute concentration, i.e., a net loss of solute and a fall in tonicity. In contrast, we found a sodium-hydrogen exchange accompanied by a net gain in solute and fluid-the concentration of this solution being hypotonic. Dr. Harry Gregor of Columbia University proposed that our observations concerning the flow of hypotonic fluid might be the result of an anomalous osmotic flux of water produced when ions of different ionic size such as sodium and hydrogen exchange across fixed-negative charge surfaces.

In order to relate the anomalous osmotic flux of solvent to ion transport the studies that follow were undertaken initially in Dr. Gregor's laboratory and have been extended in our laboratory in collaboration with Doctors Gregor and Irving Miller of the Polytechnic Institute of Brooklyn. In Figure 1 is a schematic representation of the apparatus used to study sodium-hydrogen exchange across a fixed negative charge membrane. After the sodium solution was passed over the membrane it was not recirculated. In this manner, the sodium concentration on the open side was kept constant and the hydrogen concentration on the sodium side was negligible. This maximized the gradient of both ions across the membrane. In one set of experiments the amount of solvent moved across the membrane was quantitatively determined as a function of time. The closed side of the system was filled to the upper stopcock and side-arm; as fluid entered the closed side as a result of the exchange, the excess volume of solution flowed out of the side-arm. The fluid was collected in tared vessels and the amount of solution transported determined gravimetrically. In a parallel set of experiments the change in the ionic composition of the acid side was

determined as the exchange progressed by removing solution for
assay from the stopcock in the main flow line. In Figure 2 is
depicted the water flow rate produced by sodium-hydrogen exchange
across a polyanionic membrane (polystyrene-sulfonic acid) containing
20 per cent water (transference number 0.95) (6). A net gain in
water always occurred on the closed side and this water flow produced a decline in solute concentration which is shown in Figure 3
(6) in terms of the chloride concentration and osmolality. The
decline in solute concentration was the direct result of the observed water entry produced during the sodium-hydrogen exchange
(see Figure 2) and does not reflect a loss of solute from the acid
side of the system.

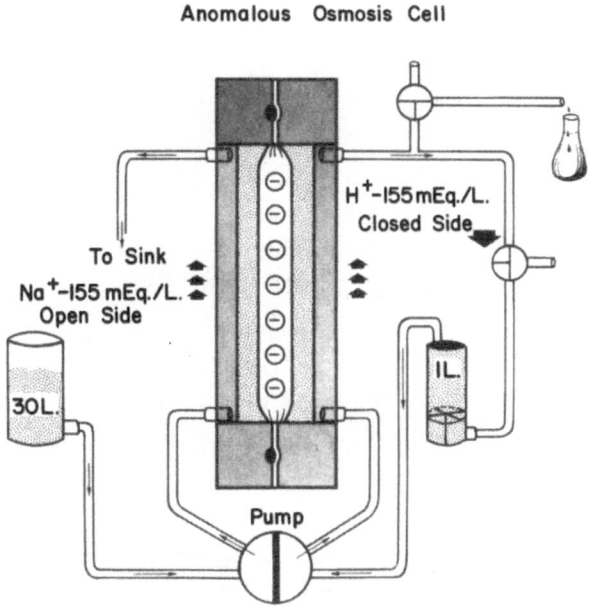

Anomalous Osmosis Cell

Figure 1.   Schematic representation of cell used to study bi-ionic
exchanges and secondary anomalous osmosis.

     Studies were then undertaken to relate ion size and membrane
hydration to the anomalous osmotic fluxes of solvent. In Figure
4 (6) are presented the anomalous osmotic fluxes resulting from
the exchange of hydrogen for a series of larger cations in two
membranes of different hydration. Water flows were two to three
times higher for the 60 per cent hydrated membrane than those
found for the 20 per cent hydrated membrane. The effect of ion
size is clear: in the exchange of hydrogen for tetramethylammonium

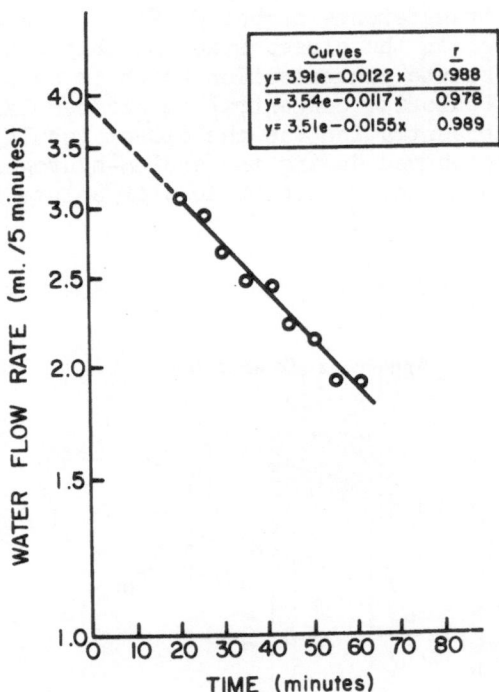

Figure 2. Anomalous osmotic water flow produced by sodium-hydrogen exchange across a sulfonic acid membrane containing 20 per cent water (6).

(crystal size 3.47Å (7)) and tetraethylammonium (crystal size 4.0Å) larger anomalous osmotic fluxes are observed than in the hydrogen-sodium (crystal size 0.95Å) exchange. The differences in water flow become much greater when the results are computed on the basis of a molar rate of ions exchanged. For both membranes the half-time for the exchange of hydrogen for the larger ions, tetramethylammonium and tetraethylammonium, is considerably greater than the half-time for the sodium-hydrogen exchange (see table at top of Figure 4). If each water curve (Figure 4) is integrated from the zero-time to the half-time of the ion flux--relating the water flows to an equal molar exchange of ions--the differences in solvent carried per ion become larger than those seen when the integration is performed as a simple function of time. These studies indicate that an anomalous osmotic flux is dependent on the membrane hydration,

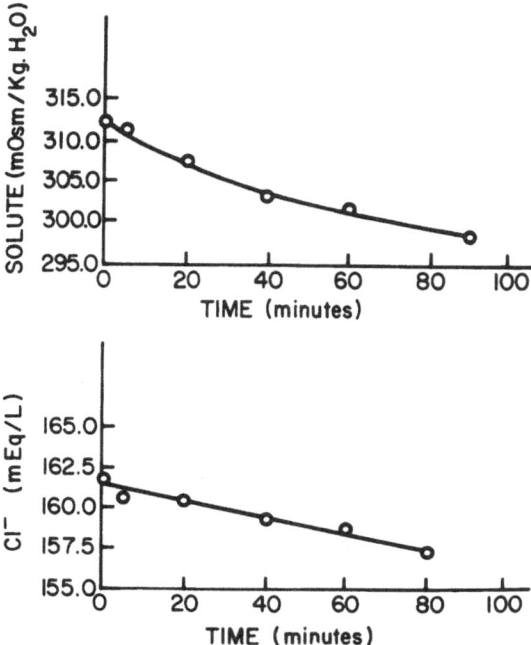

Figure 3. Decline in solute concentration produced by an anomalous flux of water shown in Figure 2. (6)

the number of ions exchanged and their ionic size.

Water flow related to ion movement has also been observed when ions pass through ion-exchange membranes under the influence of an applied electric field. The term electroosmosis (8) has been used to describe the transference of water per mole of ion moved through a highly perm-selective membrane (transference number approaching 1). It has been proposed by Tombalakian, Worsley and Graydon (9), in preliminary experiments, that anomalous osmotic fluxes of water produced by a bi-ionic exchange could be predicted from the electro-osmotic fluxes produced by the individual cations. It was shown that the greater the difference in the electroosmotic transport of the ions, the greater the anomalous flux of solvent. Breslau and Miller (10) in a recently advanced mathematical interpretation (of electroosmosis) related the velocity of the solution (in cm/sec) to the electroosmotic coefficient EO_d (expressed as moles of solution transported per faraday of current). The formulation was in terms of the hydrodynamic drag of a migrating spherical particle in a bounded medium. Based on this interpretation and the work of

Tombalakian, et al.(9) we should be able to predict the anomalous
osmotic fluxes for any system of univalent ions by:
Anomalous Osmosis (moles of solution pumped/mole of ion exchanged)=
$(EO_d)_2 - (EO_d)_1$ (moles of solution transported/faraday). (For univalent ions a faraday is equal to an equivalent). We are currently
studying the relationship between the electroosmotic fluxes produced by a variety of univalent cations and the anomalous osmotic
fluxes produced by bi-ionic exchange of these ions. Hopefully
these experiments can be extended to natural polymers.

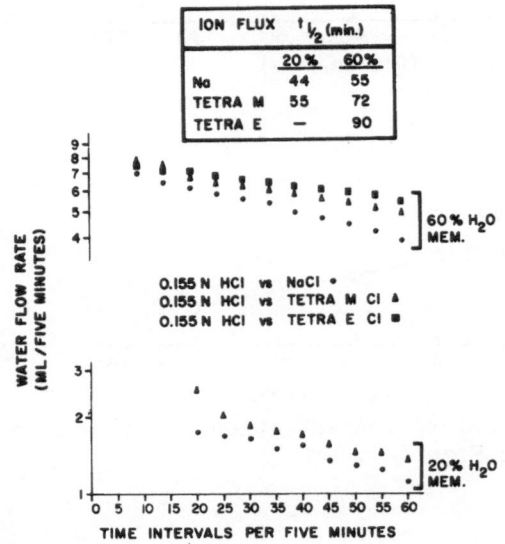

Figure 4. Summary of experimental data relating ion size and
membrane hydration to anomalous osmotic fluxes of solvent. (6)

SIGNIFICANCE OF pH AND ANTRAL PERMEABILITY IN GASTRIN RELEASE

Acid glycoproteins are the principle polymeric constituents
of the mucus coat on the surface of the stomach. Electron microscopic studies (11) reveal granular packets in the distal portions
of the surface epithelial cells and these have been shown by histochemical means to be composed of acid glycoproteins (12). The
glycoproteins are periodically released and become part of the
mucus coat of the stomach.

The mucus is seen as a vicous coat on the stomach surface in
the intact animal. The coat appears in electron micrographs as a
fuzz on the apical end of the surface epithelial cells (11). In
this symposium Doctor Brandt revealed that the surface of amoebic
cells is also covered by an amorphous coat. There is an increasing

amount of evidence that this coat, which has been called by Bennet (13) a glycocalyx, is on the surface of all cells in mammalian systems (14).

In non-secreting stomachs the mucus is a clear gel. Upon acidification of the surface it becomes opaque (15). In unrelated studies, a number of investigators have observed upon acidification of the gastric surface that radio-sodium could no longer pass from the surface fluid into the mucosa (16,17). Changes in the ionization state of the acid glycoproteins may be responsible for the pH-dependent variations observed in the physical state of the mucus and the mucosal permeability to sodium. At pH 7 the carboxyl groups of the acid glycoproteins on the lumenal side of the mucus are ionized; at pH 1 these groups are probably protonated and un-ionized leading to a loss of charge at the surface. The loss of surface charge would decrease the osmotic activity of the coat with a loss of water and secondary shrinkage. These pH-dependent physical changes could decrease the ionic permeability and increase the opacity of the mucus.

The release of gastrin, the primary humoral stimulant of gastric acid secretion is regulated by a pH-dependent servo-mechanism (18). An alteration in mucosal permeability, comparable to the pH-dependent change in sodium permeability, may be the molecular basis of this servo-mechanism. Gastrin, a peptide hormone, is produced in the distal part of the stomach known as the antrum. The surface of the gastrin containing cells is exposed to the lumen of the antral glands (19). The proximal part of the stomach known as the fundus, contains the acid and pepsinogen secreting parietal mucosa. Acid and pepsinogen are both released when the parietal mucosa is stimulated by the vagus nerve or by gastrin.

It has been known since the 1920's that when neutral solutions of glycine and other small amino acids which are normally released by peptic digestion are placed in the stomach, they stimulate acid secretion (20). We now know that the amino acids stimulate secretion by causing the release of gastrin from the antrum (21). The hormone is released into the venous blood, goes to the heart and returns to the fundic mucosa via the arterial blood. The fundic mucosa then secretes acid, the contents of the antrum decline to pH 1 and the glycine-stimulated release of gastrin is blocked (22). The mucosa at pH 1 is impermeable to the cation sodium (16,17); it should therefore be impermeable to glycine which is also a cation at this pH. We are proposing that the decline in pH prevents the permeation of the coat by glycine and leads to a diminished rate of gastrin release.

To test whether antral permeation by glycine is altered at
pH 1 antral pouches were prepared in dogs by Doctors Gerald Buetow
and Robert Cali of the Department of Surgery. 60 mM solutions of
glycine-C^{14} and ethanolamine-C^{14} were instilled into the pouches
at pH 7 and 1. The pH 7 and 1 solutions were isosmotic and contained a mixture of NaCl-KCl and HCl-KCl, respectively. At neutral
pH after two hours about 15 per cent of each compound passed through
the mucosa (Table III). There was a negligible loss of the compounds
from the acid solution (Table III) (23).

TABLE III

PERMEATION OF ANTRAL POUCHES
BY GLYCINE AND ETHANOLAMINE

PER CENT LOSS/2 HOURS

	Glycine	Ethanolamine
pH 7	16.8 (27)	14.4 (9)
pH 1	2.4 (23)	3.5 (9)

( ) = number of studies

differences significant to p<0.01 by
non-parametric tests

At pH 7, glycine is a zwitterion and ethanolamine is a cation
and the mucosa is permeable to both. The cation and zwitterion
selectivity may be the result of the carboxyl groups of the glycoproteins being ionized which imparts a negative charge to the surface. At pH 1 glycine and ethanolamine are cations and the mucosa
is impermeable to them. This may be the consequence of a reduction
in the number of unionized carboxyl groups and the accompanying
physical changes---loss of water and membrane shrinkage.

To test this hypothesis, we studied the rate of diffusion of
sodium, ethanolamine and glycine across a methacrylic acid membrane
at pH 7 and pH 1. At pH 7 the membrane had a water content of 35
per cent and was cation selective. At the lower pH it had a water
content of 5 per cent. The membrane was placed in a Lucite cell
similar to that in Figure 1. Solutions containing sodium, glycine,
and ethanolamine were placed on one side of the cell respectively,
and circulated over the membrane. In addition the solutions contained buffer or acid to maintain the pH at 7 or 1. We measured
the loss of sodium, ethanolamine and glycine from the closed side
of the system. On the open side of the system the solutions were
of identical ionic composition to the closed side, but they did

not contain the permeating species.

When the membrane is in its salt form at pH 7, 89 per cent of the sodium, 79 per cent of the ethanolamine and 40 per cent of the glycine pass through the membrane after 75 minutes of exchange (Table IV). At pH 1 when the membrane is in its acid form, only 10 per cent of the sodium, 17 per cent of the ethanolamine, and 15 per cent of the glycine pass through in the same time (Table IV) (24). These experiments indicate that a change in pH alters the cation permeability of surfaces composed of weakly acidic groups.

TABLE IV

PERMEATION OF ACID AND SALT FORMS OF METHACRYLIC
ACID MEMBRANE BY SODIUM, ETHANOLAMINE AND GLYCINE

(PER CENT LOSS)

TIME (min.)	SODIUM		ETHANOLAMINE		GLYCINE	
	ACID	SALT	ACID	SALT	ACID	SALT
15	2.2	32.1	- -	25.1	2.8	5.4
45	5.8	73.2	8.9	60.6	9.7	26.2
75	10.0	89.3	16.9	78.7	15.1	39.7

SUMMARY

The experiments with the canine fundic pouches and the sulfonic acid membranes suggest that the exchange of hydrogen for sodium across the fixed negative charges of the gastric mucus coat yield an anomalous osmotic flow of solvent. Although many of the carboxyl groups at the luminal surface are unionized when the coat is bathed with acid, the pH at the cellular side probably approaches 7 and the carboxyl groups located there are ionized. The bi-ionic exchange across the asymmetrically distributed anionic groups produces a secondary flow of solvent which transforms the primary acid secretory product into a sodium-rich hypotonic fluid.

The experiments with the antral pouches and the methacrylic acid membrane indicate that the natural stimulants of gastrin release, such as glycine, can pass through the ionized, hydrated mucus gel that lies on the surface of the antrum. On permeating the gel, these compounds diffuse into the gland and penetrate the gastrin containing cells. (How they effect gastrin release is not understood at present). The release of gastrin causes the fundic mucosa to secrete acid which diffuses over the surface of the antrum leading to the protonation of the carboxyl containing groups in the glycoproteins. With the loss of the negative charges, membrane hydration declines and the permeability to cations is altered. These changes are probably the basis of the servo-mechanism by which acid secretion diminishes the release of its primary hormonal stimulant gastrin.

Acknowledgements: This work was supported by the citizens of Nassau County, New York, and by grants from the National Cystic Fibrosis Research Foundation and the National Institutes of Health (grant # AM13674).

BIBLIOGRAPHY

1. James, A.H., THE PHYSIOLOGY OF GASTRIC DIGESTION, ch. 3 (Edward Arnold, Ltd., London, 1957).

2. Hollander, F., Fed. Proc., 11, 706 (1952).

3. Teorell, T., J. Gen. Physiol., 23, 263 (1939).

4. Teorell, T., Gastroenterology, 9, 425 (1947).

5. Berkowitz, J.M., and Janowitz, H.D., Am. J. Physiol., 210, 216 (1966).

6. Praissman, M., Miller, I.F., Gregor, H.P., and Berkowitz, J.M. Unpublished observations.

7. Nightingale, E.R., Jr., J. Phys. Chem., 63, 1381 (1959).

8. Helfferich, F., ION EXCHANGE, 402. (McGraw-Hill Book Company, Inc., New York, 1962).

9. Tombalakian, A.S., Worsley, M., and Graydon, W.F., J. Am. Chem. Soc., 88, 661 (1966).

10. Breslau, B.R., Ph.D. Dissertation, Polytechnic Institute of Brooklyn, Brooklyn, N.Y. (1969).

11. Ito, S., in HANDBOOK OF PHYSIOLOGY, Section 6: Alimentary Canal, II, 705. (Code, C.F., Ed., American Physiological Society, Washington, D.C., 1967).

12. Spicer, S.S., and Sun, D.C., Ann. N.Y. Acad. Sci. 140, 762 (1967).

13. Bennet, H.S., J. Histochem. Cytochem., 11, 14 (1963).

14. Ito, S., Fed. Proc., 28, 12 (1969).

15. Davenport, H.W., PHYSIOLOGY OF THE DIGESTIVE TRACT, 114. (Year Book Medical Publishers, Inc., Chicago, 1966).

16. Cope, O., Cohn, W.E., and Brenzier, A.G., Jr., J. Clin. Invest., 22, 103 (1943).

17. Code, C.F., Higgins, J.A., Moll, J.C., Orvis, A.L. and Scholer, J.F., J. Physiol., London, 166, 110 (1963).

18. Schofield, B., in GASTRIN, 171. (Grossman, M.I., Ed., UCLA Forum in Medical Sciences, University of California Press, Los Angeles, 1966).

19. McGuigan, J.E., Gastroenterology, 55, 315 (1968).

20. Ivy, A.C., and Javois, A.J., Am. J. Physiol., 71, 591 (1925).

21. Elwin, C.-E., and Uvnas, B., in GASTRIN, 69. (Grossman, M.I., Ed., UCLA Forum in Medical Sciences, University of California Press, Los Angeles, 1966).

22. Elwin, C.-E., Acta Physiol. Scand. 175, 36 (1960).

23. Berkowitz, J.M., Praissman, M., Beutow, G., and Cali, R. Unpublished Data.

24. Praissman, M., Miller, I.F., and Berkowitz, J.M. Unpublished Data.

PROPERTIES OF THE PLASMA MEMBRANE OF AMOEBA

Philip W. Brandt[1] and Klaus B. Hendil[2]

Columbia Univ., 630 W. 168th St., New York 10032 (1)

Carlsberg Laboratory, 10 Gl. Carlsbergvej. Valby, Dk. (2)

Pinocytosis in amoeba was described, and its brief history reviewed at an earlier ACS meeting (4). It was pointed out that pinocytosis is the descriptive term for a behavioral pattern. In amoeba this pattern is initiated by a step increase in the cation concentration at a suitable pH and pCa. This report will consider in detail the initial phase of pinocytosis, the reaction of the ameoba plasma membrane to the cationic stimulus.

The initial signal in pinocytosis is almost certainly the adsorption (2, 19, 20) or exchange of cations at the outer layer of the plasma membrane. Shortly thereafter, or simultaneously in parallel with the surface reaction, the transmembrane conductance (D.C.) increases (3) more than a decade and the impedance decreases (13). This conductance change, like pinocytosis, is dependent in magnitude upon the concentration of the stimulating cation, and upon the background pH and pCa (3). A low pCa tends to block pinocytosis (11, 14) or arrest it if the pCa is lowered after pinocytosis is initiated (3, 4). The conductance increase is time dependent, in parallel with the induction of pinocytosis channel formation (10, 12).

The plasma membranes of amoeba fixed in the high conductance state and studied with the electron microscope are thicker in comparison to control membranes. The increase in thickness, which appears to be mostly a thickening of the electron transparent zone, could be due to an increase in total lipoprotein, or more likely to an imbibition of partially hydrated ions (3, 4). Since the process

of fixation and image interpretation is not known with certainty
(15, 21), the thickening may accurately represent the in vivo
dimensions or be a propensity towards thickening brought out by
fixation. Fixation studies, which currently are being designed to
explore the area, must necessarily be indirect. However if the
thickening were due to an in vivo increase in the content of hydrated
ions, the partition coeficient of the solution: membrane system
should change for nonelectrolytes. Therefore permeability studies
on in vivo membranes will provide a direct test of the consequences
of the membrane changes.

A change in partition coeficient should alter the permeability
coeficient (Pn) for a given nonelectrolyte. The flux rates of polar
penetrants should increase as the membrane becomes relatively
richer in water and that of nonpolar penetrants should decrease.
Pn should vary with the quantities of ions in the membrane and with
the degree to which each species is hydrated. In turn these ionic
quantities will depend on the geometry, concentration, and type of
sites available in the membrane. Since changes in the pH and pCa
greatly affect the conductance changes and amount of pinocytosis
induced by a given cation, it seemed reasonable to systematically
study the effect of these variables on Pn.

After Pn was determined for a number of nonelectrolytes under
standard conditions, isopropanol was selected for intensive study.
Pn for isopropanol is about 4.5×10^{-6} cm/sec with a time constant
of about 1200 seconds in our system. Therefore the influx of ^{14}C
labeled isopropanol is sufficient in 200 seconds to be readily
separated from the background, and even when the Pn is considerably elevated by the experimental treatment the flux rates are
easily determined.

MATERIALS AND METHODS

The amoeba <u>Chaos</u> <u>chaos B</u> was grown in mass cultures by
methods quite well standardized in the Carlsberg Laboratory,
Department of Physiology. Thus large numbers of cells of a uniform
state of nutrition were available. In each experiment, several
thousand cells were separated according to size by passage through
a 2 mm ID polyethylene tube about three meters long wound around
a ring stand post 1.25 cm in diameter. The larger cells tended to
ride in the center of the flowing saline stream, while the small cells
tended to fall into the more slowly flowing saline near the tube wall;
therefore, the first cells to arrive at the lower end were the largest.
The first 1500 cells to emerge were mixed to randomize their order

then used in the experiments.

Groups of 100 cells in about 50 µl of fluid were placed in
each of 12 (approximately 1 ml capacity) pre-weighed polyethylene
capsules. The weight increase with loading was taken as the
fluid + cell volume. For each experiment, two groups were packed
by centrifugation in calibrated capillary tubes and the volume of
the cells calculated. This cell volume was confirmed by two
alternate methods, an isotope dilution method and an isotope
loading method (5).

To determine an experimental point, six capsules containing
100 cells each were shaken for five minutes in a temperature regulated water bath. After this equilibration period, a low (1 µM/L)
concentration of a ^{14}C labeled non-electrolyte dissolved in the
experimental saline was added quantitatively. A measured number of
seconds after addition of the isotope, the cells were rapidly washed
and collected by a filtration method or a centrifugation method, and
the quantity of isotope trapped in the cells was determined. In the
filtration method, the cells were transferred with a breaking pipette
from the capsule to a one inch Gelman vacuum funnel (#1112)
which contained 10 ml of wash fluid. The fluid was drawn down by
a slight vacuum to a level about 2 mm from the surface of the filter
disk and then a five ml. wash was added. This was repeated once
more, then the fluid was completely filtered off.

The filtration procedure caught the cells on the surface of
the filter only after the final wash, then the filter disk and cells
were quickly removed from the apparatus and dropped into a
scintillation vial filled with Bray's scintillation fluid. The scintillation fluid extracted the non-electrolyte from the cells and
dissolved the filter disk. The entire wash procedure took about
30 seconds. A similar washout of the same quantity of label, in
the absence of cells, provided a background count.

In the centrifuge method, brief centrifugations to pack the
cells were alternated with washes until the extracellular ^{14}C label
was minimal. It required at least 1.5 minutes but provided a check
against cell loss in the filtration method, since the surviving cells
were counted in control experiments.

The results reported here were gathered using the filtration
method; however, they suggest that amoeba in solutions in which
Pn is minimal are relatively stable during filtration while cells in
solutions which greatly increase Pn are unstable and tend to leak
or rupture during washing by filtration. Thus the curves shown in

figure 1 are probably flatter than they should be. The error is to
minimize the effects of high concentrations of cations on the
apparent permeability of the cells.

The experiments reported here are influx studies and the
following calculations were used:

$$P_n = \ln \frac{C_o}{C_o - C_i} \; \frac{V}{A\,t}$$

P = cm sec^{-1} Permeability coefficient

C_o = ^{14}C Concentration outside

C_i = ^{14}C Concentration inside

$\frac{V}{A}$ = 60 x 10^{-4} cm (9)

t = time

Each point was determined six times; therefore, the calcula-
tor was programmed to calculate and accumulate the Pn's from the
raw data and compute the mean and the standard deviation (SD).
The coefficient of variation ($\frac{SD}{mean}$ x 100) was 10% to 20% with
some higher values in conditions which maximize P_n. Therefore,
the exact magnitudes on the curves in figure 1 are not as accurate
as the direction of changes in P_n with a given variable. The
centrifugation method has an average coefficient of variation less
than 10% and this technique is currently being used to repeat and
extend the experimental curves.

The solutions all contained approximately 2 mM/L Na$^+$ but
the buffer was varied dependent on the Ca^{++} level. When Ca^{++}
was below 1.0 mM/L, a phosphate buffer was used and adjusted so
that the total Na$^+$ remained close to 2mM/L. The calcium was
simply added as a chloride between 100 µM/L and 30 mM/L.
Between 5 µM/L and 50 µM/L Ca^{++}, a 1 mM/L citrate buffer for
Ca^{++} was employed. The apparent association constant (at the
given pH) was calculated from the H$^+$ and Ca^{++} association
constants (5). The Ca^{++} buffer employed at 1 µM/L Ca^{++} was
EGTA and the apparent association constants employed were taken
from Portzehl et al.(17), or recalculated from their tables of
absolute association constants (5).

RESULTS

The Pn (isopropanol) of the amoeba plasma membrane increases
with Ca^{++} concentration and pH at 20C (Figure 1). At pH 7.0 Pn is

Figure 1

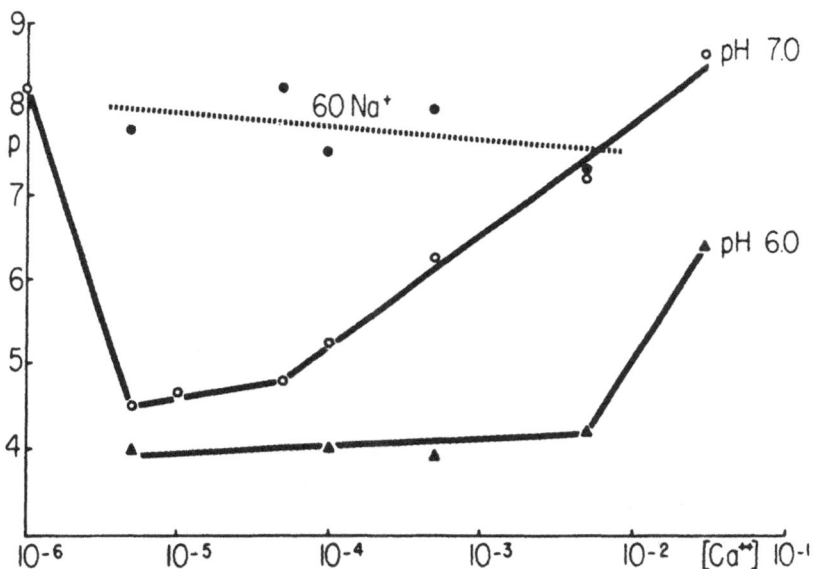

The absissa is the Ca^{++} concentration in M/L while the ordinate is the permeability coefficient in cm/sec. x 10^{-6}.   The curves indicated by pH are in a background of 2mM/L Na^+ while the curve identified as 60 Na^+ is the Pn of the membrane during the 200-400 second interval after the Na^+ was increased to this level (in mM/L) from the same Ca^{++} concentration point on the pH 7.0 line.

4.5 x 10^{-6} cm/sec. in a saline containing 5 μm/L Ca^{++} and increases progressively to 8.5 x 10^{-6} cm/sec. in 30 mM/L Ca^{++}. At pH 6.0 however, Pn is lower at all Ca^{++} concentrations than at pH 7.0 and only increase when the Ca^{++} exceeds 5 mM/L. Little further decrease in Pn is observed at pH's lower than 6 (table 1). Pn can decrease below that recorded at low pH. When the cells are cooled to 4.5 C, Pn is about 2.3 x 10^{-6} cm/sec. At very low Ca^{++} concentrations (1 μM/L) the cells are very fragile and tend to be shaken apart during the experiment, especially at pH 6.0 or lower.

At pH 8.0 Pn is higher than it is at any corresponding Ca^{++} level at a lower pH and it increases sharply with rising Ca concentration (Table 1). At 30 mM/L Ca^{++} the cells are too fragile to test, because the stirring necessary to obtain accurate Pn values homogenizes them. Thus the cells are most fragile at two extremes — low pH and low Ca^{++} and at high pH and high Ca^{++}.

Because the conductance changes induced by a step increase in the Na^+ concentration are time dependent, the time course of the change in Pn with a step increase (from 2 to 60 mM/L) in the Na^+ concentration was determined. The influx period was kept at 200 seconds, therefore, this limited the time resolution. It appeared that Pn reached a maximum during 200 to 400 second interval after the Na increase, and fell off towards control values with a time constant of about 10 minutes. This approximates the time course of the conductance changes (3, 4) and the time course of pinocytosis (9).

Pn was determined during the 200 to 400 second interval after a step increase in Na^+ at different Ca^{++} and H^+ concentrations. At pH 7.0 Pn increased to about 8.0 x 10^{-6} cm/sec. at all concentrations of Ca^{++} tested. As a consequence the high Na^+ Pn line (Figure 1) crosses the low Na^+ Pn line at about 5 mM/L Ca^{++}. We have not yet tested the effect of high Na^+ on Pn when the latter is elevated above 8.0 x 10^{-6} cm/sec. in high Ca^{++}. We have not systematically examined the effects of varying Na concentration but a few experiments using even higher Na^+ levels (than about 60 mM/L) lead us to believe that the Na^+ effect is saturated by 60 mM/L. Chapman-Andresen (10) reports that the saturation Na^+ concentration for pinocytosis is about 70 mM/L (in a phosphate buffer) which is in reasonable agreement with our data considering the accumulated errors in the two types of studies. The relative effect on Pn of 5 mM/L Ca^{++}, Sr^{++}, Ba^{++}, and Mg^{++} was tested at pH 7.0, and Pn decreased in comparisons between solutions

DATA TABLE 1

Exp. Date	pH	$Ca^{++}mM/L$	Na^+mM/L	Pn	C.V.
14/11	5.1	0.1	2	3.8	11.4
11/11	5.1	0.5	2	3.6	18.8
29/11	5.0	5.0	2	4.0	18.9
8/11	6.1	0.1	2	4.0	8.5
7/11	6.1	0.5	2	3.9	13.7
15/11	6.1	5.0	2	4.2	12.7
25/11	6.0	30.0	2	6.4	12.7
12/12	6.0	.005	~60	3.9	15.3
6/12	6.0	.01	~60	4.9	34.5
5/12	6.0	0.5	~60	5.8	20.7
9/12	6.0	5.0	~60	4.6	21.5
18/11	7.1	.001	2	8.3	38.7
25/11	7.0	.005	2	4.5	13.6
21/11	7.0	.01	2	4.7	18.2
21/11	7.0	.05	2	4.8	26.2
8/11	7.1	0.1	2	5.2	11.2
7/11	7.1	0.5	2	6.3	23.0
15/11	7.1	5.0	2	7.2	20.2
25/11	7.0	30.0	2	8.6	17.4
13/12	7.0	.005	~60	7.8	19.0
13/12	7.0	.05	~60	8.3	17.5
6/12	7.0	0.1	~60	7.5	26.6
5/12	7.0	0.5	~60	8.0	21.2
9/12	7.0	5.0	~60	7.3	20.5
14/11	8.1	0.1	2	5.5	13.3
11/11	8.0	0.5	2	6.5	16.0
28/11	8.0	5.0	2	9.7	39.0
21.7	8.0	0.1	~60	11.9	7.5

C.V. is the coefficient of variation (see text), Pn is the permeability coefficient of the membrane to isopropanol.

in the above order.

The osmotic pressure of the amoeba cytoplasm corresponds to that of a 107 to 117 milliosmolar solution (Løvtrup and Pigon, Riddick), therefore all of the solutions tested including the culture solution are hyposmotic or nearly isosmotic. Pn in low Ca^{++} and low or high Na^+ was unaffected by the addition of 120 mM/L sucrose, therefore solvent drag is not a likely factor in our system.

DISCUSSION

There are several reasons to interpret the data in table 1 and figure 1 cautiously. The coefficients of variation are roughly between 10 and 20%. It is our impression that the cells are more fragile at higher Pn's, therefore the high values are most liable to err on the low side. (Current techniques promise to reduce the coefficients of variation to about 5% and decrease the breakage of cells at high Pn).

Conceivably all our data could be generated if the cells are rapidly changing in surface area or in volume to surface ratio. There are four arguments against this possibility. First the range of Pn's measured to date are from 2.3×10^{-6}cm/sec (at 4.5^oC, 100 µM/L Ca^{++}, pH 6.75) to 13×10^{-6} cm/sec (at 20^oC, 100 µM/L Ca^{++}, 60 mM/L Cs^+, pH 8.0). This range would have to represent an overall five fold increase in surface to volume ratio, and a three fold change within 200 seconds. Second, the capsules are rapidly shaken for five minutes before an experiment begins and throughout the experiment. This causes amoeba to round up and restricts the formation of major pseudopods. Third, the high Pn values in high Na^+ are (at pH 7.0, 20^oC) associated with a ten to fifty fold increase in conductance (3, 4) over the controls. Thus an area change alone cannot simultaneously explain a two fold increase in Pn and a 50 fold increase in conductance.

Fourth, Wolpert and O'Neill (22) have measured the turnover time of amoeba membrane by tagging it with fluorescent antibody. In control, and in experiments in high Na^+, no evidence was found for a rapid turnover or formation of new membrane in less than a half time of about five hours. New membrane is thus formed at about .2% per minute. Since our experiments last five minutes, the area change and/or turnover is about 1% and can be neglected. Chapman-Andresen (8) has also argued that during pinocytosis, membranes are decreased, if anything. We conclude, therefore,

our increases in Pn's maybe too small as a consequence of a
decrease in surface area during the exposure to high concentrations
of cations. However, the volume of the cell may decrease slightly
in 60 mM/L Na^+ although this is nearly isomotic (7). Therefore,
we chose to use a constant volume/area ratio throughout all
experiments and anticipated that only small errors of variation
are introduced into the Pn values by this assumption, although
it may contain a constant error.

Thus the Pn's given are all calculated by assuming a volume
to surface ratio of 6×10^{-3} cm and may have to be adjusted on
this account. It is doubtful if the ratio is larger, but it could be
as small as 3×10^{-3} cm (9, 16). The standard errors are large if
single points (the mean of six experiments) alone are considered,
but the points fall on reasonable curves with little scatter. This
improves the reliability of the data in determining the direction
in which Pn changes with a given variable.

The data on the permeability of the amoeba to non-electrolytes was collected for two reasons. We believed, on the basis
of the earlier morphological and electrophysiological data (3, 4),
that the membrane in the high conductance state is very different
from the same membrane in the low conductance state. The present
data confirm this conclusion. Second, we hoped to establish some
link between pinocytosis and the structural and physiological
change in the membrane. The present data add little to our understanding of this link. Our current hypothesis of the process of
"excitation-pinocytosis-coupling" states that the redistribution of
ions or a specific ion in the cytoplasm as a consequence of the
permeability change, is the trigger for pinocytosis. Similar models
exist for directing amoeboid motion (1). Thus our data support the
hypothesis in so far as they demonstrate a permeability change in
the presence of pinocytosis activators (5).

In high Ca^{++} (at pH 7.0), Pn is about double that in low
Ca^{++} (figure 1). In contrast to the effect on Pn, high Ca^{++} causes
the conductance and presumably the Na^+ and K^+ permeabilities (Pe)
to decrease (3, 4, 6). In low Ca^{++} (5 - 50 µM/L), increasing
the concentration of monovalent cations (Na^+, K^+, Li^+, Cs^+, Rb^+)
is associated with about a doubling of Pn (5) and a 10 to 50 times

increase in conductance (3, 4, 13) and presumably an increase in
the permeability of the membrane towards these monovalent cations.
The cell is quite impermeable to chloride (6). From these data
it is apparent that delta Pn is not necessarily related to delta Pe's.
One possible generalization of this data is that Pn increases
whenever the instantaneous concentration in the membrane of
cations increases. Thus when Ca^{++} is low, a high membrane load
of monovalent cations causes an increase in Pn. In high Ca^{++} it
can be argued that Ca^{++} bound in the membrane diminishes Pe for
Na^+ but increases Pn.

There are several lines of evidence in support of the qualitative model just suggested.   Pn varies directly with pH.
This suggests that sites in the membrane are not available for
association with other cations when they are in the hydrogen form.
This is demonstrated by the changes in Pn at pH 6 in comparison
to pH 7 at all Ca^{++} and Na^+ levels. Presumably the sites are
more hydrated when they are in the non-hydrogen form and it is
this factor which most affects Pn.

The total number of groups available to Na^+ is determined by
the pH and the Ca^{++} concentration. This is concluded from the
horizontal line for high Na^+ in figure 1. The delta Pn is smaller
as Ca^{++} is raised until at 5 mM/L Ca^{++} there is no delta Pn. This
suggests that either all the available sites are occupied by Ca^{++},
or when Na^+ in the membrane, no net changes in the degree of
membrane hydration takes place.

It has often been questioned whether the changes in membrane
properties recorded under any experimental condition are due to
local changes in a small fraction of the total membrane, or due to
a homogeneous membrane change (21). This question resolves
formally to asking whether or not the change in property is due to
an increase in the diameter of a fixed pore, or due to an increase
in the number of transport sites. Whether the sites are confined
to a local spot is pertinant only to the second half of the question,
since the implication of "new sites" is that they form in an
undifferentiated part of the membrane. A few preliminary experiments which probed for an increase in site diameter, as opposed
to number, seemed to support the latter. Pn for slower penetrants
such as glucose, glycerol or ethylene glycol were not increased to
any greater degree by high Na^+ than Pn for isopropanol. A model
for the Pn changes based on an increase in pore area, not number,
would have to predict a relative increase in the Pn of larger
penetrants, especially those limited by pore size.

Although other models and arguments can be adduced from the data (5) the stronger ones have been presented here. We expect to be able to develop a quantitative model of the amoeba. membrane as we increase the size and quality of our data pool. The current model envisages the membrane as a collection of sites which can ionize according to the hydrogen ion concentration, and the ionic strength. The amount of water in the membrane, hence its permeability towards polar non-electrolytes, depends on the quantities of ions in the membrane at any instant. The membrane is plastic and its internal dimensions can change to accomodate different loads of hydrated ions.

SUMMARY

The permeability coefficient(Pn) of the amoeba plasma membrane for polar non-electrolytes has been found to be a function of the concentration of H^+, monovalent cations, and divalent cations. Pn increases with pH and Ca^{++} concentration. At pH 7.0 a time dependent increase in Pn to a constant value at its maximum accompanies a step increase in the Na^+ concentration at all Ca^{++} concentrations tested. A model is proposed which relates Pn to the degree of membrane hydration. The membrane is hydrated according to its total load of cations and the degree to which these ions are hydrated when bound in the membrane.

ACKNOWLEDGMENTS

Dr. Brandt wishes to acknowledge the warm hospitality of Dr. Heinz Holter, Professor of Physiology, and the other members of this department of the Carlsberg Laboratory during the year he spent in the Laboratory while this study was carried out. Dr. Brandt held a Guggenheim Fellowship over this same period, 1968-9. Supported by NIH Grant 5 RO1 - 05910.

REFERENCES

1. Bingley, M. S., and Thompson, C. M., "Bioelectric Potentials in Relation to Movement in Amoebae", J. Theoret. Biol., 2: 16-32, 1962.

2. Brandt, P. W., "A study of the Mechanism of Pinocytosis", Exp. Cell Res., 15: 300-313, 1958.

3. Brandt, P. W., and Freeman, A. R., "Plasma Membrane: Sub-
   structural Changes Correlated with Electrical Resistance and
   Pinocytosis", Science, 155: 582-585, 1967.

4. Brandt, P. W. and Freeman, A. R., "The Role of Surface Chemistry
   in the Biology of Pinocytosis", J. Colloid and Interface Science,
   25: 47-56, 1967.

5. Brandt, P. W., and Hendil, K. B., in preparation for publication in
   Comp. Rend. Trav. Lab. Carlsberg.

6. Bruce, D. L. and Marshall, J. M. "Some Ionic and Bioelectric
   Properties of the Amoeba Chaos chaos", J. Gen. Physiol. 49:151-
   178, 1965.

7. Chapman-Andresen, C., and Dick, D. A. T., "Volume Changes in
   the Amoeba Chaos Chaos L.", Comp. Rend. Trav. Lab. Carlsberg,
   32: 265-289, 1961.

8. Chapman-Andresen, C., "Factors Affecting the Duration of and the
   Utilization of Membrane During Pinocytosis", in Progress in
   Protozoology, Proceedings of the First Internat. Conf. Protozoology.
   Prague, 1961.

9. Chapman-Andresen, C., and Dick, D. A. T., "Sodium and Bromine
   Fluxes in the Amoeba Chaos Chaos L.", Comp. Rend. Trav.
   Carlsberg, 32: 445-469, 1962.

10. Chapman-Andresen, C., " Studies on Pinocytosis in Amoebae",
    Comp. Rend. Trav. Lab. Carlsberg, 33: 73-264, 1962.

11. Cooper, B. A., "Quantitative Studies of Pinocytosis Induced in
    Amoeba Proteus by Simple Cations". Comp. Rend. Trav. Lab.
    Carlsberg, 36: 385-403, 1968.

12. Holter, H., " The Induction of Pinocytosis" in Biological Approaches
    to Cancer Chemotherapy", Academic Press, Inc.,   77-88, 1960.

13. Josefsson, J. O., "Some Bioelectrical Properties of Amoeba Proteus'
    Acta Physiol. Scand., 66: 395-405, 1966.

14. Josefsson, J. O., "Induction and Inhibition of Pinocytosis in
    Amoeba Proteus", Acta Physiol. Scand., 73: 481-490, 1968.

15. Korn, E. D., "Structure of Biological Membranes", Science,
    153: 1491-1498, 1966.

16. Løvtrup, S., and Pigon, A., "Diffusion and Active Transport of Water in the Amoeba". <u>Comp. Rend. Trav. Lab. Carlsberg,</u> 28: 1-36, 1951.

17. Portzeh, L. H., Caldwell, P. C., and Ruegg, J. C., "The Dependence of Contraction and Relaxation of Muscle Fibers from the Crab <u>Maia Squinado</u> of the Internal Concentration of Free Calcium Ions". <u>Biochemica et Biophysica Acta</u>, 79: 581-591, 1964.

18. Riddick, D. H., "Contractile Vaciole in Amoeba <u>Pelomyna Carolinensis</u>". <u>Am. J. Physiol.</u>, 215: 736-740, 1968.

19. Rustad, R. C., "Molecular Orientation at the Surface of Amoeba during Pinocytosis". <u>Nature,</u> 83: 1058-1059, 1959

20. Schumaker, V. N., "Uptake of Protein from Solutions by Amoebe <u>Proteus</u>". <u>Exp. Cell Res.</u>, 15: 314-331, 1958.

21. Stoeckenius, W., and Engelman, D. M., "Current Models for the Structure of Biological Membranes", <u>J. Cell Biol.</u>, 42: 613-646, 1969.

22. Wolpert, L., and O'Neill, C. H., "Dynamics of the Membrane of <u>Amoeba Proteus</u> studied with Labelled Specific Antibody". <u>Nature</u>, 196: 1261-1266, 1962.

INDEX

Adsorption
    and electric currents, 209 ff
    of ions in amoeba, 323 ff
    of protein films at the air-water interface, 1 ff
    of proteins on colloids and cells, 217 ff
    of proteins on synthetic materials, 235 ff
    of lung surfactant, 261 ff, 275 ff

Atheromatous plaques
    composition, 55 ff

Bilayer lipid membranes
    asymmetric membranes, 155 ff
    effects of modifiers, 135 ff
    effects of pH and Ca^{2+}, 155 ff

Biological systems
    (see also Atheromatous plaques
                Gastric secretion
                Intravascular prostheses
                Lung surfactant
                Natural membranes
                Thromboresistance)
    amoeba, 323 ff
    bacterial membranes, 175 ff
    blood vessel walls, 235 ff
    lining of lung, 261 ff, 275 ff
    lining of stomach, 309 ff
    various cells including erythrocytes and tumor cells, 191 ff
    various systems including corneal stroma, cartilage, 287 ff

Calcium
    Ca^{++} dependent ATPase, 181
    effect on lecithin monolayers, 268
    effect on pinocytosis in amoeba, 323 ff

Calcium (continued)
    interactions with bilayer lipid membranes, 155 ff
    interactions with fatty acids, 23 ff

Differential thermal analysis
    phospholipids in water, 37 ff, 55 ff

Dispersions
    of phospholipids, 37 ff, 85 ff

Electrostatic effects
    and adsorption at solid-liquid interface, 209 ff
    and film pressure, 15
    and ion transport in amoeba, 323 ff
    leading to osmotic flow in stomach, 309 ff
    surface potential, 27

Experimental techniques
    adsorbed monolayers, 4
    bubble stability, 103
    differential thermal analysis, 39, 58
    immunoelectroadsorption, 211
    interfacial transport using the Schulman. chamber, 297
    microelectrophoresis, 218
    polarized light microscopy, 39
    spread monolayers, 3, 24, 103, 263
    surface potential, 24
    thin-layer chromatography, 88
    titrations of dispersed systems, 87

Gastric secretion
    control mechanism, in vivo, 309 ff

Hexadecyltrimethylammonium bromide
    interaction with bilayer lipid membranes, 141
    interaction with DNA, 119 ff

Interactions
    calcium with fatty acid monolayers, 23 ff
    choline phospholipids with sulfatide, 85 ff
    DNA with positively charged monolayers, 119 ff
    immunological reactions, 209 ff
    lipid-protein association in lung surfactant, 261 ff, 275 ff
    modifiers with bilayer lipid membranes, 135 ff
    phospholipids in water, 37 ff
    poly-L-lysine with stearic acid, 101 ff

Intravascular prostheses
    study of materials, 235 ff

Lipids
    as "carriers" in water-butanol system, 295 ff
    cholesterol and cholesteryl esters, 55 ff, 85 ff, 162, 295 ff
    fatty acids, 23 ff, 62 ff
    in lung surfactant, 261 ff
    oxidized cholesterol, 144
    phospholipids, 37 ff, 85 ff, 139, 261 ff, 295 ff
    sodium dodecyle sulfate, 93
    sulfatides, 85 ff
    triglycerides, 55 ff

Lung surfactant
    composition and surface properties, 261 ff, 275 ff

Model systems
    (see Bilayer lipid membranes
        Dispersions
        Interactions
        Lipids
        Monolayers)

Monolayers
    acyl derivatives of casein, 1 ff
    bovine serum albumin, 1 ff
    casein, 1 ff
    lung surfactant, 261 ff, 275 ff
    lysozyme, 1 ff
    fatty acids, 23 ff, 101 ff
    oleic acid, 23 ff
    phospholipids, 261 ff
    polypeptides, 101 ff
    proteins, 1 ff
    stearic acid, 23 ff, 101 ff

Natural membranes
    analogy of bilayer lipid membrane, 135 ff, 155 ff
    lipid composition, 55 ff, 76 ff, 175 ff
    phospholipid composition, 50 ff, 175 ff
    pore model, 288 ff
    red blood cells
        sheep, 217 ff
        human, 295 ff
    RNA in cell periphery, 191 ff
    surface composition of lung, 261 ff
    swelling due to water absorption, 287 ff

Nucleic acids
    DNA interaction with monolayers, 119 ff
    RNA in the cell periphery, 191 ff

Phospholipids
    as "carriers" in water-butanol system, 295 ff
    in aqueous dispersions, 37 ff, 85 ff
    in monolayers, 261 ff

Pinocytosis
    in amoeba, 323 ff

Polymers
    DNA, 119 ff
    poly-L-lysine, 101 ff
    polylysyl gelatin, 217 ff
    polypeptides (copolymers of L-lysine and L-phenylalanine), 119
    polysaccharides, 209 ff
    polystyrene latex, 217 ff
    polystyrene-sulfonic acid membranes, 309 ff
    protein monolayers, 1 ff, 209 ff, 217 ff
    RNA, 191 ff

Proteins
    enzymes in bacteria, 175 ff
    in lung surfactant, 261 ff, 275 ff
    in monolayers, 1 ff, 209 ff, 217 ff

Sugars
    kinetic and equilibrium behavior in a water-butanol-lipid
             system, 295 ff

Surface Chemistry
    (see Adsorption
         Experimental techniques
         Interactions
         Model systems
         Monolayers
         Wettability)

Thromboresistance
    "Stellite 21," 245
    surface chemical features, 235 ff

Transport
    effect of pH and Ca^{++} in amoeba, 323 ff
    of ions and water across gastric mucosa, 310 ff
    of sugars in water-butanol-lipid system, 295 ff
    of water through membranes, 287 ff

Wettability
    of blood vessel walls, 240